Fungal Pathogenesis in Humans

The Growing Threat

Special Issue Editor

Fernando Leal

MDPI • Basel • Beijing • Wuhan • Barcelona • Belgrade

Special Issue Editor
Fernando Leal
Instituto de Biología Funcional y Genómica/Universidad de Salamanca
Spain

Editorial Office
MDPI
St. Alban-Anlage 66
4052 Basel, Switzerland

This is a reprint of articles from the Special Issue published online in the open access journal *Genes* (ISSN 2073-4425) from 2018 to 2019 (available at: https://www.mdpi.com/journal/genes/special_issues/Fungal_Pathogenesis_Humans_Growing_Threat).

For citation purposes, cite each article independently as indicated on the article page online and as indicated below:

LastName, A.A.; LastName, B.B.; LastName, C.C. Article Title. *Journal Name* **Year**, *Article Number*, Page Range.

ISBN 978-3-03897-900-5 (Pbk)
ISBN 978-3-03897-901-2 (PDF)

Cover image courtesy of Fernando Leal.

Fungal Pathogenesis in Humans

Contents

About the Special Issue Editor

Fernando Leal obtained his Bachelor, Master, and Ph.D. degrees in Biology from Salamanca University (Spain). After a postdoctoral training as a Fulbright fellow at the National Cancer Institute (N.I.H., Bethesda, MD, USA) he became a faculty member (Assistant/Associate Professor/Vice Dean) of the Biology School of Salamanca University. At the same time, he became research group leader and vice director at the Institute of Microbial Biochemistry (a joint center of the Spanish Research Council and the University of Salamanca). He is currently a Full Professor at the same University. Dr. Leal has more than 30 years of research experience in different fields, such as yeast's plasma membrane, viral oncogenes, molecular biology of fungal pathogenesis, and *Aspergillus* proteomics. During a recent sabbatical stay at Edinburgh University, he was involved in a project dealing with the alteration of miRNAs export in exosomes after Vaccinia virus infection. He has published numerous scientific research papers and has collaborated as a reviewer and editor with several scientific journals. His current research interests focus on the mechanisms of gene regulation by Zn^{++} in *Aspergillus fumigatus* and their role in virulence and pathogenesis.

Editorial

Special Issue: Fungal Pathogenesis in Humans: The Growing Threat

Fernando Leal

Instituto de Biología Funcional y Genómica/Dpto. Microbiología y Genética, (CSIC/USAL), Zacarías González 2, 37007 Salamanca, Spain; fleal@usal.es

Received: 1 February 2019; Accepted: 4 February 2019; Published: 12 February 2019

Approximately 150 fungal species are considered as primary pathogens of humans and animals. The variety of infections that they may cause ranges from localized cutaneous, subcutaneous or mucosal infections to systemic and potentially fatal diseases. Many fungi are also able to cause lesions when abnormal patient susceptibility exists or after traumatic colonization of the fungus (for a comprehensive review on Medical Mycology, see Kwon-Chung and Bennet, [1]). Fungi that infect immunocompromised patients are referred to as opportunistic pathogens. The number of opportunistic fungi has recently increased due to the arrival of new and growing populations of immunocompromised hosts. In this special issue, we have attempted to compile a collection of new studies investigating the role of some virulence traits and their molecular mechanisms of action in the pathogenic outcome of fungal infections.

The term candidiasis refers to a wide clinical spectrum of infections that can be acute or chronic, superficial (cutaneous, oropharyngeal, vulvovaginal, ocular) or deep (esophageal, gastrointestinal, respiratory, urinary, etc.) and can affect either normal or immunosuppressed individuals. The major etiologic agent is *Candida albicans*, which is part of the normal human mycobiota. However, several other species are frequently encountered in certain clinical diseases (*Candida parapsilosis, Candida glabrata, Candida tropicalis, Candida lusitaniae*). Here, three different aspects of *Candida* infections are examined: the maintenance of chromosomal integrity; biofilm formation as a form of survival; and the establishment of new models of infection as an alternative to mice. Ciudad et al. [2] address the problem of repairing the alkylation base damage in the genome of *C. albicans*. After analyzing the response of three homologous recombination (HR) mutants to chromosomal damage caused by methyl methanesulfonate (MMS), these authors propose that repair takes place through a mechanism (possibly base excision repair) that does not involve homologous recombination. Biofilm formation allows *Candida* to adhere to and proliferate on medical devices and host tissues. Biofilms are constituted of a mixture of filamentous and yeast cells that surround themselves with an extracellular matrix, which provides a remarkable degree of resistance to antifungal drugs. Rodrigues et al. [3] evaluate the role of *C. parapsilosis* genes associated with the production of the biofilm matrix by monitoring their expression levels in response to treatment with antifungal drugs. They concluded that although beta-1,6-glucans and mannans are an essential part of both cells and the biofilm matrix, β-1,3-glucan seems to play a more important role in biofilm resistance to antifungal drugs. Chong et al. [4] provide a detailed review of numerous studies on *C. albicans* biofilms, which includes the analysis of the transcriptome, whole genome sequencing, functional genomic approaches to identify critical regulatory genes and comparative genomics analysis. In addition, recently discovered pathways and genes involved in the pathogenesis of the fungus are described and future directions in the development of therapeutics are suggested. Finally, Souza et al. [5] confirm the suitability of *Caenorhabditis elegans* as an alternative to the use of mouse models in pathogenesis studies, which can be infected and killed by three species of the *C. parapsilosis* complex. The progression of the infection was determined by histological examination and the immune response of *C. elegans* was monitored by analyzing gene expression. Early treatment with antifungal drugs was also found to be effective in this model.

Aspergilli are ubiquitous fungi found within our environment, which humans are continuously being exposed to. However, the diseases caused by these fungi are relatively uncommon and the severe, invasive form of these diseases is almost always confined to immunosuppressed individuals. There are three main manifestations of disease: an allergic response to inhaled aspergilli; the colonization of air spaces within the body; and tissue invasion by the fungus. *Aspergillus fumigatus* represents the most frequent etiologic agent of both noninvasive and invasive aspergillosis while *Aspergillus flavus* and *Aspergillus niger* can also provoke invasive pulmonary aspergillosis in immunosuppressed patients. Classical genetic studies on the genera soon progressed to the molecular level, which allowed for the discovery of several mechanisms involved in virulence and pathogenesis pathways. In this special issue, a review and an original article are included, which emphasize the importance of whole genome comparative studies for identifying pathogenic properties based on differences in DNA sequences. García-Rubio et al. [6] perform Whole Genome Sequencing (WGS) on more than a hundred *A. fumigatus* strains and highlight the importance of choosing the most suitable reference genome for analyzing the genetic differences between *A. fumigatus* strains, their genetic background and the development of antifungals resistance. Furthermore, Ashu and Xu [7] propose in their concept paper that these types of studies could be expanded to devise molecular epidemiology and experimental evolution methods that are useful for managing the *Aspergillus* threat. These authors provide a framework for such a purpose that implies the development of rapid and accurate diagnostic tools to genotype the infectious pathogen to the level of the species and the individual as well as drug susceptibility patterns. One of the recently described virulence mechanisms relies on the ability of *Aspergillus* to obtain essential ions (mainly Fe and Zn) from the extremely limited supply of micronutrients existing in host tissues. Vicentefranqueira et al. [8] determine how the ZafA transcription factor of *A. fumigatus* regulates zinc homeostasis and its importance for virulence. The combined use of microarrays, Electrophoretic Mobility Shift Assays (EMSA), DNAse I footprinting assays and in silico tools have been essential for obtaining a better understanding of the regulation of the homeostatic and adaptive response of this fungus to zinc starvation.

Cryptococcosis is the fourth most commonly recognized cause of life-threatening infections among AIDS patients and different types of immunosuppression are the predisposing factors influencing the rate of infection in non-AIDS patients. Cryptococcosis are infections caused by the encapsulated fungus *Cryptococcus neoformans* that occur after the spores are inhaled into the lungs. This causes pneumonia and frequently spreads hematogenously to the brain and meninges, causing meningoencephalitis. The role of small RNAs and the mechanisms by which some reach their target are addressed in two papers included within this issue. Using next-generation sequencing and bioinformatics tools, Huo et al. [9] report the existence of stable circRNAs in the genome of *Cryptococcus neoformans* for the first time. These RNAs were hosted in genes that were mainly responsible for primary metabolism and ribosomal protein production. Highly transcribed circRNAs from GTPase and RNA debranching enzyme genes were discovered. The role of these small RNAs in pathogenesis remains open for discussion. Extracellular vesicles (EVs) have been found to play important roles in crosstalk between different types of cells and tissues from the same or even different species. Many fungi use these vesicles as carriers for polysaccharides, proteins and RNAs, but their implication in pathogenesis is still not clear. Peres da Silva et al. [10] investigate if EV-mediated RNA export in *C. neoformans* was functionally connected with the Golgi reassembly and stacking protein (GRASP). The results obtained after analyzing the mutants that have defective GRASP synthesis and autophagic mechanisms suggest that GRASP, but not the autophagy regulator, is involved in the EV-mediated export of RNA. This function as a key regulator of unconventional secretion in eukaryotic cells is a new finding.

Other fungal infections that are not as widespread or as fatal as those caused by *Candida*, *Aspergillus* and *Cryptococcus* are also mentioned in this special issue. Mucormycosis are a set of infections caused by members of the order Mucorales in patients with serious underlying conditions. Vascular invasion by hyphae results in infarction and necrosis of tissues. Although *Rhizopus oryzae* is the most common causal agent of human mucormycosis, *Mucor circinelloides* isolates have been associated with outbreaks of the disease. Unfortunately, the genetic manipulation of these basal fungi is not well established, which has impeded the study of their virulence traits and pathogenesis mechanisms. Partially overcoming these challenges, Binder et al. [11] generate and functionally characterize a bioluminescent strain of *M. circinelloides* designed to be used in the monitoring of real-time and non-invasive infection in insect and murine models and in the testing of antifungal drug efficacy. Dermatophytes are fungi capable of infecting keratinized tissues, such as the epidermis, hair and nails, without affecting subcutaneous or deep tissues, whereas *Trichophyton rubrum*, the major etiologic agent of human ringworm, causes chronic lifetime infections. It is also worth mentioning the work by Petrucelli et al. [12] who describe a *T. rubrum*- HaCat keratinocyte co-culture, which is used to mimic the natural fungal–host interaction, where dual RNA-seq technology was used to evaluate the transcriptomes of both organisms. These authors found that some keratinolytic proteases and glyoxylate cycle encoding genes that may improve nutrient assimilation and fungal survival and colonization were induced in the fungus. In human keratinocytes, some genes involved in the epithelial barrier integrity were inhibited, whereas others that played a role in antimicrobial activity were induced.

A problem that is common to all fungal infections is their resistance to antifungals, with growing concern focused on how to treat most of the aforementioned diseases. Multidrug resistance transporters (MDRs) are key elements in mediating fungal resistance to pathogenesis-related stresses, a topic that was well described by Cavalheiro et al. [13]. These authors emphasize the importance of these transporters beyond the role of drug resistance and summarize their relevance in pathogenesis traits, such as resistance to host niche environments, biofilm formation, immune evasion and virulence.

I would like to express my deep appreciation for all of the hard work carried out by the investigators included in this special issue and those who were not due to different circumstances.

I hope that their enthusiasm and dedication to fungal research will encourage many young mycologists to apply the different approaches mentioned herein to the fascinating field of human fungal pathogenesis.

Conflicts of Interest: The author declare that there is no conflict of interest concerning this work.

References

1. Kwon-Chung, K.J.; Bennett, J.E. *Medical Mycology*; Lea & Febiger: Philadelphia, PA, USA, 1992; 866p.
2. Ciudad, T.; Bellido, A.; Andaluz, E.; Hermosa, B.; Larriba, G. Role of homologous recombination genes in repair of alkylation base damage by *Candida albicans*. *Genes* **2018**, *9*, 447. [CrossRef] [PubMed]
3. Rodrigues, C.F.; Henriques, M. Portrait of matrix gene expression in *Candida glabrata* biofilms with stress induced by different drugs. *Genes* **2018**, *9*, 205. [CrossRef] [PubMed]
4. Chong, P.P.; Chin, V.K.; Wong, W.F.; Madhavan, P.; Yong, V.C.; Looi, C.Y. Transcriptomic and genomic approaches for unravelling *Candida albicans* biofilm formation and drug resistance-an update. *Genes* **2018**, *9*, 540. [CrossRef] [PubMed]
5. Souza, A.C.R.; Fuchs, B.B.; Alves, V.S.; Jayamani, E.; Colombo, A.L.; Mylonakis, E. Pathogenesis of the *Candida parapsilosis* complex in the model host *Caenorhabditis elegans*. *Genes* **2018**, *9*, 401. [CrossRef] [PubMed]
6. Garcia-Rubio, R.; Monzon, S.; Alcazar-Fuoli, L.; Cuesta, I.; Mellado, E. Genome-wide comparative analysis of *Aspergillus fumigatus* strains: The reference genome as a matter of concern. *Genes* **2018**, *9*, 363. [CrossRef] [PubMed]
7. Ashu, E.E.; Xu, J. Strengthening the one health agenda: The role of molecular epidemiology in *Aspergillus* threat management. *Genes* **2018**, *9*, 359. [CrossRef] [PubMed]
8. Vicentefranqueira, R.; Amich, J.; Marin, L.; Sanchez, C.I.; Leal, F.; Calera, J.A. The transcription factor ZafA regulates the homeostatic and adaptive response to zinc starvation in *Aspergillus fumigatus*. *Genes* **2018**, *9*, 318. [CrossRef] [PubMed]
9. Huo, L.; Zhang, P.; Li, C.; Rahim, K.; Hao, X.; Xiang, B.; Zhu, X. Genome-wide identification of circRNAs in pathogenic Basidiomycetous yeast *Cryptococcus neoformans* suggests conserved circRNA host genes over kingdoms. *Genes* **2018**, *9*, 118. [CrossRef] [PubMed]

10. Peres da Silva, R.; Martins, S.T.; Rizzo, J.; Dos Reis, F.C.G.; Joffe, L.S.; Vainstein, M.; Kmetzsch, L.; Oliveira, D.L.; Puccia, R.; Goldenberg, S.; et al. Golgi reassembly and stacking protein (GRASP) participates in vesicle-mediated rna export in *Cryptococcus Neoformans*. *Genes* **2018**, *9*, 400. [CrossRef] [PubMed]

11. Binder, U.; Navarro-Mendoza, M.I.; Naschberger, V.; Bauer, I.; Nicolas, F.E.; Pallua, J.D.; Lass-Florl, C.; Garre, V. Generation of a mucor circinelloides reporter strain-A promising new tool to study antifungal drug efficacy and Mucormycosis. *Genes* **2018**, *9*, 613. [CrossRef] [PubMed]

12. Petrucelli, M.F.; Peronni, K.; Sanches, P.R.; Komoto, T.T.; Matsuda, J.B.; Silva Junior, W.A.D.; Beleboni, R.O.; Martinez-Rossi, N.M.; Marins, M.; Fachin, A.L. Dual RNA-Seq analysis of *Trichophyton rubrum* and HaCat Keratinocyte co-culture highlights important genes for fungal-host interaction. *Genes* **2018**, *9*, 362. [CrossRef] [PubMed]

13. Cavalheiro, M.; Pais, P.; Galocha, M.; Teixeira, M.C. Host-pathogen interactions mediated by MDR transporters in fungi: As pleiotropic as it gets! *Genes* **2018**, *9*, 332. [CrossRef] [PubMed]

Article

Genome-Wide Identification of circRNAs in Pathogenic Basidiomycetous Yeast *Cryptococcus neoformans* Suggests Conserved circRNA Host Genes over Kingdoms

Liang Huo [†], Ping Zhang [†], Chenxi Li, Kashif Rahim, Xiaoran Hao, Biyun Xiang and Xudong Zhu *

Beijing Key Laboratory of Genetic Engineering Drug and Biotechnology, Institute of Biochemistry and Molecular Biology, College of Life Sciences, Beijing Normal University (CLS-BNU), Beijing 100875, China; cryptoleon@gmail.com (L.H.); zp1516@163.com (P.Z.); lcx1219@163.com (C.L.); kashifbangash073@gmail.com (K.R.); hxrr_563@163.com (X.H.); xby6024@126.com (B.X.)
* Correspondence: zhu11187@bnu.edu.cn; Tel.: +86-010-5880-4722
† These authors contributed equally to this work.

Received: 9 January 2018; Accepted: 19 February 2018; Published: 26 February 2018

Abstract: Circular RNAs (circRNAs), a novel class of ubiquitous and intriguing noncoding RNA, have been found in a number of eukaryotes but not yet basidiomycetes. In this study, we identified 73 circRNAs from 39.28 million filtered RNA reads from the basidiomycete *Cryptococcus neoformans* JEC21 using next-generation sequencing (NGS) and the bioinformatics tool circular RNA identification (CIRI). Furthermore, mapping of newly found circRNAs to the genome showed that 73.97% of the circRNAs originated from exonic regions, whereas 20.55% were from intergenic regions and 5.48% were from intronic regions. Enrichment analysis of circRNA host genes was conducted based on the Gene Ontology and Kyoto Encyclopedia of Genes and Genomes pathway databases. The results reveal that host genes are mainly responsible for primary metabolism and, interestingly, ribosomal protein production. Furthermore, we uncovered a high-level circRNA that was a transcript from the guanosine triphosphate (GTP)ase gene *CNM01190* (gene ID: 3255052) in our yeast. Coincidentally, *YPT5*, *CNM01190*'s ortholog of the GTPase in *Schizosaccharomyces pombe*, protists, and humans, has already been proven to generate circRNAs. Additionally, overexpression of RNA debranching enzyme *DBR1* had varied influence on the expression of circRNAs, indicating that multiple circRNA biosynthesis pathways exist in *C. neoformans*. Our study provides evidence for the existence of stable circRNAs in the opportunistic human pathogen *C. neoformans* and raises a question regarding their role related to pathogenesis in this yeast.

Keywords: *Cryptococcus neoformans*; next-generation sequencing (NGS); circRNAs; RNA debranching enzyme; GTPase-encoding gene

1. Introduction

Circular RNAs (circRNAs), characterized by a closed loop structure, have been a hot topic of research in RNA biology since their wildly diverse and multiple functions were confirmed [1,2]. The fact that the 3′ and 5′ ends of those RNAs are joined by covalent bonds makes them lack polyadenylated tails and 5′–3′ polarity [3]. As a result, circRNAs are considered more stable than linear RNA molecules and more resistant to degradation by RNase R, which is an efficient 3′ to 5′ exoribonuclease [4].

Although the first case of circRNAs was reported in plant-based virus as early as 1976 [5], circRNAs have been discarded as "junk-RNA developed by messenger RNA (mRNA) splicing" and ignored by most research groups [6]. This situation remained for decades, until abundant circRNAs were uncovered in a variety of normal and malignant human cells in 2012 [7] and circRNAs were demonstrated as efficient "microRNA sponges" in 2013 [8]. With the development of high-throughput sequencing technology and RNA circularization prediction algorithms, such as circular RNA identification (CIRI) [9], circular (CIRC) explorer [10], and known and novel isoform explorer (KNIFE) [11], an increasing number of circRNAs have been detected in protists, yeasts, plants, flies, and mammals [12,13].

The mechanism of circRNA biogenesis is intricate, regulated by multiple factors, and varying among different species. In general, the circularization of RNAs can be accomplished through at least four disparate paths: spliceosome-dependent [14], intron-pairing-driven [15], protein factors-associated [16], and lariat-driven paths [17]. Recent studies also reveal distinctly crucial functions of circRNAs, such as microRNA (miRNA) sponge, post-transcription regulation, rolling circle translation, and creation of circRNA-derived pseudogenes [6,18,19]. However, the mechanism has not been illustrated thoroughly and it deserves to be investigated further.

The basidiomycetous yeast *Cryptococcus neoformans* is an opportunistic human pathogen that has been life-threatening to immunodeficient groups such as human immunodeficiency virus (HIV)-infected patients [20]. Efforts have been made by laboratories worldwide to understand the fundamentals of its pathogenic progress and its virulence determinants. Considering the fact that knowledge about circRNA molecules is limited, it may be necessary to define the potential role of circRNA in *C. neoformans*. Unfortunately, circRNAs have not been reported in this fungus, nor in the whole group of basidiomycetes. Thus, here we attempted to identify circRNAs from *C. neoformans*, and subsequently analyzed the features and conducted functional annotation of those circRNAs. We identified in this study 73 unique circRNAs in this basidiomycetous yeast. Interestingly, we also found the existence of small guanosine triphosphatase (GTPase)-encoding genes, which are conserved circRNA-host genes in yeasts and some other eukaryotic organisms. Finally, we demonstrate the influence of an RNA debranching enzyme, Dbr1, on the expression of circRNAs.

2. Materials and Methods

2.1. Strains and Media

The strain *C. neoformans* var. *neoformans* JEC21 (serotype D, MATα) was used for circRNA analysis in this study. Yeast extract–peptone–dextrose (YPD) medium (2% glucose, 2% peptone, 1% yeast extract, pH 6.0) was used for routine growth of *C. neoformans*.

2.2. RNA Isolation and Quality Control

JEC21 was cultured in 5 mL liquid YPD medium for 18 h at 30 °C. Fresh yeast cells were collected by centrifugation, and approximately 0.1 g of yeast cells was washed by wash buffer (0.1 M ethylenediaminetetraacetic acid (EDTA), 0.5 M sodium chloride) three times at 4 °C. Fungal capsule was broken by Bullet Blender Storm 24 (Next Advance, Troy, NY, USA) for 2 min. Total RNA was extracted using the RNAiso (Takara, Shiga, Japan) according to the protocol supplied with the reagent. RNA concentration was measured by POLARstar Omega (BMG Labtech, Offenburg, Germany), and RNA quality was tested by Agilent 2100 (Agilent Technologies, Santa Clara, CA, USA). The quality control threshold was set as follows: A260/A280 ratio > 1.8, A260/A230 ratio > 1.8, RNA integrity number value > 7.0.

2.3. Deep RNA Sequencing and In Silico Discovery of Circular RNAs

Construction of a RNA library, as well as deep RNA sequencing, was accomplished by a commercial service (Genewiz, Suzhou, China). Briefly, Ribominus™ transcriptome isolation kit (Thermo Fisher Scientific, Waltham, MA, USA) was used to remove ribosome RNA in the total RNA. RNase R (Takara) treatment was performed according to the manufacturer's protocol to remove linear RNA in the RNA samples. KAPA Stranded mRNA-seq Kit (Kapa Biosystems, Wilmington,

MA, USA) was utilized for the generation of RNA-sequencing (RNA-seq) libraries according to the manufacturer's protocol. Next-generation sequencing was then conducted on a HighSeqTM 2500 system (Illumina, San Diego, CA, USA). To remove the low-quality reads in the raw paired-end data, such as the primer/adaptor sequences and non-ATGC reads, IlluQC_PRLL.pl v 2.3.3 software [21] was used to perform quality check with the parameter set as 20. The reads with more than or equal to the specified quality score (20 in this study) are filtered as high-quality reads. Subsequently, clean data were aligned to the *C. neoformans* JEC21 genome (Cryptococcus_neoformans.GCA_000091045.1. dna.toplevel.fa, release 37) using Burrows–Wheeler aligner (BWA) (version 0.6) software with default settings [22]. The 19Mb genome sequence of *C. neoformans* JEC21 consists of 14 chromosomes with different lengths changing from 762 kilobase (kb) pairs to 2.3 megabase (Mb) pairs.

The CIRI algorithm (version 1.2) was the tool to identify circRNAs in *C. neoformans* JEC21 [9]. CIRI was performed with default options, with the computer command: CIRI_v1.2.pl -I input.sam -O output_circRNAs.txt -F genome.fa -P -A Ensembl_Cn37.gtf. Counts of identified circRNA reads were normalized by read length, and the number of reads mapping (spliced reads per billion mapping) was determined after CIRI prediction [23].

2.4. Gene Ontology Category and Kyoto Encyclopedia of Genes and Genomes Pathway Analysis

The circRNA host genes were functionally analyzed according to gene ontology (GO) by the database for annotation, visualization, and integrated discovery (DAVID) 6.7 web server (https://david.ncifcrf.gov) with the default options [24]. The KOBAS 2.0 web server was used to uncover the Kyoto Encyclopedia of Genes and Genomes (KEGG) biological pathways of circRNA host genes with the default settings [25].

2.5. Validation of Circular RNAs

To confirm the existence of certain circRNAs of interest, e.g., circCNYPT5, we adopted an approach of outward polymerase chain reaction (PCR) with a pair of primers designed for outward amplification. Briefly, total RNA was extracted with RNAiso reagent (Takara) as descripted in Section 2.2. complementary DNA (cDNA) synthesis was performed with the FastQuant RT Kit with genomic DNase (gDNase) (Tiangen Biotech, Beijing, China). The 50 μL amplification reaction system contained 0.5 μL Takara Ex Taq, 5 μL 10× Ex Taq Buffer, 4 μL deoxyribonucleotide triphosphates (dNTPs), 2 μL/2 μL forward/reverse primers, and 36.5 μL double distilled water (ddH$_2$O). The PCR program was set as follows: 98 °C for 2 min, 32 cycles at 98 °C for 10 s, 55 °C for 20 s, and 72 °C for 30 s; the final elongation step was run at 72 °C for 5 min. PCR products with expected length (~250 base pairs (bp)) were separated by 0.8% agarose gel electrophoresis and purified with TIANgel midi purification kit (Tiangen Biotech) according to the manufacturer's instruction. Sanger sequencing was employed to confirm the existence of the back-splicing junction sites (Genewiz).

2.6. Other Online Database and Software

The annotation and nomenclature of *C. neoformans* JEC21 genes in this article were referred to the Ensemble Fungi database (http://fungi.ensembl.org). The multiple alignments of amino acid sequences were conducted by the Clustal Omega web server [26] with default settings.

2.7. Construction of DBR1 Gene Overexpression Vector

To investigate the regulation of circRNAs by *DBR1*, we overexpressed the gene in the wild-type JEC21 strain. The whole *DBR1* gene, including an 800-bp flanking sequence, was obtained by PCR with the protocol described in Section 2.5, except the elongation time for each cycle was 2 min. The pBS-HYG plasmid was linearized with the restriction enzyme Hind III. Then the In-Fusion® HD cloning kit (Takara) was employed to ligate the linearized plasmid and *DBR1* fragment. Subsequently, recombinant plasmid pBS-HYG-DBR1 was linearized by Xba I enzyme and transformed into the wild-type *C. neoformans* JEC21 cells. To select positive transformants, cells were screened on YPD plates containing 100 μg/mL hygromycin. Genomic DNA of two randomly selected clones, OE-1 and OE-2, was extracted and used in subsequent experiments. PCR was performed to confirm the existence of pBS-HYG-DBR1 in the genome of selected transformants using the same protocol as described in Section 2.5.

2.8. Quantitative and Semiquantitative Reverse Transcription Polymerase Chain Reaction

Total RNA of JEC21, OE-1, and OE-2 was extracted as described in Section 2.2. Reverse transcription (RT) of total RNA was conducted by Fast Quant RT kit with gDNase (Tiangen Biotech). Briefly, 1 μL total RNA, 2 μL 5× *g* DNA buffer, and 7 μL ddH$_2$O were incubated at 42 °C for 10 min, then 2 μL 10× Fast RT Buffer, 1 μL RT enzyme mix, 2 μL Fast Quant RT primer mix, and 5 RNase-Free ddH$_2$O were added to previous tubes and incubated at 42 °C for 15 min. The reaction was stopped by incubating at 95 °C for 10 s. For *DBR1* mRNA quantification, LightCycler 480 II and corresponding LC 480 SYBR Green I Master (Roche, Basel, Switzerland) were employed. The PCR reaction system included 10 μL 2× Master Mix, 1 μL forward/reverse primers (10 μm), 1 μL cDNA, and 7 μL ddH$_2$O. Each reaction was performed in triplicate. Non-RT RNA was used as a template in negative control and actin mRNA served as reference. Specificity of primers was validated by checking the melting curves. The $2^{-\Delta\Delta Ct}$ method was employed to calculate expression levels of target genes in this study. Semiquantitative reverse transcription PCR was performed using the PCR protocol described in Section 2.5, except only 25 circles were applied.

3. Results

3.1. Genomewide Identification of Circular RNAs

To investigate circRNAs on a genome-wide level, we isolated total RNAs from the *C. neoformans* JEC21 strain. After eliminating ribosome RNAs (rRNAs) and treating with RNase R, the total RNA was utilized to construct libraries for deep sequencing by the Illumina HighSeq 2500 platform. The sequencing data reached 6.26 Giga nucleotides (Gnt) raw bases in total, covering 41.70 million paired-end individual reads sized above 150 nt. After trimming adaptors and filtering low-quality reads, we obtained 39.28 million clean reads (Table 1).

Table 1. RNA-sequencing data of *Cryptococcus neoformans* JEC1.

Sample Name	Raw Reads	Filtered Reads	Raw Base	Filtered Base	Q20 (%) [1]
Cn JEC21	41,703,834	39,280,024	6.26 Gnt	6.16 Gnt	98.52

[1] Q20 refers to the percentage of nucleotides with Phred quality score > 20, which means base accuracy is 99%. Gnt: Giga nucleotides.

Clean reads were then mapped to the *C. neoformans* JEC21 genome by BWA software. The mapped reads were input to CIRI, a published circRNA identifier, to identify the candidates of circRNAs. To reduce false-positive candidates, the circRNAs that had more than one back-splicing junction read were considered. After a two-step filtration, 73 individual circRNAs containing high-confident back-splicing junctions were obtained. The number of reads for the 73 unique circRNAs was counted to 820. Only 20 of the 73 circRNAs (27.4%) had more than four back-splicing junction reads. The 10 with the highest junction reads are listed (Table 2) and detailed information on all predicted circRNAs is available (Supplementary Table S2). The above data show that the absolute number of unique cryptococcal circRNAs is low compared to that of circRNAs in higher eukaryotes, such as animals or plants. Specifically, researchers have detected 3001 circRNAs from human cells [27] and 5372 circRNAs from soybeans [28]. However, when referring to the relative expression levels using the ratio of circRNAs number to genome size (Mb), the results changed in which the relative expression of *C. neoformans* circRNAs (~3.74) is much higher than that of human (~1.00), but a little lower than soybean (~4.88).

Table 2. Detailed information on the 10 circRNAs with the highest back-splicing reads.

circRNA ID	Chr	RNA Size	circRNA Start Loci	CircRNA End Loci	Junction Reads [1]
12:174494-175325	12	831	174494	175325	410
13:359406-359654	13	248	359406	359654	67
7:1027082-1027487	7	405	1027082	1027487	29
13:603431-604144	13	713	603431	604144	28
4:1303545-1304160	4	615	1303545	1304160	24
12:174265-175325	12	1060	174265	175325	23
12:174461-175325	12	864	174461	175325	22
11:62574-63095	11	521	62574	63095	11
13:89597-90783	13	1186	89597	90783	8
2:661847-663003	2	1156	661847	663003	7

[1] Junction reads means counts of back-splicing reads. Chr: Chromosome.

We sorted the unique circRNAs into three groups according to the positioning of their two ends on chromosomes (exonic, intronic, and intergenic regions). Among them, 54 (73.97%) of the 73 circRNAs were generated from exons of protein-coding open reading frames (ORFs) and 15 (20.55%) were intergenic circRNAs. Only four (5.48%) had intronic junctions. Besides unique circRNAs, we also calculated the total reads of each type of circRNA. Our data show that 38.17, 1.46, and 60.37% of the total 820 reads were distributed to exonic, intronic, and intergenic circRNAs, respectively (Figure 1). However, exons, introns and intergenic sequences occupy 54.14, 11.97, and 33.89%, respectively, of the whole *C. neoformans* genome [20]. Thus, these results reveal that intergenic circRNAs have higher mean reads than exonic circRNAs, although the latter consists of the majority of unique circRNAs.

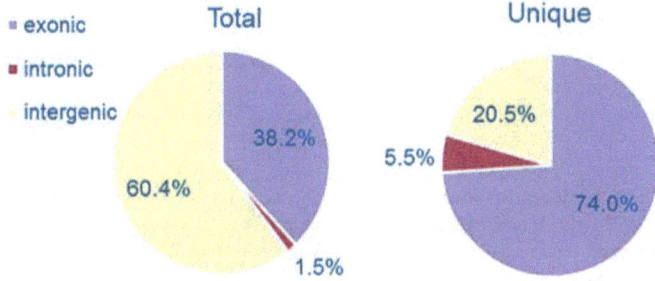

Figure 1. Percentages of three groups of circular RNAs. The circRNAs were classified as exonic, intronic, and intergenic according to the back-splicing junction position on chromosomes. Total circRNAs, calculated as back-splicing junction reads, are shown in the left panel, while unique circRNAs are shown in the right panel.

3.2. Properties of Cryptococcal Circular RNAs

In order to determine the properties of cryptococcal circRNAs, we performed a set of counting calculations for unique and total circRNAs respectively. Firstly, chromosomal distribution for unique and total circRNAs was examined. According to our analysis, 461 total reads were located on chromosome 12. The reason is simple: the highest-expressed circRNA, circ12:174494-175325 (410 reads), was found on Chr12. Correspondingly, chromosome 8 contains the least amount of total back-splicing junction reads, which is only four (Figure 2a, upper panel). The distribution of unique circRNAs among the 14 chromosomes is also displayed in Figure 2a, bottom panel. We found that chromosome 8 contains the least amount of unique circRNAs, two, and chromosomes 1 and 4 contain 10 each.

Secondly, we examined the size distribution of cryptococcal circRNAs. For unique circRNAs, the length was mostly (72.60%) between 201 and 800 nt (Figure 2b). Only a few unique circRNAs were <200 nt (1.37%) or >1400 nt (6.85%). As for total reads, the length concentrated on 801–1000 nt as the size of circ12:174494-175325, which possesses the most reads, is 831 nt. Also, only two reads in total circRNAs were found <200 nt (0.24%, totally) and 17 were >1400 nt (2.07%, totally). Finally, we normalized the expression of each unique circRNA to spliced reads per billion mapping (SRPBM), in order to analyze their expressional features. SRPBM of most circRNAs (63.01%) was <50, while only two circRNAs (2.74%) had SRPBM >500 (Figure 2c).

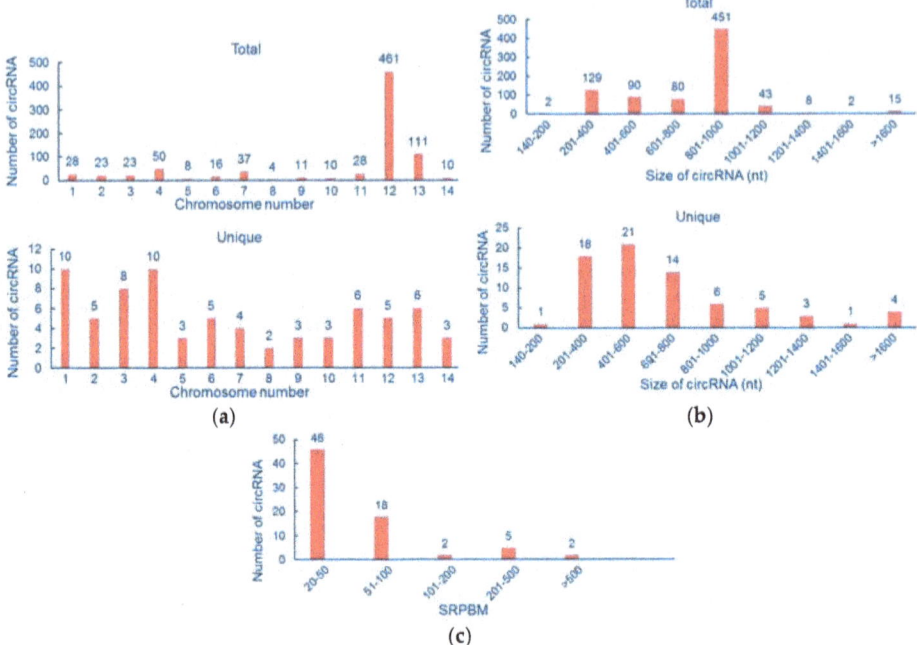

Figure 2. Counting calculations of circRNAs in *C. neoformans*. (**a**) Chromosome distribution of unique circRNAs (bottom panel) and total reads (upper panel). (**b**) Size distribution of unique circRNAs (bottom panel) and total reads (upper panel). (**c**) Expression level distribution of unique circRNAs. SRPBM: spliced reads per billion mapping.

3.3. Functional Analysis of circRNA Host Genes

To investigate the function of circRNA host genes, we performed GO analysis and KEGG pathway analysis. GO analysis suggested that circRNA host genes mainly encode proteins of the large ribosomal subunit, cell surface proteins, and plasma membrane proteins (p-value < 0.05). For GO molecular function analysis, circRNA host genes were associated with beta-glucosidase activity and structural constituents of ribosome (p-value < 0.05). For the GO biological process, those genes were enriched in translation, glucan catabolic process, arginine transport, and fungal cell wall organization (p-value < 0.05) (Figure 3). As for distribution in KEGG pathways, the results show that circRNA host genes were significantly (p-value < 0.05) enriched in two pathways: The ribosome biogenesis and starch-sucrose metabolism pathways (Table 3).

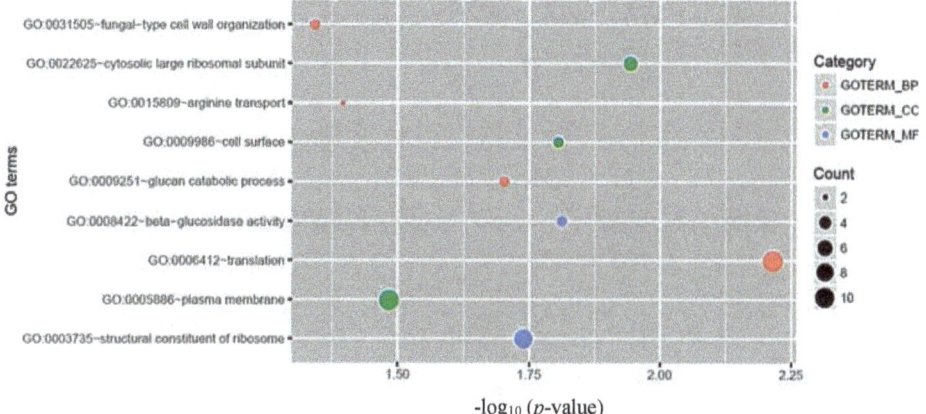

Figure 3. Gene Ontology (GO) category analysis of circRNA host genes in *C. neoformans*. GO terms with the threshold of *p*-value < 0.05 are listed. GO terms were classified in three categories: Biological process (BP), cellular component (CC), and molecular function (MF).

Table 3. Kyoto Encyclopedia of Genes and Genomes pathway enrichment analysis of circRNA-host genes in *C. neoformans*.

Pathway ID [1]	Description	Gene Count	*p*-Value
cne03010	Ribosome	6	0.023
cne00500	Starch and sucrose metabolism	3	0.031

[1] Pathway with the threshold of *p*-value < 0.05 is listed.

3.4. Small Guanosine Triphosphatase-Encoding Orthologs Are Conserved circRNA Hosts

In order to validate the existence of circRNAs, we performed verification by PCR amplification with a pair of outward-designed primers and a cDNA template. In total, primers and PCR reaction systems for six potential circRNAs were designed, including the circ13:359406-359654 whose expression was the highest among all exonic circRNAs (Supplementary Table S1). Consequently, only two specific bands were obtained, respectively, for circ13:359406-359654 and circ7:1027082-1027487. To further confirm the results, we successfully got the PCR band with expected length (248 bp) for circ13:359406-359654 (Figure 4a). Not surprisingly, we failed to observe a corresponding band at the exact level in the control reaction, in which genomic DNA was used as a template. Subsequent Sanger sequencing confirmed that it contained the predicted back-splicing junction site (Figure 4b).

According to genomic distribution data, circ13:359406-359654 was derived from the 5' end of the gene *CNM01190* (National Center for Biotechnology Information [NCBI] ID: 3255052). By virtue of splicing information stored in the Ensemble Fungi database, we found that the gene *CNM01190* has two splice variants, which encode either a short version (174 aminoacides (aa)) or a long version (252 aa) of protein (Figure 4c). Through a conserved domain basic local alignment search tool (BLAST) search [29], we found that the long variant product encoded by *CNM01190* belongs to the Rab-related GTPase family, which is exemplified by *YPT5* in *Schizosaccharomyces pombe*. Given this, we named the cryptococcal circRNA circCNYPT5. We then used Clustal Omega to compare amino acid sequences of *YPT5* and *CNM01190* long version and found that their identity rate was as high as 57.62% (Figure 4d). Surprisingly, *YPT5* has also been reported to generate circRNAs in *S. pombe*, residing at the exonic regions instead of the 5'-untranslated region (5'-UTR) [17,30]. Therefore, we also performed DNA sequence alignments between the two homologous genes with the BLAST algorithm available at the

NCBI website, but no significant similarity was found. In general, our data suggest a conservation of circRNA host genes in yeasts.

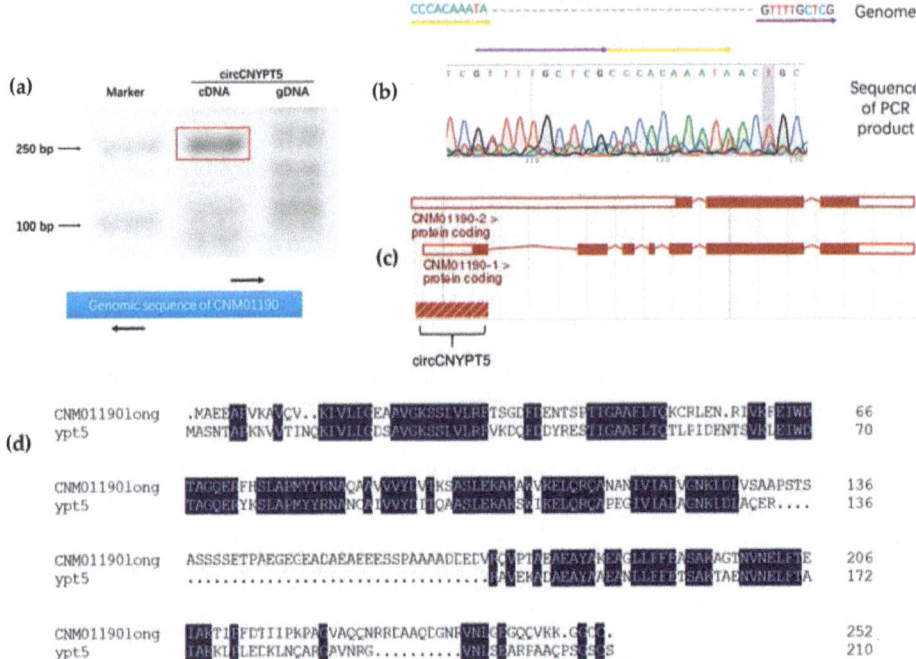

Figure 4. Small GTPase-encoding genes are conserved in circRNA origination among yeasts. (**a**) The upper panel shows that circCNYPT5 can be amplified using outward polymerase chain reactions (PCRs). By contrast, PCR using a genomic DNA template was unable to produce a corresponding band at the same level. Primers in the amplification (black arrows) are shown in the bottom panel. (**b**) Sanger sequencing confirmed the back-splicing junction in PCR products. The 5′ (yellow arrow) and 3′ (purple arrow) were found connected according to sequencing data. (**c**) circCNYPT5 is derived from the 5′ region of *CNM01190*, which has two splice variants. (**d**) Multiple alignments of amino acid sequences showed a high identity between proteins encoded by *CNM01190* in *C. neoformans* and *YPT5* in *Schizosaccharomyces pombe*. Identical amino acids between the two proteins are highlighted in dark blue.

3.5. DBR1 Expression Level Is Negatively Associated with circDPEPS but Not circCNYPT5

According to the lariat-driven model, circRNAs are derived from exon-containing lariat precursors. Inhibiting Dbr1 protein, which degrades lariat, could lead to increased circRMA levels. The expression level of global circRNAs in the *DBR1Δ* mutant strain increased by three- to four-fold over the level in the wild-type (WT) *S. pombe* [31]. To investigate the function of *DBR1* in circRNA regulation in *C. neoformans*, we tried to knock out the counterpart *DBR1* gene in *C. neoformans* JEC21 with the latest clustered regularly interspaced short palindromic repeats (CRISPR)-Cas9 editing system [32,33]. Unfortunately, we failed to get any *DBR1* knockout strain, which implies that the *DBR1* gene might be essential in serotype D strains. As a solution, we overexpressed the *DBR1* gene in JEC21. The overexpression (OE) vector pBS-HYG-DBR1 was detected in transformants OE-1 and OE-2, but not WT, through amplification of a hygromycin-resistant fragment (Figure 5a). Quantitative reverse-transcription (RT) PCR showed that *DBR1* mRNA increased up to 4.89- and 3.76-fold in OE-1 and OE-2, respectively, compared to WT (Figure 5b).

We speculated that overexpression of *DBR1* could reduce circRNA levels in *C. neoformans*. Thus, we analyzed the expression levels of circCNYPT5 in OE-1, OE-2, and WT by semiquantitative RT-PCR. To our surprise, no significant difference in circCNYPT5 expression level was found between WT and overexpression strains according to the electrophoretogram (Olympus, Tokyo, Japan). On the other hand, another highly expressed circular RNA, circDPEPS, which is derived from the whole sequence of gene *CNG03660* (putatively encoding DNA polymerase epsilon p12 subunit, gene ID: 3258898), showed a dramatically decreased level in both overexpression strains (Figure 5c). One possible explanation for these divergent results is that circDPEPS might be processed through an exon-containing lariat precursor, while the generation of circCNYPT5 might rely on a lariat-independent mechanism. On the other hand, these two circRNAs might both be regulated by the *DBR1* gene which plays an important role upstream, but the different downstream action elements may contribute to the wildly different results, which requires further investigation.

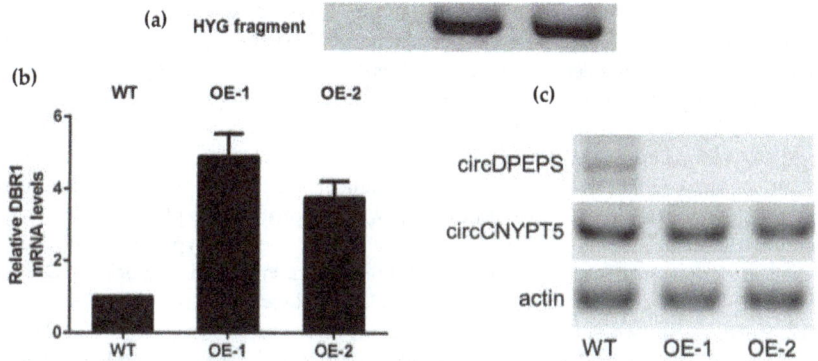

Figure 5. Overexpression (OE) of *DBR1* has a diverse influence on circDPEPS and circCNYPT5. (**a**) PCR was performed to confirm that *DBR1* overexpression of plasmid pBS-HYG-DBR1 was transformed into two transformants, OE-1 and OE-2 (lane 2 and lane 3). JEC21 (wild-type, WT) was used as control. (**b**) Quantitative reverse-transcription (RT) PCR analysis showed that *DBR1* expression level increased up to 4.89- and 3.76-fold in OE-1 and OE-2 compared to WT. Error bars show standard error of the mean. (**c**) Semiquantitative RT-PCR analysis shows that the circDPEPS level declined significantly in the OE strains, while circCNYPT5 remained at a similar level in all three strains. Actin-encoding gene *ACT1* mRNA was used as an internal control.

4. Discussion

In the present study, we report the results of a genomewide screening for circRNAs in the basidiomycetous yeast *C. neoformans* serotype D JEC21 using RNA-seq with bioinformatics analysis. A total of 73 unique circRNAs—including 54 exonic, 4 intronic, and 15 intergenic circRNAs—were identified by the CIRI algorithm. Considering the number of exonic circRNAs and intronic circRNAs, we got 58 predicted circRNA host genes, which was near the number obtained from *S. pombe* (42 genes) [30], but notably less than higher eukaryotes [27,28,34]. The reason might be that different biosynthesis paths are adopted in various organisms. For instance, neural circular RNAs were generated in a spliceosome-dependent manner [35], whereas lariat precursors, which are byproducts in an exon-skipping event, are required for the production of circRNAs in *S. pombe* [17].

Furthermore, we conducted an analysis of the functions of circRNA host genes by GO and KEGG pathways. According to the top-rank rule in GO annotation, we found that circRNA host genes are enriched in encoding structural proteins of the ribosome, plasma membrane, and cell wall, all of which are important in routine growth of the yeast. In addition, we found that the ribosome and starch–sucrose metabolism pathways are the most enriched pathways in the KEGG analysis.

However, the relationship between circRNAs and their host genes still remains elusive due to the limitations of the mutagenesis strategy in the study of circRNAs. CircRNA biosynthesis can compete with pre-mRNA splicing in some cases [36]. On the other hand, some circular intronic RNAs (ciRNAs), such as Ci-ankrd52, are able to promote the transcription of corresponding genes [37]. Additionally, circRNAs in plants like rice and tomato usually exhibit developmental specificity [38,39], while soybean circRNAs show mainly tissue specificity [28].

Interestingly, we found that Rab GTPase-encoding orthologs are conserved circRNA hosts in *C. neoformans* and *S. pombe*. In fact, previous studies reported GTPase-derived circRNAs in protists [30] and humans [40]. Given the low occurrence of unique circRNAs in yeasts, these results may have meaningful implication as to their biogenesis and the function of circRNAs. In our yeast, the host gene, *CNM01190* (NCBI ID: 3255052), has two alternative splicing products that putatively encode 174-aa and 252-aa proteins separately. The circCNYPT5 was derived from the 5′-UTR region of the gene and the first exon in the longer variant. Whether the formation of circCNYPT5 is associated with the alternative splicing process is an intriguing question. The long version of *CNM01190* encodes a member of the small GTPase family. Many of the members of the family have been shown to play vital roles in pathogenesis, thermotolerance, mating, and septin localization in *C. neoformans* [41–43]. Thus, circCNYPT5 that originated from the 5′-UTR region of the gene could presumably act as a profound regulator of the transcription of the *YPT5* gene. Biogenesis of circRNAs within the conserved host ortholog genes across the eukaryotic domain of life raises a question for further investigation.

Finally, we showed that the RNA lariat debranching enzyme (*DBR1*) has variable influence on different circRNA expression in *C. neoformans*. *DBR1* was demonstrated to play key roles in circRNA biogenesis by the lariat-driven pathway in fission yeast [17]. Whether cryptococcal *DBR1* mediates circRNAs biosynthesis in a fission yeast style is still unknown. We failed to knock out *DBR1* in the *C. neoformans* JEC21 serotype D strain, although it could be deleted in the serotype A H99 strain [44]. This phenomenon suggests diverse functions of this gene. Overexpression of *DBR1* in the JEC21 strain decreased circDPEPS levels but had no effect on circCNYPT5 levels. We also attempted to detect other potential circRNAs with PCR amplification but failed, maybe due to the background disturbance from rRNA and linear RNA which were removed from the RNA samples in the RNA-seq. Our data indicate that circCNYPT5 is not processed through an exon-containing lariat precursor. As to the biogenesis of circCNYPT5, the intron-driven model may not be applicable in this case, as no intronic secondary structure was predictable by Mfold software [45]. Whether circCNYPT5 is produced through a spliceosome-dependent path or a protein factors-associated path needs to be investigated. Consistent with our research, a recent RNA-seq study revealed an approximately three- to four-fold increase in circRNAs when *DBR1* was mutated in *S. pombe* [31]. However, researchers also found that the effect of *DBR1* deletion on particular genes seems to have statistical significance [17]. As for circDPEPS, it might be produced through a classic lariat-precursor pathway. However, there is still the possibility that this circRNA derives from other pathways. For instance, the *DBR1* gene may have an indirect influence on circDPEPS expression due to the multiple functions of *DBR1*. It is noteworthy that, according to recent work on *C. neoformans* serotype A H99, *DBR1* is indispensable in the biosynthesis of some transposon-derived small interference RNAs (siRNAs) [44]. In sum, the exact cryptococcal circRNA biosynthesis pathway, along with the comprehensive function of *DBR1* in *C. neoformans*, may need further investigation.

5. Conclusions

In the present study, we identified 73 unique circular RNAs from basidiomycetous yeast *C. neoformans* using RNA-seq and bioinformatics tools such as CIRI. The function of circRNA host genes enriches in primary metabolism, especially translation and carbohydrate metabolism. In addition, we found that the small GTPase ortholog genes are conserved hosts of circRNAs among eukaryotic organisms. Primary analysis revealed that the cryptococcal *DBR1* gene has a differential impact on the generation of different circRNAs. Our study on the identification of circRNAs opens an avenue to understanding their biological function in the pathogenesis of this pathogen.

Supplementary Materials: The following are available online at http://www.mdpi.com/2073-4425/9/3/118/s1. Table S1: Primers used in this study. Table S2: List of all predicted circRNAs in C. neoformans JEC21.

Acknowledgments: This work was partially funded by a grant from the National Science Foundation of China (#31470251).

Author Contributions: X.Z. conceived and designed the experiments; L.H., P.Z., and C.L. performed the experiments and conducted bioinformatics; K.R. and X.H. participated in analysis of the data; B.X. contributed reagents and materials; L.H. and P.Z. wrote the paper.

Conflicts of Interest: The authors declare no conflict of interest.

References

1. Salzman, J. Circular RNA expression: Its potential regulation and function. *Trends. Genet.* **2016**, *32*, 309–316. [CrossRef] [PubMed]
2. Jeck, W.R.; Sharpless, N.E. Detecting and characterizing circular RNAs. *Nat. Biotechnol.* **2014**, *32*, 453–461. [CrossRef] [PubMed]
3. Chen, L.L.; Yang, L. Regulation of circRNA biogenesis. *RNA Biol.* **2015**, *12*, 381–388. [CrossRef] [PubMed]
4. Suzuki, H.; Zuo, Y.; Wang, J.; Zhang, M.Q.; Malhotra, A.; Mayeda, A. Characterization of RNase R-digested cellular RNA source that consists of lariat and circular RNAs from pre-mRNA splicing. *Nucleic Acids Res.* **2006**, *34*, e63. [CrossRef] [PubMed]
5. Sanger, H.L.; Klotz, G.; Riesner, D.; Gross, H.J.; Kleinschmidt, A.K. Viroids are single-stranded covalently closed circular RNA molecules existing as highly base-paired rod-like structures. *Proc. Natl. Acad. Sci. USA* **1976**, *73*, 3852–3856. [CrossRef] [PubMed]
6. Liu, J.; Liu, T.; Wang, X.; He, A. Circles reshaping the RNA world: From waste to treasure. *Mol. Cancer* **2017**, *16*, 58. [CrossRef] [PubMed]
7. Salzman, J.; Gawad, C.; Wang, P.L.; Lacayo, N.; Brown, P.O. Circular RNAs are the predominant transcript isoform from hundreds of human genes in diverse cell types. *PLoS ONE* **2012**, *7*, e30733. [CrossRef] [PubMed]
8. Hansen, T.B.; Jensen, T.I.; Clausen, B.H.; Bramsen, J.B.; Finsen, B.; Damgaard, C.K.; Kjems, J. Natural RNA circles function as efficient microRNA sponges. *Nature* **2013**, *495*, 384–388. [CrossRef] [PubMed]
9. Gao, Y.; Wang, J.; Zhao, F. Ciri: An efficient and unbiased algorithm for de novo circular RNA identification. *Genome Biol.* **2015**, *16*, 4. [CrossRef] [PubMed]
10. Zhang, X.O.; Wang, H.B.; Zhang, Y.; Lu, X.; Chen, L.L.; Yang, L. Complementary sequence-mediated exon circularization. *Cell* **2014**, *159*, 134–147. [CrossRef] [PubMed]
11. Szabo, L.; Morey, R.; Palpant, N.J.; Wang, P.L.; Afari, N.; Jiang, C.; Parast, M.M.; Murry, C.E.; Laurent, L.C.; Salzman, J. Statistically based splicing detection reveals neural enrichment and tissue-specific induction of circular RNA during human fetal development. *Genome Biol.* **2015**, *16*, 126. [CrossRef] [PubMed]
12. Zhao, Y.; Alexandrov, P.N.; Jaber, V.; Lukiw, W.J. Deficiency in the ubiquitin conjugating enzyme UBE2A in Alzheimer's Disease (AD) is linked to deficits in a natural circular miRNA-7 sponge (circRNA; ciRS-7). *Genes* **2016**, *7*, 116. [CrossRef] [PubMed]
13. Chen, L.L. The biogenesis and emerging roles of circular RNAs. *Nat. Rev. Mol. Cell Biol.* **2016**, *17*, 205–211. [CrossRef] [PubMed]
14. Lasda, E.; Parker, R. Circular RNAs: Diversity of form and function. *RNA* **2014**, *20*, 1829–1842. [CrossRef] [PubMed]
15. Liang, D.; Wilusz, J.E. Short intronic repeat sequences facilitate circular RNA production. *Genes Dev.* **2014**, *28*, 2233–2247. [CrossRef] [PubMed]
16. Conn, S.J.; Pillman, K.A.; Toubia, J.; Conn, V.M.; Salmanidis, M.; Phillips, C.A.; Roslan, S.; Schreiber, A.W.; Gregory, P.A.; Goodall, G.J. The RNA binding protein quaking regulates formation of circRNAs. *Cell* **2015**, *160*, 1125–1134. [CrossRef] [PubMed]
17. Barrett, S.P.; Wang, P.L.; Salzman, J. Circular RNA biosynthesis can proceed through an exon-containing lariat precursor. *eLife* **2015**, *4*, e07540. [CrossRef] [PubMed]

18. Huang, S.; Yang, B.; Chen, B.J.; Bliim, N.; Ueberham, U.; Arendt, T.; Janitz, M. The emerging role of circular RNAs in transcriptome regulation. *Genomics* **2017**, *109*, 401–407. [CrossRef] [PubMed]

19. Granados-Riveron, J.T.; Aquino-Jarquin, G. The complexity of the translation ability of circRNAs. *Biochim. Biophys. Acta* **2016**, *1859*, 1245–1251. [CrossRef] [PubMed]

20. Loftus, B.J.; Fung, E.; Roncaglia, P.; Rowley, D.; Amedeo, P.; Bruno, D.; Vamathevan, J.; Miranda, M.; Anderson, I.J.; Fraser, J.A.; et al. The genome of the basidiomycetous yeast and human pathogen *Cryptococcus neoformans*. *Science* **2005**, *307*, 1321–1324. [CrossRef] [PubMed]

21. Patel, R.K.; Jain, M. NGS QC toolkit: a toolkit for quality control of next generation sequencing data. *PLoS ONE* **2012**, *7*, e30619. [CrossRef] [PubMed]

22. Li, H.; Durbin, R. Fast and accurate long-read alignment with burrows-wheeler transform. *Bioinformatics* **2010**, *26*, 589–595. [CrossRef] [PubMed]

23. Li, Y.; Zheng, Q.; Bao, C.; Li, S.; Guo, W.; Zhao, J.; Chen, D.; Gu, J.; He, X.; Huang, S. Circular RNA is enriched and stable in exosomes: A promising biomarker for cancer diagnosis. *Cell Res.* **2015**, *25*, 981–984. [CrossRef] [PubMed]

24. Huang da, W.; Sherman, B.T.; Lempicki, R.A. Systematic and integrative analysis of large gene lists using DAVID bioinformatics resources. *Nat. Protoc.* **2009**, *4*, 44–57. [CrossRef] [PubMed]

25. Xie, C.; Mao, X.; Huang, J.; Ding, Y.; Wu, J.; Dong, S.; Kong, L.; Gao, G.; Li, C.Y.; Wei, L. KOABS 2.0: A web server for annotation and identification of enriched pathways and diseases. *Nucleic Acids Res.* **2011**, *39*, W316–W322. [CrossRef] [PubMed]

26. Sievers, F.; Wilm, A.; Dineen, D.; Gibson, T.J.; Karplus, K.; Li, W.; Lopez, R.; McWilliam, H.; Remmert, M.; Soding, J.; et al. Fast, scalable generation of high-quality protein multiple sequence alignments using clustal omega. *Mol. Syst. Biol.* **2011**, *7*, 539. [CrossRef] [PubMed]

27. Shen, Y.; Guo, X.; Wang, W. Identification and characterization of circular RNAs in zebrafish. *FEBS Lett.* **2017**, *591*, 213–220. [CrossRef] [PubMed]

28. Zhao, W.; Cheng, Y.; Zhang, C.; You, Q.; Shen, X.; Guo, W.; Jiao, Y. Genome-wide identification and characterization of circular RNAs by high throughput sequencing in soybean. *Sci. Rep.* **2017**, *7*, 5636. [CrossRef] [PubMed]

29. Johnson, M.; Zaretskaya, I.; Raytselis, Y.; Merezhuk, Y.; McGinnis, S.; Madden, T.L. NCBI BLAST: A better web interface. *Nucleic Acids Res.* **2008**, *36*, W5–W9. [CrossRef] [PubMed]

30. Wang, P.L.; Bao, Y.; Yee, M.C.; Barrett, S.P.; Hogan, G.J.; Olsen, M.N.; Dinneny, J.R.; Brown, P.O.; Salzman, J. Circular RNA is expressed across the eukaryotic tree of life. *PLoS ONE* **2014**, *9*, e90859. [CrossRef] [PubMed]

31. Bitton, D.A.; Atkinson, S.R.; Rallis, C.; Smith, G.C.; Ellis, D.A.; Chen, Y.Y.; Malecki, M.; Codlin, S.; Lemay, J.F.; Cotobal, C.; et al. Widespread exon skipping triggers degradation by nuclear RNA surveillance in fission yeast. *Genome Res.* **2015**, *25*, 884–896. [CrossRef] [PubMed]

32. Wang, Y.; Wei, D.; Zhu, X.; Pan, J.; Zhang, P.; Huo, L.; Zhu, X. A 'suicide' CRISPR-Cas9 system to promote gene deletion and restoration by electroporation in *Cryptococcus neoformans*. *Sci. Rep.* **2016**, *6*, 31145. [CrossRef] [PubMed]

33. Rahim, K.; Huo, L.; Li, C.; Zhang, P.; Basit, A.; Xiang, B.; Ting, B.; Hao, X.; Zhu, X. Identification of a basidiomycete-specific Vilse-like GTPase activating proteins (GAPs) and its roles in the production of virulence factors in *Cryptococcus neoformans*. *FEMS Yeast Res.* **2017**, *17*. [CrossRef] [PubMed]

34. Lu, T.; Cui, L.; Zhou, Y.; Zhu, C.; Fan, D.; Gong, H.; Zhao, Q.; Zhou, C.; Zhao, Y.; Lu, D.; et al. Transcriptome-wide investigation of circular RNAs in rice. *RNA* **2015**, *21*, 2076–2087. [CrossRef] [PubMed]

35. You, X.; Vlatkovic, I.; Babic, A.; Will, T.; Epstein, I.; Tushev, G.; Akbalik, G.; Wang, M.; Glock, C.; Quedenau, C.; et al. Neural circular RNAs are derived from synaptic genes and regulated by development and plasticity. *Nat. Neurosci.* **2015**, *18*, 603–610. [CrossRef] [PubMed]

36. Ashwal-Fluss, R.; Meyer, M.; Pamudurti, N.R.; Ivanov, A.; Bartok, O.; Hanan, M.; Evantal, N.; Memczak, S.; Rajewsky, N.; Kadener, S. CircRNA biosynthesis competes with pre-mRNA splicing. *Mol. Cell* **2014**, *56*, 55–66. [CrossRef] [PubMed]

37. Zhang, Y.; Zhang, X.O.; Chen, T.; Xiang, J.F.; Yin, Q.F.; Xing, Y.H.; Zhu, S.; Yang, L.; Chen, L.L. Circular intronic long noncoding RNAs. *Mol. Cell* **2013**, *51*, 792–806. [CrossRef] [PubMed]

38. Ye, C.Y.; Chen, L.; Liu, C.; Zhu, Q.H.; Fan, L. Widespread noncoding circular RNAs in plants. *New Phytol.* **2015**, *208*, 88–95. [CrossRef] [PubMed]

39. Tan, J.; Zhou, Z.; Niu, Y.; Sun, X.; Deng, Z. Identification and functional characterization of tomato circRNAs derived from genes involved in fruit pigment accumulation. *Sci. Rep.* **2017**, *7*, 8594. [CrossRef] [PubMed]
40. Siede, D.; Rapti, K.; Gorska, A.A.; Katus, H.A.; Altmuller, J.; Boeckel, J.N.; Meder, B.; Maack, C.; Volkers, M.; Muller, O.J.; et al. Identification of circular RNAs with host gene-independent expression in human model systems for cardiac differentiation and disease. *J. Mol. Cell Cardiol.* **2017**, *109*, 48–56. [CrossRef] [PubMed]
41. Ballou, E.R.; Selvig, K.; Narloch, J.L.; Nichols, C.B.; Alspaugh, J.A. Two Rac paralogs regulate polarized growth in the human fungal pathogen *Cryptococcus neoformans*. *Fungal Genet. Biol.* **2013**, *57*, 58–75. [CrossRef] [PubMed]
42. Ballou, E.R.; Nichols, C.B.; Miglia, K.J.; Kozubowski, L.; Alspaugh, J.A. Two *CDC42* paralogues modulate *Cryptococcus neoformans* thermotolerance and morphogenesis under host physiological conditions. *Mol. Microbiol.* **2010**, *75*, 763–780. [CrossRef] [PubMed]
43. Ballou, E.R.; Kozubowski, L.; Nichols, C.B.; Alspaugh, J.A. Ras1 acts through duplicated Cdc42 and Rac proteins to regulate morphogenesis and pathogenesis in the human fungal pathogen *Cryptococcus neoformans*. *PLoS Genet.* **2013**, *9*, e1003687. [CrossRef] [PubMed]
44. Dumesic, P.A.; Natarajan, P.; Chen, C.; Drinnenberg, I.A.; Schiller, B.J.; Thompson, J.; Moresco, J.J.; Yates, J.R., 3rd; Bartel, D.P.; Madhani, H.D. Stalled spliceosomes are a signal for RNAi-mediated genome defense. *Cell* **2013**, *152*, 957–968. [CrossRef] [PubMed]
45. Zuker, M. Mfold web server for nucleic acid folding and hybridization prediction. *Nucleic Acids Res.* **2003**, *31*, 3406–3415. [CrossRef] [PubMed]

Article

Portrait of Matrix Gene Expression in *Candida glabrata* Biofilms with Stress Induced by Different Drugs

Célia F. Rodrigues [ID] and Mariana Henriques * [ID]

Laboratório de Investigação em Biofilmes Rosário Oliveira (LIBRO), Centre of Biological Engineering, University of Minho, 4710-057 Braga, Portugal; rodriguescf@ceb.uminho.pt
* Correspondence: mcrh@deb.uminho.pt; Tel.: +351-253-604401; Fax: +351-253-604429

Received: 6 March 2018; Accepted: 5 April 2018; Published: 10 April 2018

Abstract: (1) Background: *Candida glabrata* is one of the most significant *Candida* species associated with severe cases of candidiasis. Biofilm formation is an important feature, closely associated with antifungal resistance, involving alterations of gene expression or mutations, which can result in the failure of antifungal treatments. Hence, the main goal of this work was to evaluate the role of a set of genes, associated with matrix production, in the resistance of *C. glabrata* biofilms to antifungal drugs. (2) Methods: the determination of the expression of *BGL2*, *XOG1*, *FKS1*, *FKS2*, *GAS2*, *KNH1*, *UGP1*, and *MNN2* genes in 48-h biofilm's cells of three *C. glabrata* strains was performed through quantitative real-time PCR (RT-qPCR), after contact with Fluconazole (Flu), Amphotericin B (AmB), Caspofungin (Csf), or Micafungin (Mcf). (3) Results: Mcf induced a general overexpression of the selected genes. It was verified that the genes related to the production of β-1,3-glucans (*BGL2*, *XOG1*, *GAS2*) had the highest expressions. (4) Conclusion: though β-1,6-glucans and mannans are an essential part of the cell and biofilm matrix, *C. glabrata* biofilm cells seem to contribute more to the replacement of β-1,3-glucans. Thus, these biopolymers seem to have a greater impact on the biofilm matrix composition and, consequently, a role in the biofilm resistance to antifungal drugs.

Keywords: *Candida*; biofilms; matrix; drug resistance; gene expression; *Candida glabrata*

1. Introduction

Fungal infections continue to increase worldwide, particularly among immunosuppressed patients, individuals under prolonged hospitalization, catheterization, or continued antimicrobial treatments [1–3]. *Candida* spp. are the commonest fungal species involved in these diseases. *Candida albicans* is the most isolated species, but *Candida glabrata* and *Candida parapsilosis* are the second most isolated species in the United States of America and Europe, respectively [1,4,5]. Though *C. glabrata* does not have the capacity to form hyphae and pseudohyphae or to secret proteases, this species has other virulence factors, such as the ability to secrete phospholipases, lipases, and haemolysins and, importantly, the capacity to form biofilms [6–8]. These factors highly contribute to a high aggressiveness, resulting in a low therapeutic response and severe cases of recurrent candidiasis [8,9]. Biofilms are communities of microorganisms that colonize tissues and indwelling medical devices, embedded in an extracellular matrix [10,11]. These heterogeneous structures provide high resistance to antifungal therapy and strong host immune responses [7,8,12]. *C. glabrata* has shown to form a compact biofilm structure in different multilayers [6,7], with proteins, carbohydrates, and ergosterol into their matrices [6,7,13].

Various reports have shown the presence of β-1,3 glucans in the biofilm matrices of *C. albicans* [14–17]. Interestingly, it has been demonstrated that an increase in cell wall glucan was associated with biofilm growth [14] and, more recently, β-1,3 glucans were shown to be also present in the matrices of *C. glabrata* biofilms [13,18,19]. This specific carbohydrate has been associated with a general increase of extracellular matrix delivery, which is critical for securing biofilm cells to a surface and crucial to develop an antifungal drug resistance phenotype [14,19–23]. Several genes are involved in the delivery and accumulation of extracellular matrix. It is recognized that, in *C. albicans*, the major β-1,3 glucan synthases are encoded mainly by *FKS1* but also by *FKS2* [24]. The *BGL2* and *XOG1* genes also have important roles in glucan matrix delivery by encoding glucanosyltransferases and β-1,3 exoglucanase, respectively [25,26]. These genes play an important part in cell wall remodeling, however, the influence of the corresponding enzymes in matrix glucan delivery does not appear to affect cell wall ultrastructure or β-1,3 glucan concentration, suggesting that these enzymes function specifically for matrix delivery [17,19,26–28]. Identical to *Saccharomyces cerevisiae*, in *C. glabrata*, the *GAS* gene family is a regulator of the production of β-1,3 glucan [29]. *Gas2*, a glycosylphosphatidylinositol (GPI)-anchored cell surface protein [30,31], is a putative carbohydrate-active enzyme that may change cell wall polysaccharides [29,32].

Another carbohydrate of *C. glabrata* cell wall is β-1,6-glucan, present as a polymer covalently attached to glycoproteins [33–36], β-1,3-glucan, and chitin [37]. Nagahashi et al. [36] reported the isolation of *KNH1* homologs (genes encoding cell surface *O*-glycoproteins), suggesting the evolutionary conservation of these molecules as essential components of β-1,6-glucan synthesis in *C. glabrata*, which was also discussed before [35,38]. Additionally, the *UGP1* gene is a putative uridine diphosphate (UDP)-glucose pyrophosphorylase related to the general β-1,6-D-glucan biosynthetic process [39,40]. During stress conditions, several *S. cerevisiae* orthologous genes are induced in *C. glabrata*. In glucose starvation stress, *UGP1* is induced [39].

The external layer cell wall of *Candida* spp. also consists of highly glycosylated mannoproteins [41–43], which play a major role in host recognition, adhesion, and cell wall integrity [44–56]. These proteins have both *N*- and *O*-linked sugars, predominantly mannans, which are also known to be present in the biofilm matrices of *C. albicans* [57–59]. The *MNN2* gene is one putative element of *N*-linked glycosylation, directly responsible for mannans production for both cell and biofilm matrices of *C. glabrata* [57–59].

The goal of this work was to determine the expression profile of selected genes (Tables 1 and 2) related to the production of biofilm matrix components in response to stress caused by drugs from the most important antifungal classes: azoles (Fluconazole, Flu), polyenes (Amphotericin B, AmB), and echinocandins (Caspofungin, Csf, and Micafungin, Mcf).

2. Materials and Methods

2.1. Organisms

Three strains of *C. glabrata* were used in the course of this study: One reference strain (*C. glabrata* ATCC 2001) from the American Type Culture Collection (Manassas, VA, USA), one strain recovered from the urinary tract (*C. glabrata* 562123) of a patient, and one strain recovered from the vaginal tract of a patient (*C. glabrata* 534784) in the Hospital Escala, Braga, Portugal. The identity of all isolates was confirmed using CHROMagar™ *Candida* (CHROMagar™, Paris, France) and by PCR-based sequencing using specific primers (*ITS1* and *ITS4*) against the 5.8 s subunit gene reference [60]. The PCR products were sequenced using the ABI-PRISM Big Dye terminator cycle sequencing kit (Perkin Elmer, Applied Biosystems, Warrington, UK).

2.2. Growth Conditions

For each experiment, *C. glabrata* ATCC2001, *C. glabrata* 534784, and *C. glabrata* 562123 strains were subcultured on Sabouraud dextrose agar (SDA) (Merck, Darmstadt, Germany) for 24 h at 37 °C. The cells were then inoculated in Sabouraud dextrose broth (SDB) (Merck) and incubated for 18 h at 37 °C under agitation at 120 rpm. After incubation, the cells were harvested by centrifugation at $3000 \times g$ (Thermo Scientific, CL10, Hampton, NH, USA) for 10 min at 4 °C and washed twice with phosphate buffered saline (PBS, pH = 7.5). The cell pellets were then suspended in Roswell Park memorial institute (RPMI), and the cellular density was adjusted to 1×10^5 cells/mL, using a Neubauer counting chamber.

2.3. Antifungal Drugs

Flu, Csf, and Mcf were kindly provided by Pfizer® (New York, NY, USA), MSD® (Kenilworth, NJ, USA) and Astellas Pharma, Ltd., (Tokyo, Japan), respectively, in their pure form. AmB was purchased from Sigma® (Sigma-Aldrich, Buffalo, NY, USA). Aliquots of 5000 mg/L were prepared using dimethyl sulfoxide (DMSO). The final concentrations used were prepared in RPMI-1640 (Sigma-Aldrich).

2.4. Biofilm Formation

The minimum biofilm eradicatory concentration (MBEC) values were previously determined by the group, according to the European committee on antimicrobial susceptibility testing (EUCAST) guidelines [61,62]. For biofilm formation, standardized cell suspensions (1000 µL) were placed into selected wells of 24-wells polystyrene microtiter plates (Orange Scientific, Braine-l'Alleud, Belgium). At 24 h, 500 µL of RPMI-1640 was removed, and an equal volume of fresh RPMI-1640 plus the antifungal solution was added, on the basis of the MBEC values determined and indicated in bold in Table 1 ($2\times$ concentrated). The plates were incubated at 37 °C for additional 24 h at 120 rpm. RPMI-1640 containing only the antifungal agent was used as a negative control. As a positive control, cell suspensions were tested in the absence of the antifungal agent [18].

Table 1. Minimum biofilm eradicatory concentrations (MBEC) for the *Candida glabrata* strains of fluconazole (Flu), amphotericin B (AmB), caspofungin (Csf), and micafungin (Mcf) (mg/L).

Origin	Strain	Flu	AmB	Csf	Mcf
Reference (Wild Type)	ATCC2001	>1250	4	2.5–3	16–17
Urinary Tract	562123	625	2	0.5–1	16–17
Vaginal Tract	534784	>1250	2	2.5–3	5.5–6

Bold: concentrations applied to the pre-formed biofilms.

2.5. Gene Expression Analysis

2.5.1. Gene Selection and Primer Design for Quantitative Real-Time PCR

Genes related to the production of biofilm matrix components (β-1,3, β-1,6 glucans, and mannans)—*BGL2, FKS1, FKS2, GAS2, KNH1, UGP1, XOG1 and MNN2*—were selected for this study. The gene sequences of interest were obtained from *Candida* Genome Database [63] and the primers for quantitative real-time PCR (RT-qPCR) were designed using Primer 3 [64] web-based software and are listed in Table 2. *ACT1* was chosen as a housekeeping gene. In order to verify the specificity of each primer pair for its corresponding target gene, the PCR products were first amplified from *C. glabrata* ATCC2001.

2.5.2. Preparation of Biofilm Cells for RNA Extraction

After biofilm formation, the medium was eliminated, and the wells were washed with sterile water to remove non-adherent cells. The biofilms were scraped from the wells with 1 mL of sterile water and sonicated (Ultrasonic Processor, Cole-Parmer, IL, USA) for 30 s at 30 W to separate the cells from the biofilm matrix. The cells were harvested by centrifugation at $8000\times g$ for 5 min at 4 °C [18].

2.5.3. RNA Extraction

RNA extraction was performed using PureLink RNA Mini Kit (Invitrogen, Carlsbad, CA, USA). Prior to RNA extraction, a lysis buffer from PureLink RNA Mini kit was prepared by adding 1% of β-mercaptoethanol to the supplied buffer solution. Then, 500 µL of lysis buffer containing glass beads (0.5 mm diameter) was added to each pellet. The cell suspensions were homogenized twice for 30 s using a Mini-Bead-Beater-8 (Stratech Scientific, Soham, UK). After cell disruption, the PureLink RNA Mini Kit (Invitrogen) was used for total RNA extraction according to the manufacturer's recommended protocol. To avoid potential DNA contamination, the samples were treated with RNase-Free DNase I (Invitrogen) [18].

Table 2. Primers, targets used, and specific function of the genes used for the expression analysis.

Sequence (5′ → 3′)	Primer	Target	Properties and Proposed Function [a]
5′-GGC AAG AAA CTG GAC AGA GC-3′ 5′-GGA AAA CTT GGG TCC TGC TG-3′	F R	BGL2	β-1,3-glucanosyltransferase activity; glucan endo-β-1,3-D-glucosidase activity
5′-GTC CTA ACC TTG CAC ACC AG-3′ 5′-CTA CGC CCA AAC ATC AGC-3′	F R	FKS1	β-1,3-D-glucan synthase activity
5′-GGG TCA CTG TGA AAT GTT-3′ 5′-GTA GAC GGG TTC GGA TT-3′	F R	FKS2	β-1,3-D-glucan synthase activity
5′-ACC AGT CGT ACC ATT ACC GG-3′ 5′-CCT GCC CAA CTT CTA ACA GC-3′	F R	GAS2	β-1,3-glucanosyltransferase activity
5′-CGG TGC CAA CGG TTA CTA-3′ 5′-GTG ACA CGG GTT TCA GGA-3′	F R	KNH1	β-1,6-D-glucan biosynthetic process
5′-AAT CGC ACA AGG CAG AGA-3′ 5′-ACT TGG GCG ACT TCC AAT-3′	F R	UGP1	β-1,6-D-glucan biosynthetic process
5′-GGT GAG TTG CAA CGT GAC AT-3′ 5′-ATT CGG TTA AAG CGG CAC TC-3′	F R	XOG1	Glucan endo-β-1,6 and 1,3-glucosidase activity
5′-GAA GCC TGA TGG TGG TGA-3′ 5′-ATT GGG CGA TGA CCT TCT-3′	F R	MNN2	α-mannosyltransferase biosynthetic process
5′-GTT GAC CGA GGC TCC AAT GA-3′ 5′-CAC CGT CAC CAG AGT CCA AA-3′	F R	ACT1	Housekeeping gene

[a] CGD: Candida Genome Database [63]; F: forward; R: reverse.

2.5.4. Synthesis of Complementary DNA

To synthesize complementary DNA (cDNA), the iScript cDNA Synthesis Kit (Bio-Rad, Hercules, CA, USA) was used according to the manufacturer's instructions. For each sample, 10 µL of the extracted RNA was used in a final reaction volume of 50 µL. cDNA synthesis was performed firstly at 70 °C for 5 min and then at 42 °C for 1 h. The reaction was stopped by heating for 5 min at 95 °C [18].

2.5.5. Quantitative Real-Time PCR

RT-qPCR (CFX96 Real-Time PCR System, Bio-Rad) was performed to determine the relative levels of all genes mRNA transcripts in the RNA samples, with *ACT1* used as a reference *Candida* housekeeping gene. Each reaction mixture consisted of a working concentration of SoFast EvaGreen Supermix (Bio-Rad), 50 µM forward and reverse primers, and 4 µL cDNA, in a final reaction volume of 20 µL. Negative controls (water) as well as non-transcriptase reverse controls (NRT) were included in each run [18]. The relative quantification of gene expression was performed by the $2^{-\Delta C_T}$ method [65]. Each reaction was performed in triplicate, and mean values of relative expression were determined for each gene. The results are presented after calculation of $2^{-\Delta C_T}$.

2.6. Statistical Analysis

All experiments were repeated three times in independent assays. The results were compared using one-way analysis of variance (ANOVA), Dunnett's multiple comparisons tests, using GraphPad™ Prism™ 7 software (GraphPad Software, San Diego, CA, USA). All tests were performed with a confidence level of 95%. In order to determine the similarity of the strains' gene profiles, the Pearson Correlation Coefficient (r) was also applied.

3. Results and Discussion

Candidaemia related to *C. glabrata* has been increasing in the last years in parallel with its high drug resistance, particularly to the azole antifungal class [1,20,66]. Biofilms of *C. glabrata* are highly recalcitrant to treatments with antifungal agents as a consequence of multiple resistance mechanisms, such as those linked to the presence of a strong net of exopolysaccharydes and other biopolymers that protect the cells and hinder the diffusion of the drugs [1,15,67–69]. In order to stress *C. glabrata* biofilm cells, four antifungals were applied (at concentrations based on MBECs values, Table 1) in pre-formed biofilms, and then an evaluation of biofilms' matrix gene expression was performed and compared with the expression of a housekeeping gene.

Figure 1 shows the heatmap with the results of the RT-qPCR expression profiling of biofilm cells of *C. glabrata* ATCC2001 (A), *C. glabrata* 562123 (B), and *C. glabrata* 534784 (C) in the presence of antifungal drugs. The final data are presented in fold-change (FC) in comparison to the expression of the housekeeping gene ($2^{-\Delta C_T}$) [70].

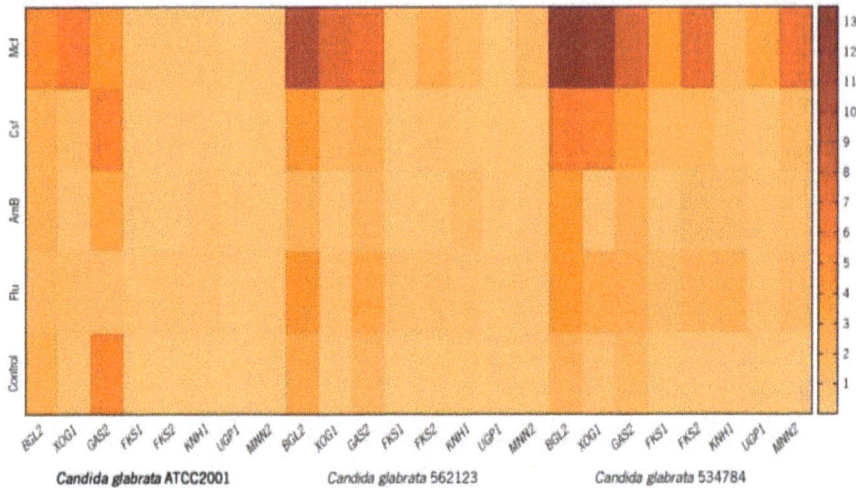

Figure 1. Real-time PCR expression profiling of *BGL2, XOG1, GAS2, FKS1, FKS2, KNH1, UGP1,* and *MNN2* genes in biofilm cells of *Candida glabrata* ATCC2001 (reference strain, in bold) (a), *C. glabrata* 562123 (b), and *C. glabrata* 534784 (c) in the presence of antifungal drugs. The heatmap was generated by a log transformation of the RT-qPCR data and the fold-change (FC) expression determined through $2^{-\Delta C_T}$. The numerical scale on the right represents the FC. (Control: non-treated cells).

Generally, *BGL2, FKS1, FKS2, GAS2,* and *XOG1* displayed higher expression levels in biofilm cells in response to the drugs and, by contrast, *KNH1, UPG1,* and *MNN2* displayed minor expression changes (Figure 1 and Table 3).

BGL2 showed similar expression in the control groups, and its FC expression decreased in the reference strain when Flu and Csf were present (FC: 1.40 and 2.00 respectively) and in the urinary strain when AmB was present (FC: 1.71). In all the other cases, *BGL2* FC expression increased, particularly when the biofilms were treated with Mcf (FC: *C. glabrata* ATCC2001: 5.13; *C. glabrata* 562123: 10.58; *C. glabrata* 534784: 13.49). All changes in *BGL2* expression were statistically significant compared to the untreated cells ($p < 0.0001$). Compared to the controls, *XOG1* gene revealed a statistically significant downregulation after contact with AmB (FC: *C. glabrata* ATCC2001 0.37; *C. glabrata* 562123 0.10; *C. glabrata* 534784: 0.27) and overexpression in the presence of all the other antifungals, in all strains. For this gene, the most noteworthy overexpression was observed in Flu-treated *C. glabrata* 534784 (FC: 2.35, $p < 0.0001$), Csf-treated *C. glabrata* 562123 and *C. glabrata* 534784 (FC: 1.42 and 5.45, $p < 0.0001$, respectively), and in all strains after contact with Mcf (FC: *C. glabrata* ATCC2001: 6.54, *C. glabrata* 562123: 7.89, *C. glabrata* 534784: 12.38, all $p < 0.0001$).

Table 3. Real-time PCR expression profiling of *BGL2*, *FKS1*, *FKS2*, *GAS2*, *KNH1*, *UGP1*, *XOG1*, and *MNN2* genes in biofilm cells of *C. glabrata* ATCC2001, *C. glabrata* 562123, *C. glabrata* 534784 with and without antifungal treatment (FC: $2^{-\Delta C_T}$). The significance of the FC results was determined by comparing the treated groups with the non-treated ones (controls) (* $p < 0.05$; ** $p < 0.001$; *** $p < 0.0005$; **** $p < 0.0001$).

Gene		*Candida glabrata* ATCC2001	*Candida glabrata* 562123	*Candida glabrata* 534784
			Fold-Change	
BGL2	Non treated	2.23	2.77	1.90
	Flu	1.41 ****	4.01 ****	3.66 ****
	AmB	2.04 ****	1.71 ****	3.85 ****
	Csf	2.00 ****	3.82 ****	5.98 ****
	Mcf	5.13 ****	10.58 ****	13.49 ****
XOG1	Non treated	0.57	0.15	0.46
	Flu	0.96 ****	0.54 ****	2.35 ****
	AmB	0.37 ****	0.10 ***	0.27 ****
	Csf	1.08 ****	1.42 ****	5.45 ****
	Mcf	6.54 ****	7.89 ****	12.38 ****
GAS2	Non treated	5.34	1.67	1.39
	Flu	0.76 ****	2.73 ****	2.26 ****
	AmB	3.02 ****	1.56 ****	1.95 ****
	Csf	5.72 ****	2.08 ****	3.39 ****
	Mcf	3.99 ****	7.43 ****	8.18 ****
FKS1	Non treated	0.11	0.11	0.17
	Flu	0.22 ****	0.65 ****	1.07 ****
	AmB	0.16 ****	0.08 ****	0.19 (ns)
	Csf	0.07 ***	0.20 ****	1.27 ****
	Mcf	0.49 ****	0.94 ****	3.55 ****
FKS2	Non treated	0.14	0.20	0.27
	Flu	0.61 ****	0.64 ****	1.77 ****
	AmB	0.06 ***	0.19 *	1.05 ****
	Csf	0.28 ****	0.55 ****	1.66 ****
	Mcf	0.43 ****	2.29 ****	7.50 ****
KNH1	Non treated	0.06	0.20	0.22
	Flu	0.50 ****	0.41 ****	2.08 ****
	AmB	0.51 ****	0.87 ****	0.94 ***
	Csf	0.24 ****	0.21 (ns)	0.72 ****
	Mcf	0.45 ****	1.33 ****	1.43 ****
UGP1	Non treated	0.002	0.07	0.20
	Flu	0.10 ****	0.15 ****	0.55 ****
	AmB	0.17 ****	0.05 ****	0.46 ****
	Csf	0.01 (ns)	0.06 **	0.33 ****
	Mcf	0.04 **	0.21 ****	3.17 ****
MNN2	Non treated	0.02	0.18	0.13
	Flu	0.19 ****	0.34 ****	1.17 ****
	AmB	0.13 ****	0.24 ****	0.71 ****
	Csf	0.18 ****	0.13 ***	1.40 ****
	Mcf	0.31 ****	1.21 ****	7.03 ****

(ns, non-significant; Non-treated, controls).

In an important report, Taff et al. [19] concluded that mutants of *C. albicans* unable to produce Bgl2 and Xog1 enzymes did not show perturbations in the cell wall glucan composition of biofilm cells, and that these enzymes were not necessary for filamentation or biofilm formation. However, the biofilms had a reduced matrix glucan content, reduced total matrix biomass accumulation, and improved susceptibility to antifungal drug therapy [19]. Similarly, Li et al. [71] showed that, in *C. albicans*' persister cells (frequent in biofilms [72,73]), there was an increased expression of cell wall integrity proteins such as Xog1 and Bgl2. These studies recognized a biofilm-specific pathway involving Bgl2 and Xog1 (and Phr1) enzymes and affecting matrix delivery, by which these enzymes release and modify cell wall glucan for deposition in the extracellular space; however, an alternative explanation is that these enzymes act in the extracellular space, being crucial for mature matrix organization and function [19]. These enzymes have been localized in the cell wall, supporting the hypothesis of their activity in the cell wall, but have also secretion sequences that support an extracellular function. As seen earlier, *BGL2* is one of the glucan modifying genes for glucan delivery, and *XOG1* is a glucanase [19], necessary for modification and delivery of carbohydrates to the mature biofilm matrix. Without delivery and accumulation of matrix glucan, the biofilms exhibit enhanced susceptibility to antifungal drugs [19]. The change in the regulation of *BGL2* and *XOG1* in the biofilm cells of *C. glabrata* after drug treatment that we observed is interpreted as a response of the biofilm cells to the reduction of biofilm matrix, specifically of β-1,3-glucans, and it has been described before [7,13,18].

The *GAS* gene family is also a regulator in the production of β-1,3-glucan, and Gas2 is a glycosylphosphatidylinositol (GPI)-anchored cell surface protein [31] involved in the production of β-1,3-glucan in *C. glabrata* [29,30]. Gas2 is a documented putative carbohydrate-active enzyme and consequently it can alter the cell wall polysaccharides in order to build and remodel the cell wall glycan network during growth in *C. glabrata* [29]. In *C. glabrata* ATCC2001, *GAS2* was highly expressed in the non-treated group and after Csf contact (FC: 5.72), while Flu, AmB, and Mcf led to its downregulation. The clinical isolates upregulated the gene in all conditions, except for AmB treatment of *C. glabrata* 562123 (Figure 1 and Table 3, all $p < 0.0001$). Hence, when analyzing the results of *C. glabrata* 562123 and *C. glabrata* 534784, the *GAS2* network seems to be activated also after glycan's loss following drug treatment, in order to replace the lack of 1,3-β-glucans and re-establish biofilm cells' homeostasis. All results compared with those from untreated cells (controls) were statistically significant ($p < 0.0001$).

The resistance to echinocandins increased from 4.9% to 12.3% between 2001 and 2010 [74] with a rapid development of *FKS* mutations in *Candida* spp., especially in *C. glabrata* [75,76]. The amino acid substitutions occurring in *FKS1* [30,77–79] and *FKS2* [30,80] are directly related to the resistance to this class of drugs: acquired *FKS* mutations [81] are reported to confer low β-(1,3)-D-glucan synthase sensitivity and to increase the minimum inhibitory concentration (MIC) values, which are related to clinical failure [82]; intrinsic *FKS* mutations, also lead to elevated MIC levels but have a weaker effect on the reduction of β-(1,3)-D-glucan synthase sensitivity [82–85]. Generally, in the *C. glabrata* strains, a 24 h contact with both echinocandins upregulated *FKS1* and *FKS2* genes and, in the reference strain, the presence of Mcf upregulated *FKS1*. More specifically, the results showed that all strains upregulated the expression of *FKS1* after drug exposure (statistically significant), with the exception of Csf in *C. glabrata* ATCC2001 (FC: 0.07 $p < 0.001$) and Csf in *C. glabrata* 562123 (FC: 0.08; $p < 0.0001$). For *FKS2*, its overexpression was observed for almost all treatments in the three strains, excluding following AmB treatment in *C. glabrata* ATCC2001 (FC: 0.28; $p < 0.001$) and *C. glabrata* 562123 (FC: 0.55; $p < 0.05$). *C. glabrata* 534784 revealed, again, to have the highest capacity to overexpress both genes in response to drug stress. These differences among the strains may be related to the described *Candida* spp. intra-strains variations [62]. Bizerra et al. [76] reported the occurrence of a mutation associated with the resistance phenotype against echinocandins in *C. glabrata* isolated from a single cancer patient with candidemia exposed to antifungal prophylaxis with Mcf. Arendrup et al. [86] revealed that Mcf MICs of *C. glabrata* *FKS* hot spot mutant isolates were less raised than those obtained for the other echinocandins, showing that the efficacy of Mcf could be differentially dependent on specific *FKS* genes mutations. These reports mention singularities regarding the *FKS* gene and Mcf, which can also be observed in our results (Figure 1 and Table 3). Interestingly, up and downregulations of *FKS1* and *FKS2* were similar in the clinical isolates and parallel to those observed for *BGL2*, which makes sense, since this gene has shown to perform, with *XOG1* (and *PHR1*), in a complementary manner in order to distribute the matrix downstream of the primary β-1,3 glucan synthase encoded by *FKS1* [19]. Previous investigations also found elevated transcript levels of *FKS1*, *BGL2*, and *XOG1* during in vivo *C. albicans* biofilm growth when compared to planktonic growth, which is consistent with our results and with a role in a biofilm-specific function, such as matrix formation [87,88].

The overexpressed values obtained for *BGL1*, *XOG1*, *FKS1*, *FKS2*, and *GAS2* after the stress conditions induced by most antifungals endorse the impact of β-1,3-glucans in the maintenance of the cell and biofilm matrix structure.

Sequencing studies have shown that *C. glabrata* is more closely related to *S. cerevisiae* than to *C. albicans* [89], with some genes functionally interchangeable among the two species [90,91]. An important component of the cell wall and the biofilm matrix is β-1,6-glucan, which is regulated by several genes, such as *KNH1*. Preceding studies have demonstrated that the *KNH1* homologs are essential components of β-1,6-glucan synthesis in *C. glabrata* [35,36,38]. In *S. cerevisiae*, many genes involved in β-1,6-glucan synthesis were isolated through mutations (*kre* [killer resistant] mutations) that are responsible for the resistance to the K1 killer toxin, which kills sensitive yeast cells after binding to β-1,6-glucan [35,38,92]. Dijkgraaf and colleagues [35] reported that the disruption of both *KRE9* and *KNH1* was synthetically lethal for *C. glabrata*, demonstrating the importance of these genes in the maintenance of cell structure. In the present study, after a drug stress, all *C. glabrata* strains upregulated this gene (Figure 1 and Table 3), indicating an effort to replace these β-1,6-glucans after losses due to the aggression of the antifungals, confirming also a certain degree of relevance of these elements in the cell wall and biofilm matrix of *C. glabrata* [35,36,38]. *C. glabrata* ATCC2001 showed to upregulate the *KNH1* gene in the presence of antifungal drugs, and *C. glabrata* 562123 indicated an identical pattern by marginally increasing *KNH1* gene expression in these conditions. Compared to the other two strains, with the exception of Flu treatment (FC: 2.08; $p < 0.0001$), *KNH1* showed a different regulation in the vaginal tract strain (Figure 1 and Table 3). *C. glabrata* 534784 demonstrated to have the highest upregulation capacity, presenting overexpression almost for all genes (Figure 1 and Table 3).

During glucose starvation, a set of genes orthologous to *S. cerevisiae* is induced in *C. glabrata*, including *UGP1*, related to the β-1,6-D-glucan biosynthetic process [39,40], which shows that the environmental stress response is conserved between *S. cerevisiae* and *C. glabrata* [39]. *UPG1* showed to have the lowest expression, compared with other genes and controls. Nonetheless, except for one condition and strain, in the presence of antifungal drugs, several overexpression states were observed (Figure 1 and Table 3). The reference strain displayed overexpression in all conditions, with the highest gene upregulation occurring in the presence of AmB (FC: 0.17; $p < 0.0001$) and the lowest in the presence of Csf (FC: 0.01; non-significant); the urinary tract strain also revealed limited gene upregulation in the presence of Csf and AmB and the highest expression in the presence of Mcf (0.15; $p < 0.0001$). Generally, *C. glabrata* 534784 demonstrated the highest FC expression in all conditions. The lowest gene upregulation was observed in biofilm cells stressed by Csf (FC: 0.33; $p < 0.0001$). Srikantha and colleagues [91] identified a set of genes that are upregulated by the transcription factor Bcr1, involved in impermeability, impenetrability, and drug resistance of *C. albicans'* biofilms. The authors concluded that the induction of Bcr1 overexpression in weak biofilms of *C. albicans* conferred those three characteristics and, in these cases, *UGP1* gene was downregulated [91]. This result supports the FC expression we obtained: since *C. glabrata* biofilms were weakened by the drugs, *UGP1* expression was increased in order to balance this defect (as seen with *KNH1*). The overexpression values we obtained for both *KNH1* and *UPG1* point to the relevance of β-1,6-glucans in the maintenance of a good cell and matrix structure.

Regarding mannans regulation, all strains showed a low or moderate expression of *MNN2* in the controls (non-treated cells), but relevant expression changes arose in the presence of all drugs, particularly when Mcf was added (Figure 1 and Table 3). *C. glabrata* ATCC2001 demonstrated the lowest expression in the control group, among all strains (Figure 1 and Table 3). The urinary strain presented the lowest gene expression, when compared to the untreated group (Figure 1 and Table 3). Flu and Mcf induced the highest *MNN2* values (FC: 0.34 and 1.21, respectively, both $p < 0.0001$), while AmB and Csf were associated with the lowest expressions (FC: 0.24, $p < 0.0001$ and 0.13, $p < 0.0005$). For *C. glabrata* ATCC2001 and *C. glabrata* 534784, the weakest effects were associated with the biofilm cells that were stressed by AmB (FC: 0.13 and 0.71 respectively, both $p < 0.0001$). When Mcf was applied, the biofilm cells of the vaginal strain showed a strong response to the stress, compared to the other two strains (FC: 7.03; $p < 0.0001$). Our team has also found that *C. glabrata* ATCC2001 increased the amounts of mannans on its cell walls in the presence of these drugs (data not shown), revealing a possible adaptation of the cells to the stress caused by the antifungal drugs. Other studies reported analogous adjustments of the cell walls after environmental drug stress, which has been related to high antifungal resistance events [1,2,19,92–94], supporting these results.

Interestingly, and when compared to the rest of the genes, the present results demonstrate that *KNH1*, *UGP1*, and *MNN2* had the lowest values of FC expression. This seems to indicate that, although β-1,6-glucans are an important part of the cell and biofilm matrix, the cells appear to invest more in replacing the lost β-1,3-glucans, leading to consider that these components have a greater significance in the maintenance of the homeostasis of the biofilm matrix and the biofilm cells. In fact, in studies developed in our group [13,18], the total polysaccharides and β-1,3-glucans concentrations increased significantly in *C. glabrata* biofilm matrices after Flu, AmB, and Mcf contact. These higher concentrations in β-1,3 glucans content might explain part of the main biofilm resistance to the drugs that was formerly described [7,95–97].

Finally, downregulation of most genes and strains happened in the presence of AmB whereas, in opposition, Mcf induced the main overexpression alterations (Figure 1 and Table 3). AmB is a fungicidal drug and the most important antifungal polyene used for the treatment of systemic fungal infections [98,99]. This drug binds to the ergosterol of the cell membrane but also induces oxidative stress. This explains the existence of a still low reported rate of resistance and the good effectiveness of AmB [1,13,100–104]. Also, this low resistance may be associated with the lower gene expression effects that we detected after AmB exposure in *C. glabrata* (Figure 1 and Table 3). In opposition, the most acute upregulations occurred in the presence of echinocandins and, particularly, when Mcf was applied. This class of antifungals act by inhibiting β-1,3-glucan synthesis [1,100,105], which affects cell wall and matrix composition. By overexpressing the genes related to β-1,3-glucan synthesis (*BGL2*, *FKS1*, *FKS2*, *GAS2*, *XOG1*), the cells were attempting to compensate and replace the β-1,3-glucan losses in their matrices induced by the drugs and, thus, protect and decrease their susceptibility to the antifungals [19]. This general increase in total carbohydrates and specifically in β-1,3-glucans in *Candida* spp. biofilm matrices has already been described [7,13,18].

Regarding the correlation between the gene expression profiles in *C. glabrata*, the results based on the r are displayed in Table 4.

Table 4. Pearson Correlation Coefficient (r) determined for the expression profiles of *BGL2*, *FKS1*, *FKS2*, *GAS2*, *KNH1*, *UGP1*, *XOG1*, and *MNN2* genes in biofilm cells of *C. glabrata* ATCC2001, *C. glabrata* 562123, *C. glabrata* 534784, in the presence or absence of antifungal drugs.

Gene	ATCC2001 vs. 562123	ATCC2001 vs. 534784	562123 vs. 534784
BGL2	0.9100	0.9145	0.946
XOG1	0.9965	0.9459	0.9646
FKS1	0.8947	0.8723	0.8778
FKS2	0.5074	0.4481	0.9937
GAS2	−0.0514	0.1091	0.9663
KNH1	0.6427	0.8107	0.3697
UGP1	−0.1091	-0.1122	0.8519
MNN2	0.7924	0.8618	0.9728

The results showed a strong positive correlation (r near 1) between the response profiles of *BGL2*, *XOG1*, *FKS1*, and *MNN2* gene expression in the three strains, which means that up and downregulation had a high tendency to occur similarly in all strains. The scores of the r for the profile of the *FKS2* gene revealed a moderate positive correlation between the reference strain (*C. glabrata* ATCC2001) and the isolates (*C. glabrata* 562123 and *C. glabrata* 534784). This indicates that, although the correlation was positive, it was weak, and the profiles of the gene response were variable in the three strains. On the other hand, the clinical isolates showed strong positive correlation between the expression profiles of this gene. *C. glabrata* ATCC2001 demonstrated a moderate positive correlation between the expression profiles of the *GAS2* gene. *C. glabrata* 562123 and *C. glabrata* 534784 had a strong positive correlation between the expression profiles of the *GAS2* gene. *KNH1* gene was the most variable and difficult gene to correlate between the strains. The reference and the 562123 strain showed a moderate correlation, whereas the reference and the 534784 strain showed a strong correlation, and the clinical isolates showed the only weak correlation detected in this study. As for the *UPG1* gene, although there was a negative correlation, the association between its expression in ATCC2001 and in the clinical isolates can be considered weak. Between the isolates, it was determined that *UPG1* up and downregulation had a high tendency to occur similarly in all strains (thus showing strong correlation). In summary, *BGL2*, *XOG1*, *FKS1*, and *MNN2* appeared to be the genes presenting the most similar responses to antifungal drugs within the transcriptome of the three strains; also, the clinical isolates appeared to be nearer each other than to the reference strain. Once more, β-1,3-glucan synthesis was identified as important in *C. glabrata* (three of the four genes affected are responsible for β-1,3-glucan production). These similarities among the two clinical strains may be due to the fact that both were derived from a hospital environment, and it is probable that they had already been challenged by several drugs, so their responses were prompter compared to the reference strain that is a wild type strain.

4. Conclusions

The in vitro high-dose paradox associated with *Candida* spp. isolates is being increasingly reported and connected to slightly elevated MICs, potentially contributing to clinical resistance and failure of antifungal treatments. These drug tolerance and adaptive mechanisms are highly related to *Candida* spp. biofilm forms. *C. glabrata* extracellular matrix is crucial for mature biofilm formation, not only contributing to the adhesive nature of the biofilm cells, but also protecting the cells from antifungal agents and from the host immune system. Understanding the production of the biofilm matrix components and the associated delivery processes is important for the development of effective biofilm therapies. All stakeholders in this process represent potentially attractive targets for detection of and therapeutic interventions against candidiasis.

Acknowledgments: This study was supported by the Portuguese Foundation for Science and Technology (FCT) under the scope of the strategic funding of UID/BIO/04469/2013 unit and COMPETE 2020 (POCI-01-0145-FEDER-006684) and BioTecNorte operation (NORTE-01-0145-FEDER-000004) funded by the European Regional Development Fund under the scope of Norte2020—Programa Operacional Regional do Norte and Célia F. Rodrigues' [SFRH/BD/93078/2013] Ph.D. grant. We also would like to acknowledge MSD® and Astellas® for the kind donation of Caspofungin and Micafungin, respectively.

Author Contributions: Célia F. Rodrigues conceived, designed, performed the experiments, analyzed the data, and wrote the paper; Mariana Henriques conceived, designed the experiments, analyzed the data, and wrote the paper.

Conflicts of Interest: The authors declare no conflict of interest.

References

1. Rodrigues, C.F.; Rodrigues, M.E.; Silva, S.; Henriques, M. *Candida glabrata* biofilms: How far have we come? *J. Fungi* **2017**, *3*, 11. [CrossRef] [PubMed]
2. Silva, S.; Rodrigues, C.F.; Araújo, D.; Rodrigues, M.E.; Henriques, M. *Candida* species biofilms' antifungal resistance. *J. Fungi* **2017**, *3*, 8. [CrossRef] [PubMed]
3. Costa-Orlandi, C.; Sardi, J.; Pitangui, N.; de Oliveira, H.; Scorzoni, L.; Galeane, M.; Medina-Alarcón, K.; Melo, W.; Marcelino, M.; Braz, J.; et al. Fungal biofilms and polymicrobial diseases. *J. Fungi* **2017**, *3*, 22. [CrossRef] [PubMed]
4. McCall, A.; Edgerton, M. Real-time approach to flow cell imaging of *Candida albicans* biofilm development. *J. Fungi* **2017**, *3*, 13. [CrossRef] [PubMed]
5. Nobile, C.J.; Johnson, A.D. *Candida albicans* biofilms and human disease. *Annu. Rev. Microbiol.* **2015**, *69*, 71–92. [CrossRef] [PubMed]
6. Silva, S.; Henriques, M.; Martins, A.; Oliveira, R.; Williams, D.; Azeredo, J. Biofilms of non-*Candida albicans* Candida species: Quantification, structure and matrix composition. *Med. Mycol.* **2009**, *47*, 681–689. [CrossRef] [PubMed]

7. Fonseca, E.; Silva, S.; Rodrigues, C.F.; Alves, C.T.; Azeredo, J.; Henriques, M. Effects of fluconazole on *Candida glabrata* biofilms and its relationship with ABC transporter gene expression. *Biofouling* **2014**, *30*, 447–457. [CrossRef] [PubMed]
8. Rodrigues, C.F.; Silva, S.; Henriques, M. *Candida glabrata*: A review of its features and resistance. *Eur. J. Clin. Microbiol. Infect. Dis.* **2014**, *33*, 673–688. [CrossRef] [PubMed]
9. Negri, M.; Silva, S.; Henriques, M.; Oliveira, R. Insights into *Candida tropicalis* nosocomial infections and virulence factors. *Eur. J. Clin. Microbiol. Infect. Dis.* **2012**, *31*, 1399–1412. [CrossRef] [PubMed]
10. Costerton, J.W.; Lewandowski, Z.; Caldwell, D.E.; Korber, D.R.; Lappin-Scott, H.M. Microbial Biofilms. *Annu. Rev. Microbiol.* **1995**, *49*, 711–745. [CrossRef] [PubMed]
11. Donlan, R.; Costerton, J. Biofilms: Survival mechanisms of clinically relevant microorganisms. *Clin. Microbiol. Rev.* **2002**, *15*, 167–193. [CrossRef] [PubMed]
12. Silva, S.; Negri, M.; Henriques, M.; Oliveira, R.; Williams, D.W.; Azeredo, J. *Candida glabrata*, *Candida parapsilosis* and *Candida tropicalis*: Biology, epidemiology, pathogenicity and antifungal resistance. *FEMS Microbiol. Rev.* **2012**, *36*, 288–305. [CrossRef] [PubMed]
13. Rodrigues, C.F.; Silva, S.; Azeredo, J.; Henriques, M. *Candida glabrata*'s recurrent infections: Biofilm formation during Amphotericin B treatment. *Lett. Appl. Microbiol.* **2016**, *63*, 77–81. [CrossRef] [PubMed]
14. Nett, J.; Lincoln, L.; Marchillo, K.; Massey, R.; Holoyda, K.; Hoff, B.; VanHandel, M.; Andes, D. Putative role of β-1,3 glucans in *Candida albicans* biofilm resistance. *Antimicrob. Agents Chemother.* **2007**, *51*, 510–520. [CrossRef] [PubMed]
15. Nett, J.E.; Sanchez, H.; Cain, M.T.; Andes, D.R. Genetic basis of *Candida* biofilm resistance due to drug-sequestering matrix glucan. *J. Infect. Dis.* **2010**, *202*, 171–175. [CrossRef] [PubMed]
16. Nett, J.E.; Crawford, K.; Marchillo, K.; Andes, D.R.; Nett, J.E.; Crawford, K.; Marchillo, K.; Andes, D.R. Role of Fks1p and matrix glucan in *Candida albicans* biofilm resistance to an echinocandin, pyrimidine, and polyene. *Antimicrob. Agents Chemother.* **2010**, *54*, 3505–3508. [CrossRef] [PubMed]
17. Nett, J.E.; Sanchez, H.; Cain, M.T.; Ross, K.M.; Andes, D.R. Interface of *Candida albicans* biofilm matrix-associated drug resistance and cell wall integrity regulation. *Eukaryot. Cell* **2011**, *10*, 1660–1669. [CrossRef] [PubMed]
18. Rodrigues, C.F.; Gonçalves, B.; Rodrigues, M.E.; Silva, S.; Azeredo, J.; Henriques, M. The effectiveness of voriconazole in therapy of *Candida glabrata*'s biofilms oral infections and its influence on the matrix composition and gene expression. *Mycopathologia* **2017**, *182*, 653–664. [CrossRef] [PubMed]
19. Taff, H.T.; Nett, J.E.; Zarnowski, R.; Ross, K.M.; Sanchez, H.; Cain, M.T.; Hamaker, J.; Mitchell, A.P.; Andes, D.R. A *Candida* biofilm-induced pathway for matrix glucan delivery: Implications for drug resistance. *PLoS Pathog.* **2012**, *8*. [CrossRef] [PubMed]
20. Rodrigues, C.F.; Henriques, M. Oral mucositis caused by *Candida glabrata* biofilms: Failure of the concomitant use of fluconazole and ascorbic acid. *Ther. Adv. Infect. Dis.* **2017**, *1*. [CrossRef] [PubMed]
21. Lopez-Ribot, J.L. Large-scale biochemical profiling of the *Candida albicans* biofilm matrix: New compositional, structural, and functional insights. *MBio* **2014**, *5*, e01314–e01333. [CrossRef] [PubMed]
22. Ramage, G.; Rajendran, R.; Sherry, L.; Williams, C. Fungal biofilm resistance. *Int. J. Microbiol.* **2012**, *2012*. [CrossRef] [PubMed]
23. Mitchell, K.F.; Zarnowski, R.; Andes, D.R. Fungal super glue: The biofilm matrix and its composition, assembly, and functions. *PLoS Pathog.* **2016**, *12*. [CrossRef] [PubMed]
24. Douglas, C. Fungal beta (1,3)-D-glucan synthesis. *Med. Mycol.* **2001**, *39*, 55–66. [CrossRef] [PubMed]
25. Del Mar Gonzalez, M.; Diez-Orejas, R.; Molero, G.; Alvarez, A.M.; Pla, J.; Nombela, C.; Sanchez-Perez, M. Phenotypic characterization of a *Candida albicans* strain deficient in its major exoglucanase. *Microbiology* **1997**, *143*, 3023–3032. [CrossRef] [PubMed]
26. Sarthy, A.V.; McGonigal, T.; Coen, M.; Frost, D.J.; Meulbroek, J.A.; Goldman, R.C. Phenotype in *Candida albicans* of a disruption of the *BGL2* gene encoding a 1,3-β-glucosyltransferase. *Microbiology* **1997**, *143*, 367–376. [CrossRef] [PubMed]
27. Mouyna, I.; Fontaine, T.; Vai, M.; Monod, M.; Fonzi, W.A.; Diaquin, M.; Popolo, L.; Hartland, R.P.; Latgé, J.P. Glycosylphosphatidylinositol-anchored glucanosyltransferases play an active role in the biosynthesis of the fungal cell wall. *J. Biol. Chem.* **2000**, *275*, 14882–14889. [CrossRef] [PubMed]

28. Goldman, R.C.; Sullivan, P.A.; Zakula, D.; Capobianco, J.O. Kinetics of β-1,3 glucan interaction at the donor and acceptor sites of the fungal glucosyltransferase encoded by the *BGL2* gene. *Eur. J. Biochem.* **1995**, *227*, 372–378. [CrossRef] [PubMed]

29. De Groot, P.W.J.; Kraneveld, E.A.; Yin, Q.Y.; Dekker, H.L.; Gross, U.; Crielaard, W.; de Koster, C.G.; Bader, O.; Klis, F.M.; Weig, M. The cell wall of the human pathogen *Candida glabrata*: Differential incorporation of novel adhesin-like wall proteins. *Eukaryot. Cell* **2008**, *7*, 1951–1964. [CrossRef] [PubMed]

30. Garcia-Effron, G.; Lee, S.; Park, S.; Cleary, J.D.; Perlin, D.S. Effect of *Candida glabrata* *FKS1* and *FKS2* mutations on echinocandin sensitivity and kinetics of 1,3-β-D-glucan synthase: Implication for the existing susceptibility breakpoint. *Antimicrob. Agents Chemother.* **2009**, *53*, 3690–3699. [CrossRef] [PubMed]

31. Vai, M.; Orlandi, I.; Cavadini, P.; Alberghina, L.; Popolo, L. *Candida albicans* homologue of *GGP1/GAS1* gene is functional in *Saccharomyces cerevisiae* and contains the determinants for glycosylphosphatidylinositol attachment. *Yeast* **1996**, *12*, 361–368. [CrossRef]

32. Miyazaki, T.; Nakayama, H.; Nagayoshi, Y.; Kakeya, H.; Kohno, S. Dissection of Ire1 functions reveals stress response mechanisms uniquely evolved in *Candida glabrata*. *PLoS Pathog.* **2013**, *9*. [CrossRef] [PubMed]

33. Montijn, R.C.; van Rinsum, J.; van Schagen, F.A.; Klis, F.M. Glucomannoproteins in the cell wall of *Saccharomyces cerevisiae* contain a novel type of carbohydrate side chain. *J. Biol. Chem.* **1994**, *269*, 19338–19342. [PubMed]

34. Kapteyn, J.C.; Montijn, R.C.; Vink, E.; de la Cruz, J.; Llobell, A.; Douwes, J.E.; Shimoi, H.; Lipke, P.N.; Klis, F.M. Retention of *Saccharomyces cerevisiae* cell wall proteins through a phosphodiester-linked β-1,3-/β-1,6-glucan heteropolymer. *Glycobiology* **1996**, *6*, 337–345. [CrossRef] [PubMed]

35. Dijkgraaf, G.J.; Brown, J.L.; Bussey, H. The *KNH1* gene of *Saccharomyces cerevisiae* is a functional homolog of *KRE9*. *Yeast* **1996**, *12*, 683–692. [CrossRef]

36. Nagahashi, S.; Lussier, M.; Bussey, H. Isolation of *Candida glabrata* homologs of the *Saccharomyces cerevisiae* *KRE9* and *KNH1* genes and their involvement in cell wall β-1,6-glucan synthesis. *J. Bacteriol.* **1998**, *180*, 5020–5029. [PubMed]

37. Cid, V.J.; Durán, A.; del Rey, F.; Snyder, M.P.; Nombela, C.; Sánchez, M. Molecular basis of cell integrity and morphogenesis in *Saccharomyces cerevisiae*. *Microbiol. Rev.* **1995**, *59*, 345–386. [PubMed]

38. Brown, J.L.; Kossaczka, Z.; Jiang, B.; Bussey, H. A mutational analysis of killer toxin resistance in *Saccharomyces cerevisiae* identifies new genes involved in cell wall (1–6)-β-glucan synthesis. *Genetics* **1993**, *133*, 837–849. [PubMed]

39. Roetzer, A.; Gregori, C.; Jennings, A.M.; Quintin, J.; Ferrandon, D.; Butler, G.; Kuchler, K.; Ammerer, G.; Schüller, C. *Candida glabrata* environmental stress response involves *Saccharomyces cerevisiae* Msn2/4 orthologous transcription factors. *Mol. Microbiol.* **2008**, *69*, 603–620. [CrossRef] [PubMed]

40. Dodgson, A.R.; Pujol, C.; Denning, D.W.; Soll, D.R.; Fox, A.J. Multilocus sequence typing of *Candida glabrata* reveals geographically enriched clades. *J. Clin. Microbiol.* **2003**, *41*, 5709–5717. [CrossRef] [PubMed]

41. Ruiz-Herrera, J.; Victoria Elorza, M.; Valentin, E.; Sentandreu, R. Molecular organization of the cell wall of *Candida albicans* and its relation to pathogenicity. *FEMS Yeast Res.* **2006**, *6*, 14–29. [CrossRef] [PubMed]

42. Latgé, J.-P. The cell wall: A carbohydrate armour for the fungal cell. *Mol. Microbiol.* **2007**, *66*, 279–290. [CrossRef] [PubMed]

43. Netea, M.G.; Brown, G.D.; Kullberg, B.J.; Gow, N.A.R. An integrated model of the recognition of *Candida albicans* by the innate immune system. *Nat. Rev. Microbiol.* **2008**, *6*, 67–78. [CrossRef] [PubMed]

44. Netea, M.G.; Mardi, L. Innate immune mechanisms for recognition and uptake of *Candida* species. *Trends Immunol.* **2010**, *31*, 346–353. [CrossRef] [PubMed]

45. Gow, N.A.R.; Netea, M.G.; Munro, C.A.; Ferwerda, G.; Bates, S.; Mora-Montes, H.M.; Walker, L.; Jansen, T.; Jacobs, L.; Tsoni, V.; et al. Immune recognition of *Candida albicans* β-glucan by Dectin-1. *J. Infect. Dis.* **2007**, *196*, 1565–1571. [CrossRef] [PubMed]

46. Van de Veerdonk, F.L.; Kullberg, B.J.; van der Meer, J.W.; Gow, N.A.; Netea, M.G. Host-microbe interactions: Innate pattern recognition of fungal pathogens. *Curr. Opin. Microbiol.* **2008**, *11*, 305–312. [CrossRef] [PubMed]

47. Reid, D.M.; Gow, N.A.; Brown, G.D. Pattern recognition: Recent insights from Dectin-1. *Curr. Opin. Immunol.* **2009**, *21*, 30–37. [CrossRef] [PubMed]

48. Mora-Montes, H.M.; Bates, S.; Netea, M.G.; Castillo, L.; Brand, A.; Buurman, E.T.; Diaz-Jimenez, D.F.; Jan Kullberg, B.; Brown, A.J.P.; Odds, F.C.; et al. A multifunctional mannosyltransferase family in *Candida albicans* determines cell wall mannan structure and host-fungus interactions. *J. Biol. Chem.* **2010**, *285*, 12087–12095. [CrossRef] [PubMed]

49. Taylor, P.R.; Tsoni, S.V.; Willment, J.A.; Dennehy, K.M.; Rosas, M.; Findon, H.; Haynes, K.; Steele, C.; Botto, M.; Gordon, S.; et al. Dectin-1 is required for β-glucan recognition and control of fungal infection. *Nat. Immunol.* **2007**, *8*, 31–38. [CrossRef] [PubMed]

50. Murciano, C.; Moyes, D.L.; Runglall, M.; Islam, A.; Mille, C.; Fradin, C.; Poulain, D.; Gow, N.A.R.; Naglik, J.R. *Candida albicans* cell wall glycosylation may be indirectly required for activation of epithelial cell proinflammatory responses. *Infect. Immun.* **2011**, *79*, 4902–4911. [CrossRef] [PubMed]

51. Gow, N.A.; Hube, B. Importance of the *Candida albicans* cell wall during commensalism and infection. *Curr. Opin. Microbiol.* **2012**, *15*, 406–412. [CrossRef] [PubMed]

52. Bates, S.; Hughes, H.B.; Munro, C.A.; Thomas, W.P.H.; MacCallum, D.M.; Bertram, G.; Atrih, A.; Ferguson, M.A.J.; Brown, A.J.P.; Odds, F.C.; et al. Outer chain N-glycans are required for cell wall integrity and virulence of *Candida albicans*. *J. Biol. Chem.* **2006**, *281*, 90–98. [CrossRef] [PubMed]

53. Bates, S.; MacCallum, D.M.; Bertram, G.; Munro, C.A.; Hughes, H.B.; Buurman, E.T.; Brown, A.J.P.; Odds, F.C.; Gow, N.A.R. *Candida albicans* Pmr1p, a secretory pathway P-type Ca²⁺/Mn²⁺-ATPase, is required for glycosylation and virulence. *J. Biol. Chem.* **2005**, *280*, 23408–23415. [CrossRef] [PubMed]

54. Mora-Montes, H.M.; Bates, S.; Netea, M.G.; Diaz-Jimenez, D.F.; Lopez-Romero, E.; Zinker, S.; Ponce-Noyola, P.; Kullberg, B.J.; Brown, A.J.P.; Odds, F.C.; et al. Endoplasmic reticulum -glycosidases of *Candida albicans* are required for N glycosylation, cell wall integrity, and normal host-fungus interaction. *Eukaryot. Cell* **2007**, *6*, 2184–2193. [CrossRef] [PubMed]

55. Munro, C.A.; Bates, S.; Buurman, E.T.; Hughes, H.B.; MacCallum, D.M.; Bertram, G.; Atrih, A.; Ferguson, M.A.J.; Bain, J.M.; Brand, A.; et al. Mnt1p and Mnt2p of *Candida albicans* are partially redundant α-1,2-mannosyltransferases that participate in O-linked mannosylation and are required for adhesion and virulence. *J. Biol. Chem.* **2005**, *280*, 1051–1060. [CrossRef] [PubMed]

56. Saijo, S.; Ikeda, S.; Yamabe, K.; Kakuta, S.; Ishigame, H.; Akitsu, A.; Fujikado, N.; Kusaka, T.; Kubo, S.; Chung, S.; et al. Dectin-2 recognition of α-mannans and induction of Th17 cell differentiation is essential for host defense against *Candida albicans*. *Immunity* **2010**, *32*, 681–691. [CrossRef] [PubMed]

57. Lal, P.; Sharma, D.; Pruthi, P.; Pruthi, V. Exopolysaccharide analysis of biofilm-forming *Candida albicans*. *J. Appl. Microbiol.* **2010**, *109*, 128–136. [CrossRef] [PubMed]

58. Correia, I.A. *Role of Secreted Aspartyl Proteases in Candida Albicans Virulence, Host Immune Response and Immunoprotection in Murine Disseminated Candidiasis*; Universidade do Minho: Braga, Portugal, 2012.

59. Johnson, C.J.; Cabezas-Olcoz, J.; Kernien, J.F.; Wang, S.X.; Beebe, D.J.; Huttenlocher, A.; Ansari, H.; Nett, J.E. The extracellular matrix of *Candida albicans* biofilms impairs formation of neutrophil extracellular traps. *PLoS Pathog.* **2016**, *12*. [CrossRef] [PubMed]

60. Williams, D.W.; Wilson, M.J.; Lewis, M.A.O.; Potts, A.J.C. Identification of *Candida* species by PCR and restriction fragment length polymorphism analysis of intergenic spacer regions of ribosomal DNA. *J. Clin. Microbiol.* **1995**, *33*, 2476–2479. [PubMed]

61. Arendrup, M.C.; Arikan, S.; Barchiesi, F.; Bille, J.; Dannaoui, E.; Denning, D.W.; Donnelly, J.P.; Fegeler, W.; Moore, C.; Richardson, M.; et al. EUCAST Technical note on the method for the determination of broth dilution minimum inhibitory concentrations of antifungal agents for conidia—Forming moulds. *ESCMID Tech. Notes* **2008**, *14*, 982–984.

62. *European Committee on Antimicrobial Susceptibility Testing, EUCAST Breakpoint Tables for Interpretation of MICs*, Version 8.1; Available online: http://www.eucast.org.

63. Skrzypek, M.S.; Binkley, J.; Binkley, G.; Miyasato, S.R.; Simison, M.; Sherlock, G. The Candida Genome Database (CGD): Incorporation of Assembly 22, systematic identifiers and visualization of high throughput sequencing data. *Nucleic Acids Res.* **2017**, *45*, D592–D596. [CrossRef] [PubMed]

64. Untergasser, A.; Nijveen, H.; Rao, X.; Bisseling, T.; Geurts, R.; Leunissen, J.A.M. Primer3Plus, an enhanced web interface to Primer3. *Nucleic Acids Res.* **2007**, *35*, W71–W74. [CrossRef] [PubMed]

65. Schmittgen, T.D.; Livak, K.J. Analyzing real-time PCR data by the comparative C_T method. *Nat. Protoc.* **2008**, *3*, 1101–1108. [CrossRef] [PubMed]

66. Cho, E.-J.; Shin, J.H.; Kim, S.H.; Kim, H.-K.; Park, J.S.; Sung, H.; Kim, M.N.; Im, H.J. Emergence of multiple resistance profiles involving azoles, echinocandins and Amphotericin B in *Candida glabrata* isolates from a neutropenia patient with prolonged fungaemia. *J. Antimicrob. Chemother.* **2015**, *70*, 1268–1270. [CrossRef] [PubMed]

67. Douglas, L.J. *Candida* biofilms and their role in infection. *Trends Microbiol.* **2003**, *11*, 30–36. [CrossRef]

68. Zarnowski, R.; Westler, W.M.; Lacmbouh, G.A.; Marita, J.M.; Bothe, J.R.; Bernhardt, J.; Sahraoui, A.L.H.; Fontainei, J.; Sanchez, H.; Hatfeld, R.D.; et al. Novel entries in a fungal biofilm matrix encyclopedia. *MBio* **2014**, *5*, 1–13. [CrossRef] [PubMed]

69. Mukherjee, P.K.; Chandra, J. *Candida* biofilm resistance. *Drug. Resist. Updat.* **2004**, *7*, 301–309. [CrossRef] [PubMed]

70. Livak, K.J. *User Bulletin #2 ABI P RISM 7700 Sequence Detection System SUBJECT: Relative Quantitation of Gene Expression—Updated 2001*; Applied Biosystems: Foster City, CA, USA, 1997.

71. Li, P.; Seneviratne, C.; Alpi, E.; Vizcaino, J.; Jin, L. Delicate metabolic control and coordinated stress response critically determine antifungal tolerance of *Candida albicans* biofilm persisters. *Antimicrob. Agents Chemother.* **2015**, *59*, 6101–6112. [CrossRef] [PubMed]

72. Al-Dhaheri, R.S.; Douglas, L.J. Absence of Amphotericin B-tolerant persister cells in biofilms of some *Candida* species. *Antimicrob. Agents Chemother.* **2008**, *52*, 1884–1887. [CrossRef] [PubMed]

73. Sun, J.; Li, Z.; Chu, H.; Guo, J.; Jiang, G.; Qi, Q. *Candida albicans* Amphotericin B-tolerant persister formation is closely related to surface adhesion. *Mycopathologia* **2016**, *181*, 41–49. [CrossRef] [PubMed]

74. Alexander, B.D.; Johnson, M.D.; Pfeiffer, C.D.; Jimenez-Ortigosa, C.; Catania, J.; Booker, R.; Castanheira, M.; Messer, S.A.; Perlin, D.S.; Pfaller, M.A. Increasing echinocandin resistance in *Candida glabrata*: Clinical failure correlates with presence of *FKS* mutations and elevated minimum inhibitory concentrations. *Clin. Infect. Dis.* **2013**, *56*, 1724–1732. [CrossRef] [PubMed]

75. Pinhati, H.M.S.; Casulari, L.A.; Souza, A.C.R.; Siqueira, R.A.; Damasceno, C.M.G.; Colombo, A.L. Outbreak of candidemia caused by fluconazole resistant *Candida parapsilosis* strains in an intensive care unit. *BMC Infect. Dis.* **2016**, *16*. [CrossRef] [PubMed]

76. Bizerra, F.C.; Jimenez-Ortigosa, C.; Souza, A.C.R.; Breda, G.L.; Queiroz-Telles, F.; Perlin, D.S.; Colombo, A.L. Breakthrough candidemia due to multidrug-resistant *Candida glabrata* during prophylaxis with a low dose of micafungin. *Antimicrob. Agents Chemother.* **2014**, *58*, 2438–2440. [CrossRef] [PubMed]

77. Park, S.; Kelly, R.; Kahn, J.N.N.; Robles, J.; Hsu, M.-J.; Register, E.; Li, W.; Vyas, V.; Fan, H.; Abruzzo, G.; et al. Specific substitutions in the echinocandin target Fks1p account for reduced susceptibility of rare laboratory and clinical *Candida* sp. isolates. *Antimicrob. Agents Chemother.* **2005**, *49*, 3264–3273. [CrossRef] [PubMed]

78. Desnos-Ollivier, M.; Moquet, O.; Chouaki, T.; Guerin, A.M.; Dromer, F. Development of echinocandin resistance in *Clavispora lusitaniae* during caspofungin treatment. *J. Clin. Microbiol.* **2011**, *49*, 2304–2306. [CrossRef] [PubMed]

79. Jensen, R.H.; Johansen, H.K.; Arendrup, M.C. Stepwise development of a homozygous S80P substitution in Fks1p, conferring echinocandin resistance in *Candida tropicalis*. *Antimicrob. Agents Chemother.* **2013**, *57*, 614–617. [CrossRef] [PubMed]

80. Lewis, J.S.; Wiederhold, N.P.; Wickes, B.L.; Patterson, T.F.; Jorgensen, J.H. Rapid emergence of echinocandin resistance in *Candida glabrata* resulting in clinical and microbiologic failure. *Antimicrob. Agents Chemother.* **2013**, *57*, 4559–4561. [CrossRef] [PubMed]

81. Shields, R.K.; Nguyen, M.H.; Press, E.G.; Kwa, A.L.; Cheng, S.; Du, C.; Clancy, C.J. The presence of an *FKS* mutation rather than MIC is an independent risk factor for failure of echinocandin therapy among patients with invasive candidiasis due to *Candida glabrata*. *Antimicrob. Agents Chemother.* **2012**, *56*, 4862–4869. [CrossRef] [PubMed]

82. Beyda, N.D.; Lewis, R.E.; Garey, K.W. Echinocandin resistance in *Candida* species: Mechanisms of reduced susceptibility and therapeutic approaches. *Ann. Pharmacother.* **2012**, *46*, 1086–1096. [CrossRef] [PubMed]

83. Barchiesi, F.; Spreghini, E.; Tomassetti, S.; Della Vittoria, A.; Arzeni, D.; Manso, E.; Scalise, G. Effects of caspofungin against *Candida guilliermondii* and *Candida parapsilosis*. *Antimicrob. Agents Chemother.* **2006**, *50*, 2719–2727. [CrossRef] [PubMed]

84. Garcia-Effron, G.; Katiyar, S.K.; Park, S.; Edlind, T.D.; Perlin, D.S. A naturally occurring proline-to-alanine amino acid change in Fks1p in *Candida parapsilosis*, *Candida orthopsilosis*, and *Candida metapsilosis* accounts for reduced echinocandin susceptibility. *Antimicrob. Agents Chemother.* **2008**, *52*, 2305–2312. [CrossRef] [PubMed]

85. Forastiero, A.; Garcia-Gil, V.; Rivero-Menendez, O.; Garcia-Rubio, R.; Monteiro, M.C.; Alastruey-Izquierdo, A.; Jordan, R.; Agorio, I.; Mellado, E. Rapid development of *Candida krusei* echinocandin resistance during caspofungin therapy. *Antimicrob. Agents Chemother.* **2015**, *59*, 6975–6982. [CrossRef] [PubMed]

86. Arendrup, M.; Perlin, D.; Jensen, R.; Howard, S.; Goodwin, J.; Hopec, W. Differential in vivo activities of anidulafungin, caspofungin, and micafungin against *Candida glabrata* isolates with and without *FSK* resistance mutations. *Antim Agents Chemoter.* **2012**, *56*, 2435–2442. [CrossRef] [PubMed]

87. Fanning, S.; Mitchell, A.P. Fungal Biofilms. *PLoS Pathog.* **2012**, *8*. [CrossRef] [PubMed]

88. Nett, J.; Lepak, A.; Marchillo, K.; Andes, D. Time course global gene expression analysis of an in vivo *Candida* biofilm. *J. Infect. Dis.* **2009**, *200*, 307–313. [CrossRef] [PubMed]

89. Barns, S.M.; Lane, D.J.; Sogin, M.L.; Bibeau, C.; Weisburg, W.G. Evolutionary relationships among pathogenic *Candida* species and relatives. *J. Bacteriol.* **1991**, *173*, 2250–2255. [CrossRef] [PubMed]

90. Kitada, K.; Yamaguchi, E.; Arisawa, M. Cloning of the *Candida glabrata TRP1* and *HIS3* genes, and construction of their disruptant strains by sequential integrative transformation. *Gene* **1995**, *165*, 203–206. [CrossRef]

91. Nakayama, H.; Ueno, K.; Uno, J.; Nagi, M.; Tanabe, K.; Aoyama, T.; Chibana, H.; Bard, M. Growth defects resulting from inhibiting *ERG20* and *RAM2* in *Candida glabrata*. *FEMS Microbiol. Lett.* **2011**, *317*, 27–33. [CrossRef] [PubMed]

92. Boone, C.; Sommer, S.S.; Hensel, A.; Bussey, H. Yeast KRE genes provide evidence for a pathway of cell wall beta-glucan assembly. *J. Cell Biol.* **1990**, *110*, 1833–1843. [CrossRef] [PubMed]

93. Srikantha, T.; Daniels, K.J.; Pujol, C.; Kim, E.; Soll, D.R. Identification of genes upregulated by the transcription factor Bcr1 that are involved in impermeability, impenetrability, and drug resistance of *Candida albicans* a/α biofilms. *Eukaryot. Cell* **2013**, *12*, 875–888. [CrossRef] [PubMed]

94. Chen, K.-H.; Miyazaki, T.; Tsai, H.-F.; Bennett, J.E. The bZip transcription factor Cgap1p is involved in multidrug resistance and required for activation of multidrug transporter gene *CgFLR1* in *Candida glabrata*. *Gene* **2007**, *386*, 63–72. [CrossRef] [PubMed]

95. Ferrari, S.; Sanguinetti, M.; De Bernardis, F.; Torelli, R.; Posteraro, B.; Vandeputte, P.; Sanglard, D. Loss of mitochondrial functions associated with azole resistance in *Candida glabrata* results in enhanced virulence in mice. *Antimicrob. Agents Chemother.* **2011**, *55*, 1852–1860. [CrossRef] [PubMed]

96. Mathé, L.; Van Dijck, P. Recent insights into *Candida albicans* biofilm resistance mechanisms. *Curr. Genet.* **2013**, *59*, 251–264. [CrossRef] [PubMed]

97. Marco, F.; Pfaller, M.A.; Messer, S.A.; Jones, R.N. Activity of MK-0991 (L-743,872), a new echinocandin, compared with those of LY303366 and four other antifungal agents tested against blood stream isolates of *Candida* spp. *Diagn. Microbiol. Infect. Dis.* **1998**, *32*, 33–37. [CrossRef]

98. Kuhn, D.M.; George, T.; Chandra, J.; Mukherjee, P.K.; Ghannoum, M.A. Antifungal susceptibility of *Candida* biofilms: Unique efficacy of amphotericin B lipid formulations and echinocandins. *Antimicrob. Agents Chemother.* **2002**, *46*, 1773–1780. [CrossRef] [PubMed]

99. Scorzoni, L.; de Paula e Silva, A.C.A.; Marcos, C.M.; Assato, P.A.; de Melo, W.C.; de Oliveira, H.C.; Costa-Orlandi, C.B.; Mendes-Giannini, M.J.; Fusco-Almeida, A.M. Antifungal therapy: New advances in the understanding and treatment of mycosis. *Front. Microbiol.* **2017**, *8*, 1–23. [CrossRef] [PubMed]

100. Pierce, C.G.; Srinivasan, A.; Uppuluri, P.; Ramasubramanian, A.K.; López-Ribot, J.L. Antifungal therapy with an emphasis on biofilms. *Curr. Opin. Pharmacol.* **2013**, *13*. [CrossRef] [PubMed]

101. Pappas, P.G.; Kauffman, C.A.; Andes, D.R.; Clancy, C.J.; Marr, K.A.; Ostrosky-Zeichner, L.; Reboli, A.C.; Schuster, M.G.; Vazquez, J.A.; Walsh, T.J.; et al. Clinical practice guideline for the management of Candidiasis: 2016 update by the infectious diseases society of America. *Clin. Infect. Dis.* **2015**, *62*, e1–e50. [CrossRef] [PubMed]

102. Canuto, M.M.; Rodero, F.G. Antifungal drug resistance to azoles and polyenes. *Lancet Infect. Dis.* **2002**, *2*, 550–563. [CrossRef]

103. Rex, J.H.J.; Walsh, T.J.; Sobel, J.D.J.; Filler, S.G.; Pappas, P.G.; Dismukes, W.E.; Edwards, J.E. Practice guidelines for the treatment of candidiasis. *Clin. Infect. Dis.* **2000**, *30*, 662–678. [CrossRef] [PubMed]

104. Schmalreck, A.F.; Willinger, B.; Haase, G.; Blum, G.; Lass-Flörl, C.; Fegeler, W.; Becker, K.; Antifungal susceptibility testing-AFST study group. Species and susceptibility distribution of 1062 clinical yeast isolates to azoles, echinocandins, flucytosine and amphotericin B from a multi-centre study. *Mycoses* **2012**, *55*, e124–e137. [CrossRef] [PubMed]
105. Perlin, D.S. Mechanisms of echinocandin antifungal drug resistance. *Ann. N. Y. Acad. Sci.* **2015**, *1354*, 1–11. [CrossRef] [PubMed]

Article

The Transcription Factor ZafA Regulates the Homeostatic and Adaptive Response to Zinc Starvation in *Aspergillus fumigatus*

Rocío Vicentefranqueira [1], Jorge Amich [2] , Laura Marín [1], Clara Inés Sánchez [1], Fernando Leal [1] and José Antonio Calera [1,*]

[1] Instituto de Biología Funcional y Genómica (IBFG-CSIC), Departamento de Microbiología y Genética, Universidad de Salamanca, 37007 Salamanca, Spain; rvrdew@usal.es (R.V.); lauramarin@usal.es (L.M.); cisanchezs@usal.es (C.I.S.); fleal@usal.es (F.L.)
[2] Manchester Fungal Infection Group (MFIG), Faculty of Biology, Medicine and Health, University of Manchester, Manchester M13 9NT, UK; jorge.amichelias@manchester.ac.uk
* Correspondence: jacalera@usal.es; Tel.: +34-923-294891; Fax: +34-923-224876

Received: 24 May 2018; Accepted: 20 June 2018; Published: 26 June 2018

Abstract: One of the most important features that enables *Aspergillus fumigatus* to grow within a susceptible individual and to cause disease is its ability to obtain Zn^{2+} ions from the extremely zinc-limited environment provided by host tissues. Zinc uptake from this source in *A. fumigatus* relies on ZIP transporters encoded by the *zrfA*, *zrfB* and *zrfC* genes. The expression of these genes is tightly regulated by the ZafA transcription factor that regulates zinc homeostasis and is essential for *A. fumigatus* virulence. We combined the use of microarrays, Electrophoretic Mobility Shift Assays (EMSA) analyses, DNase I footprinting assays and in silico tools to better understand the regulation of the homeostatic and adaptive response of *A. fumigatus* to zinc starvation. We found that under zinc-limiting conditions, ZafA functions mainly as a transcriptional activator through binding to a zinc response sequence located in the regulatory regions of its target genes, although it could also function as a repressor of a limited number of genes. In addition to genes involved in the homeostatic response to zinc deficiency, ZafA also influenced, either directly or indirectly, the expression of many other genes. It is remarkable that the expression of many genes involved in iron uptake and ergosterol biosynthesis is strongly reduced under zinc starvation, even though only the expression of some of these genes appeared to be influenced directly or indirectly by ZafA. In addition, it appears to exist in *A. fumigatus* a zinc/iron cross-homeostatic network to allow the adaptation of the fungus to grow in media containing unbalanced Zn:Fe ratios. The adaptive response to oxidative stress typically linked to zinc starvation was also mediated by ZafA, as was the strong induction of genes involved in gliotoxin biosynthesis and self-protection against endogenous gliotoxin. This study has expanded our knowledge about the regulatory and metabolic changes displayed by *A. fumigatus* in response to zinc starvation and has helped us to pinpoint new ZafA target genes that could be important for fungal pathogens to survive and grow within host tissues and, hence, for virulence.

Keywords: *Aspergillus fumigatus*; zinc; transcription; regulation

1. Introduction

Zinc is an essential nutrient element that plays a critical role in many different biological processes as a structural and/or as a catalytic component of many different enzymes [1,2]. Indeed, zinc is required, in both pathogens and their hosts, for the normal functioning of hundreds of enzymes and regulatory proteins in all cells. For this reason, zinc uptake and distribution are subjected to a strict homeostatic control. This is critical to ensure an appropriate steady supply of zinc and to prevent the noxious effects that both excess and deficiency of zinc may have on cell growth and differentiation.

Free-living, mutualist and commensal microorganisms obtain zinc from non-living matter such as water, soil or organic compounds present in their surrounding environment. In contrast, microbial pathogens have to obtain zinc from the tissues of the host they parasitize. However, zinc in living tissues is subjected to a strict homeostatic control that restricts the access of pathogens to it. Indeed, intracellular eukaryotic zinc proteins bind zinc tightly with pK_d stability constants (i.e., $-\log K_d$) of 10–12 [3]. As a consequence, although the concentration of total zinc within most eukaryotic cells is in the micromolar range (100–500 μM) [4,5], the amount of intracellular free Zn^{2+} ions is commensurately about six to seven orders of magnitude lower than the overall cellular zinc concentration. It has been estimated that the concentration of free Zn^{2+} ions in different types of mammalian cells and tissues range from tens to hundreds of picomoles per liter [6]. For instance, the concentration of free Zn^{2+} ions reported in human red blood cells is 24 pM [7], which indicates that a single red blood cell (90 fL volume) should have 1–2 atoms of free Zn^{2+} ions [8].

Aspergillus fumigatus is a saprophyte filamentous fungus that usually grows on organic decaying matter. However, it is also able to grow as a parasite in the lungs of immunosuppressed individuals and cause invasive pulmonary aspergillosis (IPA), which is one of the fungal infectious diseases with the highest mortality rate among immunosuppressed patients [9]. One of the most important features that enables *A. fumigatus* to grow within a susceptible host and causes disease is its ability to obtain Zn^{2+} ions from the hostile environment provided by host tissues [10]. *A. fumigatus* zinc uptake from host tissues relies on transporters of the ZIP family encoded by the *zrfA*, *zrfB* and *zrfC* genes [11,12]. The ZrfC transporter seems to be well adapted to function under the alkaline zinc-limiting conditions of mammals' bodies and plays a major role in zinc uptake from host tissues. Interestingly, the ZrfA and ZrfB zinc transporters, despite their expression are higher in acidic than in alkaline zinc-limiting media, also play a relevant role in virulence. Thus, the virulence of a Δ*zrfA*Δ*zrfB*Δ*zrfC* mutant is fully abrogated in a murine model of invasive pulmonary aspergillosis [13], as is that of a Δ*zafA* mutant [14], which lacks the gene encoding the zinc-responsiveness transcriptional factor ZafA that regulates zinc homeostasis in *A. fumigatus*. Hence, the ZafA-mediated induction of the zinc uptake system under zinc-limiting conditions is a key homeostatic function required for optimal fungal growth within host tissues.

The most investigated ZafA orthologue is the Zap1 transcription factor of *Saccharomyces cerevisiae* [15]. Several genome-wide transcriptional profiling studies in *S. cerevisiae* have shown that Zap1 influences the expression of many genes required to implement properly the adaptive and homeostatic responses to zinc deficiency [16–18]. Similar studies on Zap1-regulated genes have been also performed in the pathogenic yeasts *Candida albicans* and *Cryptococcus gattii* [19,20]. These Zap1-like yeast transcription factors have 3–4 canonical zinc fingers domains (ZFDs) of the C2H2-type and four ZFDs of the CWCH2-type. The latter are typically clustered in pairs forming two tandem motifs (tCWCH2) [21] (Figure 1). Interestingly, in transcription factors carrying tCWCH2 motifs, only the CWCH2 zinc finger of the tCWCH2 motif closer to a canonical C2H2 zinc finger interacts with the DNA backbone, whereas its partner CWCH2 zinc finger is dedicated to protein-protein interactions and does not interact with the DNA [22–24]. Actually, it has been proposed that the tCWCH2 motifs are involved in inter-ZFD interactions and influence both the recognition and binding capacities of their neighboring canonical C2H2 domains to DNA [21]. Since the activity of the transcription factors having tCWCH2 motifs appear to be largely influenced by this type of ZFDs, it is very likely that the transcriptional profile under zinc starvation in yeast carrying Zap1-like factors with two tCWCH2 motifs differs to a certain extent from that observed in *A. fumigatus* and other filamentous fungi carrying ZafA-like factors with a unique tCWCH2 motif [14] (Figure 1). To discover new ZafA target genes essential for fungal virulence other than those encoding ZIP transporters, we performed a genome-wide transcription profiling in *A. fumigatus* grown in vitro under alkaline zinc-limiting conditions to mimic those found by the fungus in the lungs of a susceptible host. In addition, we analyzed the promoter regions of the genes that exhibited the highest differential expression levels between zinc-replete and zinc-limiting conditions. To this purpose, we used a combination of in vitro and in silico procedures to identify the consensus DNA zinc response motif to which ZafA binds under zinc-limiting conditions.

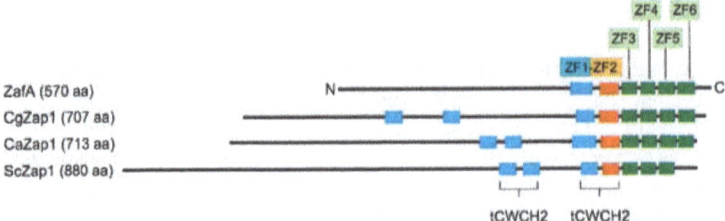

Figure 1. Schematic representation of the transcriptional regulators of zinc homeostasis from *Aspergillus fumigatus* (ZafA), *Cryptococcus gattii* (CgZap1), *Candida albicans* (CaZap1) and *Saccharomyces cerevisiae* (ScZap1). Green boxes indicate canonical zinc finger (ZF) domains of the C2H2-type. Blue and orange boxes indicate ZFs of the CWCH2-type. These ZFs are typically arranged in pairs forming tandem CWCH2 motifs (tCWCH2). The ZF of the CWCH2-type that is presumably able to bind DNA is indicated as an orange box. Protein sequences were aligned taking as reference the ZF of the C2H2-type located towards the N-terminus in ZafA (i.e., ZF3).

2. Materials and Methods

2.1. Strains and Culture Media

The *A. fumigatus* strains used in this study are listed in Table 1. These fungal strains were routinely grown in either the PDA complex medium (20 g/L potato dextrose agar, 20 g/L sucrose, 2.5 g/L $MgSO_4$-$7H_2O$) or in the liquid synthetic dextrose nitrate zinc-limiting medium (SDN–Zn, pH 7.2) (1.7 g/L YNB without amino acids, without ammonium sulphate and without zinc [CYN2401, Formedium], 20 g/L Dextrose, 3 g/L $NaNO_3$, 12 μM $FeSO_4$-$7 H_2O$, 6 μM $CuSO_4$-$5 H_2O$, 10 μM Na_2MoO_4-$2 H_2O$), which was supplemented with zinc as specified using a sterile stock solution of 1 mM $ZnSO_4$-$7H_2O$ in ultrapure water.

Table 1. *Aspergillus fumigatus* strains used in this study. Genes in brackets were reintroduced into the *A. fumigatus* genome by targeting them at the intergenic region between the AFUA_2G08360 (*pyrG*) and AFUA_2G08350 genes.

Strain	Detailed Genotype	Reference
CEA17	*pyrG1* (auxotrophic PyrG⁻)	[11]
AF14	Wild-type (isogenic to CEA17)	[12]
AF171	Δ*zafA::hisG* (isogenic to CEA17)	[14]
AFZR0	Wild-type [*luc* ← Pz*rfC^{wt}* → *gfp*] (isogenic to CEA17)	This study
AFZR1	Wild-type [*luc* ← Pz*rfC^{ZR123}* → *gfp*] (isogenic to CEA17)	This study

2.2. Culture of Aspergillus fumigatus for RNA Isolation

To obtain RNA under zinc-limiting culture conditions to be used for both microarray analyses and Real Time quantitative PCR (RT-qPCR) experiments, the wild-type and Δ*zafA* mutant strain were inoculated to a density of 1.5×10^6 spores/mL in 200 mL of the SDN–Zn zinc limiting medium and incubated at 37 °C with shaking at 200 rpm for 20 h. Similarly, to obtain RNA under zinc-replete culture conditions, the wild-type strain was inoculated and incubated as described previously in the SDN–Zn zinc limiting medium supplemented with 100 μM zinc. In all cases, spores grown in PDA were used as inoculum.

The mycelia were harvested by filtration on filter paper, washed twice with sterile water and snap-frozen in liquid nitrogen. After grinding the mycelia in the presence of liquid nitrogen, the RNA was extracted using the RNeasy Plant Mini Kit (74904, QIAGEN, Hilden, Germany) according to manufacturer's instructions. RNA was eluted in 50 µL of RNase-free water. The concentration and quality of RNA was determined by UV spectrometry (Nanodrop ND1000 spectrophotometer, Thermo Fisher Scientific, Waltham, MA, USA) and checked on 0.8% agarose gels stained with ethidium bromide. RNA samples were stored at −80 °C until use.

2.3. Microarray Experiments

The *A. fumigatus* Af293 DNA microarray slides (version 4) with 2× replicate 70-mer oligonucleotide printings used in this study were obtained (April 2010) through the Pathogen Functional Genomics Resource Center of the National Institute of the Allergy and Infectious Diseases/National Institute of Health (NIAID/NIH), managed and funded by the Division of Microbiology and Infectious Diseases (DMID) and operated by the J. Craig Venter Institute.

Total RNA was treated with RNase DNase-free and 2 µg of total RNA DNA-free was used for cDNA synthesis, which was performed using SuperScript III RT and a dNTP/5-(3-aminoallyl)-dUTP labelling mix. Upon elimination of unincorporated aa-dUTP, either the Cy5 or Cy3 dye was coupled to the aminoallyl-labelled cDNA by incubation for 1 h at room temperature in 0.1 M sodium carbonate (pH 9.3). Free dyes were removed using the QIAGEN PCR purification kit and were calculated for the total picomoles of dyes incorporation and the pmol cDNA/pmol Cy dye ratios upon measuring spectrophotometrically the cDNA concentration, $OD_{650\,nm}$ for Cy5 and $OD_{550\,nm}$ for Cy3. Microarray experiments were performed in duplicate with dye labels reversed. Each slide was hybridized with 1:1 mix of two differentially labeled probes Cy3 and Cy5 (1100 pmoles of each one) in a HS 4800 Pro™ Hybridization Station (Tecan Group Ltd., Männedorf, Switzerland) and scanned using the Axon GenePix 4000B microarray scanner. The TIFF images generated were analyzed using Spotfinder 321a to obtain relative transcript levels. Data were normalized with MIDAS using LOWESS (Locally Weighted Scatterplot Smoothing) and the dye-swap procedure (all softwares were developed by the J. Craig Venter Institute to be used for DNA microarray slides provided by the NIAID/NIH). The resulting data were averaged from duplicated genes on each array and from duplicate flip-dye arrays for each experiment, taking a total of four intensity data points for each gene. A volcano-plot algorithm was used to identify genes that exhibited statistical significance ($p < 0.05$) with a change in expression levels ≥ 1.5-fold. Differentially expressed genes at the 95% confidence level were determined using T-test implemented in the MEV program of the TM4 microarray software suite (http://www.tm4.org). The microarray data have been deposited in the GEO database under accession number GSE109389.

2.4. RT-qPCR

The concentration of total RNA in all samples were brought to a final concentration of 150 ng/µL. To remove genomic DNA, 1.5 µg of total RNA were treated with RQ1 DNase I (M610, Promega, Madison, WI, USA) and checked by conventional PCR for the complete absence of gDNA. Next, 1 µg of DNase-treated RNA was reversed transcribed using the SuperScript II Reverse Transcriptase (18064-014, Invitrogen, ThermoFisher Scientific) and oligo (dT)$_{15}$ (C1101, Promega) or random hexamers (SO142, ThermoFisher Scientific) as primers. Prior to the qPCR reactions, the cDNA samples were diluted 1:3 in water (except for reactions against the 18S rRNA that were diluted 1:1200 in water). Quantitative real time PCR (qPCR) reactions were performed in a BioRad CFX96 equipment. A typical qPCR reaction mixture (10 µL) contained 13.5 ng cDNA (32 pg when the qPCR was for 18S rRNA), a specific pair of primers (150 nM final concentration) and the SYBR Premix ExTaq (Tli RNaseH Plus) (RR420A, Takara). The amount of cDNA used per reaction was calculated on the basis that all RNA had been reversed transcribed into cDNA, including the rRNAs. Primers used for qPCR are listed in Table S1. PCRs were carried out for 40 cycles, denaturation at 95 °C for 10 s, annealing at 59 °C for 20 s and extension at 72 °C for 20 s. The relative expression level with respect to 18S rRNA (REL/18S) was calculated by the $2^{-\Delta Ct}$ method. The relative expression ratio (rER) was calculated by the $2^{-\Delta\Delta Ct}$ method using the expression level of the 18S rRNA as internal reference.

2.5. Expression and Purification of the Recombinant ZafA Protein

An *Eco*RI-flanked DNA fragment (482 bp) carrying the coding sequence for the C-terminal 140 amino acids of ZafA (residues 434–570) was obtained by high-fidelity PCR using as template the plasmid pZAF14, which carried the complete cDNA for the ZafA coding sequence [14], and the pair of oligonucleotides JA181/JA182 (Table S2). This fragment was digested with *Eco*RI and ligated to the *Eco*RI site of the pGEX-5X-3 plasmid (GE Healthcare, Little Chalfont, UK) to generate the pGEX-5X-ZafA plasmid. The ZafA coding sequence inserted into the pGEX-5X-ZafA plasmid was confirmed by sequencing. The pGEX-5X-ZafA plasmid was used to transform *Escherichia coli* BL21-DE3. The resulting strain was able to synthesize high amounts of the four C-terminal zinc fingers of ZafA (ZF3-6) fused to the C-terminus of GST upon induction with IPTG (0.25 mM). Although the GST-ZafA^{ZF3-6} protein aggregated into inclusion bodies, it was easily solubilized using a slightly acidic lysis buffer (20 mM Tris-HCl, 150 mM NaCl, 100 mM ZnCl$_2$, 0.1% Triton X-100, 5 mM DTT [pH 6.7]), instead of the most standard lysis buffer at pH 7.5. The GST-ZafA^{ZF3-6} protein was purified using Glutathione Sepharose 4B (GE Healthcare, Cat. 17075601) and eluted with elution buffer (50 mM Tris-ClH [pH 8.0], 150 mM NaCl, 0.1% Triton X-100, 10 mM reduced glutathione). The eluted GST-ZafA^{ZF3-6} protein was concentrated with an Amicon Ultra-15 3K device (Millipore, Burlington, MA, USA) and dialyzed in a D-Tube Dialyzer Maxi unit (MWCO 3.5 kDa) (Merck-Millipore, Novagen, Temecula, CA, USA) against the factor Xa buffer (20 mM Tris-Cl, pH 6.5, 50 mM NaCl, 1 mM CaCl$_2$). Upon optimizing the cleavage conditions of the GST-ZafA^{ZF3-6} protein with factor Xa, the purified GST-ZafA^{ZF3-6} protein was treated with 25 ng factor Xa (NEB, P8010, Ipswich, MA, USA) per 260 ng of protein for 6 h at room temperature to achieve 100% cleavage of GST-ZafA^{ZF3-6} protein, as judged in SDS-PAGE gels stained with coomassie blue. The factor Xa was removed from the cleavage reaction with Xarrest™ Agarose (Merck-Millipore, cat. 69038). Following the capture of Factor Xa, the agarose was removed by spin-filtration. The cleavage reaction without factor Xa, which contained equimolar amounts of GST and ZafA^{ZF3-6}, was dialyzed against binding buffer (25 mM Tris-HCl, pH 8.0, 50 mM KCl, 10 µM ZnCl$_2$, 1 mM MgCl$_2$, 1 mM DTT). Proteins were stored at −20 °C until use.

The GST protein that was used as control was expressed from plasmid pGEX-5X-3, purified, digested with factor Xa and dialyzed as described for the GST-ZafA^{ZF3-6} protein.

Protein concentrations were estimated by measuring A280 nm in a NanoDrop spectrophotometer (ThermoFisher) and taking into consideration that the extinction molar coefficients for the GST and ZafA^{ZF3-6} moieties of the GST-ZafA^{ZF3-6} protein were 1.632 and 0.1811 mg^{-1} × mL × cm^{-1}, respectively.

2.6. Construction of Plasmids to Obtain Probes for Electrophoretic Mobility Shift Assays and DNase I Footprinting Experiments

Two overlapping DNA fragments of the *zafA* promoter (P*zafA*) of 397 bp and 233 bp were obtained by PCR using as template gDNA from the wild-type AF14 strain and the pair of oligonucleotides JA204/JA394 and JA375/JA204, respectively (Table S2). The PCR products were ligated to the pGEM-T vector (Promega, cat. A1360) to generate the pZAF151 and pZAF91 plasmids.

The complete sequence of the bidirectional promoter P*aspf2-zrfC* that drives the expression of the *aspf2* and *zrfC* genes (abbreviated as P*zrfC*, 884 bp) was amplified by PCR using as template gDNA from the wild-type AF14 strain and the pair of oligonucleotides JA194/JA196. The PCR product was ligated to the pGEM-T easy vector (A1360, Promega) to generate the pASFP2-ZRF31 plasmid. Similarly, a 215-bp DNA fragment of P*zrfC* right next to the *zrfC* coding sequence was amplified using the pair of oligonucleotides JA376/JA377 and the PCR product was ligated to the pGEM-T vector to generate the pASFP2-ZRF311 plasmid. Finally, 539 bp of P*zrfC* right next to the *zrfC* coding sequence were excised from the pASFP2-ZRF31 plasmid following an *Eco*RV/*Msc*I digestion-religation to generate the pASFP2-ZRF312 plasmid, which only carried 349 bp of P*zrfC* right next to the *aspf2* coding sequence.

A 222-bp DNA fragment of the *zrfA* promoter (P*zrfA*) was amplified by PCR using as template gDNA from the wild-type AF14 strain and the pair of oligonucleotides JA378/JA379. The PCR product was ligated to the pGEM-T vector to generate the pZRF10 plasmid.

A 256-bp DNA fragment of the *zrfB* promoter (P*zrfB*) was amplified by PCR using as template gDNA from the wild-type AF14 strain and the pair of oligonucleotides JA380/JA381. The PCR product was ligated to the pGEM-T vector to generate the pZRF261 plasmid.

A 191 bp-DNA fragment of the *alcA* promoter of *Aspergillus nidulans* (P*alcA*) was amplified by high-fidelity PCR using the pair of oligonucleotides JA261/JA388 and as template gDNA from the *A. nidulans* strain G1059 [25]. The PCR product was ligated to the pGEM-T vector to generate plasmid pALC3.

All subcloned PCR products were sequenced to verify the absence of any mutation that could have been introduced during PCR.

2.7. Preparation of Probes for Electrophoretic Mobility Shift Assays

DNA probes for different promoters (P*zafA*, P*zrfC*, P*zrfA*, P*zrfB* and P*alcA*) were amplified by high-fidelity PCR using Pfu polymerase and purified using the QIAquick PCR Purification Kit (QIAGEN). The oligonucleotides used as primers (Table S2) and plasmids used as templates for DNA amplification by PCR were the following: For P*zafA* EMSA assays we used a 233-bp DNA fragment amplified by PCR using a dilution of the pZAF91 plasmid as template and the pair of oligonucleotides JA375/JA204 as primers; For P*zrfC* EMSA assays we used a 221-bp DNA fragment amplified by PCR using a dilution of the pASFP2-ZRF311 plasmid as template and the pair of oligonucleotides JA376/JA377; For P*zrfA* EMSA assays we used a 222-bp DNA fragment amplified by PCR using a dilution of the pZRF10 plasmid as template and the pair of oligonucleotides JA378/JA379 as primers; For P*zrfB* EMSA assays we used a 256-bp DNA fragment amplified by PCR using a dilution of the pZRF261 plasmid as template and the pair of oligonucleotides JA380/JA381 as primers; For P*alcA* EMSA assays we used a 213-bp DNA fragment amplified by PCR using a dilution of the pALC3 plasmid as template and the pair of oligonucleotides JA261/JA388 as primers.

2.8. Preparation of Probes for DNase I Footprinting Assays

6-FAM-5′-labelled DNA probes for different promoters (P*zafA*, P*zrfC*, P*zrfA*, P*zrfB* and P*alcA*) were amplified by high-fidelity PCR using Pfu polymerase and purified using the QIAquick PCR Purification Kit (QIAGEN). The oligonucleotides used as primers (Table S2) and plasmids used as templates in DNA amplification by PCR are described next.

Four different probes were used for the P*zafA* DNase I footprinting assays: (i) Two 432 bp DNA fragments amplified by PCR using the pZAF91 plasmid as template and the pair of oligonucleotides FAM-Rv/Fw and Rv/FAM-Fw as primers to get them labelled in 5′-ends of the sense and antisense strands respectively (these probes carried 229 bp of P*zafA* plus two flanking sequences corresponding to the MCS of the vector); (ii) A 536 bp DNA fragment amplified by PCR using the pZAF151 plasmid as template and the pair of oligonucleotides FAM-Rv/JA204 as primers to get it labelled in the 5′-end of the sense strand (this probe carried 396 bp of P*zafA* plus one flanking sequence corresponding to the MCS of the vector); (iii) A 435 bp DNA fragment amplified by PCR using the pZAF151 plasmid as template and the pair of oligonucleotides Rv/FAM-JA439 as primers to get it labelled in the 5′-end of the antisense strand (this probe carried 295 bp of P*zafA*).

Four different probes were used for the P*zrfC* DNase I footprinting assays: (i) A 322 bp DNA fragment amplified by PCR using the pASPF2-ZRF311 plasmid as template and the pair of oligonucleotides FAM-Fw/JA377 as primers to get it labelled in the 5′-end of the sense strand for *zrfC* (this probe carried 215 bp of P*zrfC* plus one flanking sequence corresponding to the MCS of the vector); (ii) A 347 bp DNA fragment amplified by PCR using the pASPF2-ZRF311 plasmid as a template and the pair of oligonucleotides JA376/FAM-Rv as primers to get it labelled in the 5′-end of the antisense strand for *zrfC* (this probe carried 215 bp of P*zrfC* plus one flanking sequence corresponding to the MCS of the vector); (iii) A 393 bp DNA fragment amplified by PCR using the pASPF2-ZRF312 plasmid as template and the pair of oligonucleotides FAM-Fw/JA7 as primers to get it labelled in the 5′-end of the sense strand for *zrfC* (this probe carried 349 bp of P*zrfC* plus one flanking sequence corresponding to the MCS of the vector); (iv) A 504 bp DNA fragment amplified by PCR using the pASPF2-ZRF312 plasmid as template and the pair of oligonucleotides JA194/FAM-Rv as primers to get it labelled in the 5′-end of the antisense strand for *zrfC* (this probe carried 349 bp of P*zrfC* plus one flanking sequence corresponding to the MCS of the vector).

Two different probes were used for the P*zrfB* DNase I footprinting assays: (i) A 456 bp DNA fragment amplified by PCR using the pZRF261 plasmid as template and the pair of oligonucleotides FAM-Fw/Rv as primers to get it labelled in the 5′-end of the sense strand; (ii) A 456 bp DNA fragment amplified by PCR using the pZRF261 plasmid as template and the pair of oligonucleotides Fw/FAM-Rv as primers to get it labelled in the 5′-end of the antisense strand. Both probes carried 256 bp of P*zrfB* plus two flanking sequences corresponding to the MCS of the vector.

One probe was used for the P*zrfA* DNase I footprinting assays. A 350 bp DNA fragment amplified by PCR using the pZRF10 as template and the pair of oligonucleotides JA378/FAM-Rv as primers to get it labelled in the 5′-end of the antisense strand (this probe carried 217 bp of P*zrfA* plus one flanking sequence corresponding to the MCS of the vector).

One probe was used for the P*alcA* DNase I footprinting assays. A 413 bp DNA fragment of the pALC3 plasmid carrying 213 bp of P*alcA* plus two flanking sequences corresponding to the MCS of the vector were amplified by high-fidelity PCR and the pair of oligonucleotides FAM-Fw/Rv or Fw/FAM-Rv.

2.9. Electrophoretic Mobility Shift Assays and DNase I Footprinting Assays Reactions and Analyses

Reactions for EMSA experiments (20 µL) were assembled in 0.5-mL microtubes and carried out in EMSA buffer (25 mM Tris-HCl [pH 8], 50 mM KCl, 1 mM MgCl$_2$, 12.5% glycerol, 1 mM DTT), containing 1 µL BSA (0.2 mg/mL), 0.25–0.75 pmoles of the target DNA and the amount of the equimolar mixture of GST and ZafA^{ZF3-6} required to add 0.125–24 pmoles ZafA^{ZF3-6}. Reactions were incubated for 1 h at 30 °C before adding 4 µL of loading buffer (20% sucrose, 0.002% bromophenol blue). Reactions were loaded in a polyacrylamide 5% EMSA gel that had been pre-run in TBE 1× for 10 min at 4 °C in a cold room. Gels were run at 100 V for approximately 75 min, stained with a solution of ethidium bromide (5 µg/mL) for 20 min at room temperature, washed for 10 min with distilled water and photographed on a UV transilluminator in a gel documentation system.

Reactions for footprinting assays (60 µL) were assembled in 0.5-mL microtubes and carried out in an EMSA buffer containing 3 µL BSA (0.2 mg/mL), 0.5 pmoles of the desired 6-carboxyfluorescein (6-FAM)-labeled DNA fragment and 12 pmoles of ZafA^{ZF3-6}. Control reactions were assembled identically but omitting ZafA^{ZF3-6}. All DNA fragments used for footprinting assays were obtained by high-fidelity PCR using as primers both a 6-FAM-5′-labelled oligonucleotide and a non-labeled oligonucleotide, as described above. After incubating the footprinting reactions for 45 min at 30 °C, DNA was partially digested with 0.01 U of DNase I by adding 4 µL of a DNase I stock solution (0.0025 U/µL) per reaction and incubated at 25 °C for 1 min. The reaction was stopped immediately by adding 340 µL of stop buffer (9 mM Tris-HCl [pH 8], 40 mM EDTA). DNA was extracted with one volume of phenol:chloroform:isoamyl alcohol (25:24:1) followed by an extraction with one volume of chloroform:isoamyl alcohol (24:1). DNA in the aqueous phase was precipitated with 0.1 volumes of a non-buffered 3.0 M AcNa solution plus 2.5 volumes of ethanol. Precipitation of DNA was left to stand to proceed overnight at −20 °C. After centrifugation, the DNA pellet was washed twice with 70% ethanol, allowed to air dry at room temperature, suspended in 10 µL of Hi-Di formamide, denatured at 95 °C for 3 min and cooled on ice. All DNase I footprinting reactions were analyzed electrophoretically on an automated capillary DNA sequencer using the GeneScan™ 500 LIZ (Thermofisher) dye Size Standard as a calibrator. Each footprinting reaction was performed in parallel to four dideoxynucleotide-based sequencing reactions of the corresponding DNA probe, using as a primer the same 6-FAM 5′-labelled oligonucleotides utilized to amplify each DNA fragment by PCR.

2.10. Construction of Plasmids Used for Aspergillus fumigatus Transformation

The strains AFZR0 and AFZR1 of *A. fumigatus* used in this study were able to express the coding sequences of both the firefly luciferase (*luc*; as a reporter of *aspf2* expression) and the green fluorescent protein (*gfp*; as a reporter of *zrfC* expression) under control of either a wild-type version of the bidirectional *aspf2-zrfC* promoter (PzrfC^wt) or a mutant version of this promoter whose ZR motifs had been inactivated by site directed mutagenesis (PzrfC^ZR123).

The CEA17 uridine-uracil-auxotrophic *pyrG1* strain was transformed with two different EcoRI-EcoRI DNA fragments of 7286 bp, excised respectively from plasmids pPYRGQ191 and pPYRGQ193 to generate respectively the AFZR0 and AFZR1 strains (Figure S1).

The plasmids pPYRGQ191 and pPYRGQ193 were generated by ligating XbaI-FspI DNA fragments (4105 bp), which carried respectively the [*luc* ← PzrfC^wt → *gfp*] and [*luc* ← PzrfC^ZR123 → *gfp*] constructs, to the pPYRGQ31 plasmid digested with XbaI/SmaI. The pPYRGQ31 plasmid was an improved version of the pPYRGQ3 plasmid that had been designed previously in our laboratory to revert specifically the *pyrG1* mutation (C756T) in the *A. fumigatus* CEA17 or in any PyrG⁻ CEA17 derivative strain and select PyrG⁺ prototrophic strains bearing the DNA fragment of interest inserted between its AFUA_2G08360 (*pyrG*) and AFUA_2G08350 loci [11].

The plasmids used to transform *A. fumigatus* were linearized by digestion with *EcoRI*, extracted with phenol:chloroform:IAA, precipitated with acetate/isopropanol, washed in 70% ethanol, dissolved in 50% (*v/v*) of KC solution (0.6 M KCl, 50 mM CaCl; pH 6.0–6.5) and used for transformation as described below.

2.11. Generation of Protoplasts and Transformation of Aspergillus fumigatus

Protoplasts of *A. fumigatus* were prepared using a new protocol developed in our laboratory to obtain high-quality protoplasts that were very susceptible to transformation. In brief, to obtain protoplasts of the CEA17 PyrG⁻ strain we inoculated 5×10^8 conidia of this strain in the SDN medium supplemented with 0.05% (*w/v*) uracil, 0.12% (*w/v*) uridine and 20% (*v/v*) of a sterile conditioned medium that was produced by ourselves and contained a highly active α-glucanase. The culture was incubated at 37 °C for 14 h. Germlings were collected by filtration through a cell strainer unit (40 μm), suspended in 10 mL of protoplasting buffer (0.2 g VinoTaste, 0.75 M KCl, 25 mM citrate/phosphate; pH 5.8) (it had been previously sterilized by filtration through a 0.22 μm filter unit) and incubated at 35 °C with shaking at 120 rpm for 2 h. The protoplasting suspension was filtered through a miracloth (Merck-Millipore, Calbiochem Cat. 475855; 22–25 μm) and centrifuged at 1200× *g* for 10 min a 4 °C. The protoplast pellet was washed in 15 mL of a cold KC solution (0.6 M KCl, 50 mM CaCl; pH 6.0–6.5) by mixing gently and centrifuged at 1200× *g* for 10 min at 4 °C. The protoplast pellet was suspended into 0.4 mL of KC solution and used for transformation.

Aliquots of 0.2 mL of protoplast suspension were dispensed in 15-mL conical tubes to which were added sequentially the transforming DNA (~10 μg) and 0.5 mL of PEG buffer (30% PEG-6000 [*w/v*], 0.6 M KCl, 50 mM CaCl₂, 10 mM Tris-HCl; pH 7.5). The protoplast-DNA-PEG mixture was incubated on ice for 30 min and then at 25 °C for 30 min in a water bath. Next, 9.3 mL of KC solution was added to each protoplast-DNA-PEG solution, mixed by inversion, centrifuged at 1200× *g* for 10 min at 4 °C, suspended in 0.5 mL of KC solution, mixed with 6 mL of soft agar (AMM with 0.6 M KCl plus 0.4% [*w/v*] agarose) and poured onto two selective AMM plates. Plates were incubated at 37 °C until PyrG⁺ fungal transformants had grown. Several independent transformants were isolated and re-isolated on AMM agar plates. The transformants that had undergone homologous recombination at the expected locus were confirmed by PCR. A sample of conidia scrapped with a cotton swab from a colony of each isolated was inoculated into 1 mL of AMM and incubated overnight at 37 °C to obtain minipreps of genomic DNA suitable for PCR analysis.

2.12. Detection of SOD Activity on PAGE Gels

After 20 h of incubation of a wild-type strain in liquid SDN–Zn medium, mycelium was harvested through filtration, washed with cold sterile water and snap-frozen on liquid N_2. Mycelium was ground in a mortar with a pestle in presence of liquid N_2 until it became a fine dust, suspended into 0.4 mL of native protein extraction buffer (36 mM KH_2PO_4/K_2HPO_4 [pH 7.8]; 1 mM EDTA; 0.1% Triton X-100; 5% [v/v] glycerol; 0.5% [v/v] Protease inhibitor cocktail-EDTA [Thermo Scientific, cat. 87785]) and clarified by centrifugation at 14,000 × g for 10 min at 4 °C. Proteins were separated by native PAGE (T = 8%) at 120 V at 4 °C and gels were washed twice with distilled water (10 min each). Gels were incubated for 20 min in 50 mL of an Nitro Blue Tetrazolium Chloride (NBT) solution (2.45 mM [N6876, Sigma-Aldrich, St. Louis, MO, USA] in 36 mM KH_2PO_4/K_2HPO_4 [pH 7.8]), transferred to the developing solution (28 µM riboflavin, 28 mM TEMED in 36 mM KH_2PO_4/K_2HPO_4 [pH 7.8]) and illuminated with white light until transparent bands were readily observed in a dark-blue background due to formazan generated after reduction of NBT with O_2^- formed by reduction of riboflavin in the presence of O_2.

3. Results

3.1. Under Zinc-Limiting Conditions ZafA Influences the Expression of Genes Related to Many Different Metabolic and Biosynthetic Processes in Addition to Those Required to Maintain Zinc Homeostasis

When *A. fumigatus* grows under zinc-limiting conditions, a significant change in the transcription level of the most direct ZafA-target genes is expected. In addition, other genes not targeted by ZafA may also change their expression levels as a side effect of the adaptation to the zinc-limiting conditions triggered by ZafA. To identify all the putative ZafA-target genes we designed an experimental approach based in the following theoretical assumptions: (i) Some genes that changed their expression level in a wild-type strain under zinc-limiting conditions would be regulated either directly or indirectly by ZafA, whereas other genes would change their expression level in response to zinc starvation in a ZafA-independent manner. We referred to the former genes as putative directly and indirectly ZafA-regulated (DZR and IZR) genes, whereas the later ones were referred to as non-ZafA-regulated (NZR) genes; (ii) The NZR genes could be either up-regulated or down-regulated both in the wild-type strain and Δ*zafA* mutant grown under zinc-limiting with respect to zinc-replete conditions; (iii) The expression level of the DZR genes in a Δ*zafA* strain should be similar under both zinc-limiting and zinc-replete conditions; (iv) The intrinsic lack of the appropriate homeostatic and adaptive responses to zinc starvation in the Δ*zafA* strain would result in a strain-specific adaptive response to relieve the lack of *zafA* in this mutant that might involve changes in the expression level of many genes that in a wild-type strain would not change under zinc starvation, and that we referred to as strain-specific adaptive (SSA) genes.

Therefore, by keeping these assumptions in mind we performed two independent microarray experiments (Exp#1 and Exp#2) (Figure 2). Exp#1 was aimed to determine genes expressed differentially in a wild-type strain grown under zinc-limiting conditions versus zinc-replete conditions (Results in Table S3). Exp#2 was aimed to determine the genes expressed differentially in a Δ*zafA* strain with respect to a wild-type strain, both grown under zinc-limiting conditions (Table S4). Upon comparing the differentially regulated genes detected in Exp#1 and Exp#2, 112 putative IZR genes (19 up-regulated and 93 down-regulated in both the wild-type strain and Δ*zafA* mutant) and 153 putative DZR genes were identified (Figure 2). More specifically, 118 DZR genes that were up-regulated in the wild-type strain according to Exp#1 emerged as down-regulated in the Δ*zafA* strain according to Exp#2, whereas 35 DZR genes that were down-regulated in the wild-type strain according to Exp#1 emerged as up-regulated in the Δ*zafA* strain according to Exp#2 (Table S5). It would be expected that the direct targets of ZafA were included among the DZR genes. It would be expected that the genes targeted indirectly by ZafA were included among the IZR genes.

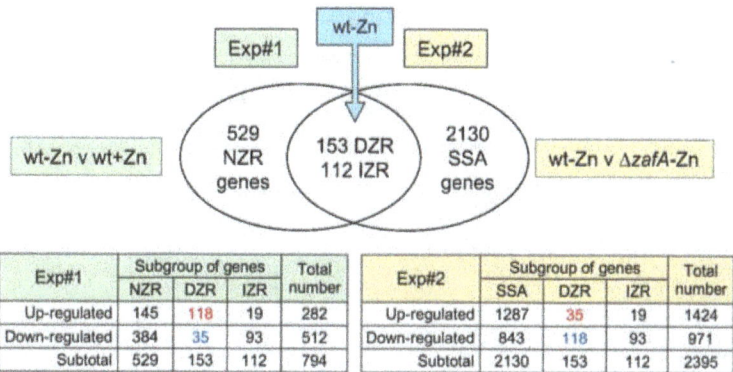

Figure 2. Quantitative summary of different types of genes detected in microarray experiments. SSA: strain-specific adapted (genes); DZR: directly ZafA regulated; IZR: indirectly ZafA regulated ; NZR: non-ZafA regulated.

To validate the microarray analysis, we measured by RT-qPCR the relative expression ratio of 56 genes whose expression changed significantly in a wild-type strain grown under zinc-limiting versus zinc-replete conditions. More precisely, we measured the relative expression of 10 IZR, 22 NZR and 24 DZR genes (Table 2). The expression levels of the *zafA*, *zrfA*, *zrfB* and *zrfC* genes, whose expression profiles had been well defined previously [11,12,14], were measured as positive controls. The expression of the actin gene (AFUA_6G04740/*actA*), whose expression did not change significantly under zinc starvation neither in a wild-type nor in a Δ*zafA* mutant [11,14], was measured as a negative control. In addition, the expression levels of six additional genes selected at random, whose expression did not change also under zinc starvation neither in a wild-type nor in a Δ*zafA* mutant, were also measured as negative controls (Table 2). The results obtained by both techniques were coincident and the correlation between microarray data and those measured by RT-qPCR were highly significant (Pearson's $r = 0.69$, $P < 0.0001$) in spite of the high difference in sensitivity and accuracy between these experimental procedures.

Table 2. Measurement by RT-qPCR of the expression levels of several genes of each group detected in microarrays, both in a wild-type strain grown in zinc-limiting versus zinc-replete conditions (as in Exp#1) and in a Δ*zafA* strain grown under zinc-limiting conditions with respect to a wild-type strain grown in the same culture conditions (i.e., as in Exp#2). ZafA can be formally considered as an inductor or repressor of the DZR genes, as indicated. The indicated genes (*) should be reallocated into the specified group according to RT-qPCR data, provided that the genes were considered (arbitrarily) to be expressed differentially when their transcription levels changed > 1.5-fold. Expression of genes in red and blue show induction or repression respectively.

Gene ID	Group of Genes	Gene Name	wt − Zn/wt + Zn rER ± SD	Δ*zafA* − Zn/wt − Zn rER ± SD	Predicted Effect of ZafA on Gene Expression	Reallocating Subgroup of Genes *
AFUA_1G01550	DZR	*zrfA*	231.3 ± 27.8	−207.7 ± 18.9	Induction	
AFUA_1G02150	Control	*gfdA*	−1.25 ± 0.15	1.17 ± 0.11		
AFUA_1G03150	DZR	*erg24A*	−49.9 ± 7.0	2.0 ± 0.2	Repression	
AFUA_1G04620	IZR	-	−4.9 ± 0.4	−1.9 ± 0.2		
AFUA_1G07480	NZR	*hem13*	−6.0 ± 0.8	−1.6 ± 0.1		IZR
AFUA_1G10060	NZR	-	3.1 ± 0.2	−1.2 ± 0.1		

Table 2. *Cont.*

Gene ID	Group of Genes	Gene Name	wt − Zn/wt + Zn rER ± SD	ΔzafA − Zn/wt − Zn rER ± SD	Predicted Effect of ZafA on Gene Expression	Reallocating Subgroup of Genes *
AFUA_1G10080	DZR	zafA	31.5 ± 3.0	−21582 ± 4316	Induction	
AFUA_1G10130	NZR	sahA	1.9 ± 0.1	1.1 ± 0.1		
AFUA_1G12830	IZR	niaD	−1.6 ± 0.1	−5.2 ± 0.7		
AFUA_1G13510	NZR	facB	4.6 ± 0.6	1.0 ± 0.1		
AFUA_1G14550	DZR	sodC/sod3	1698.7 ± 198.2	−1176.5 ± 142.4	Induction	
AFUA_1G15590	NZR	cybS/sdh4	2.9 ± 0.4	1.1 ± 0.1		
AFUA_1G17190	NZR	sidI	−92.5 ± 8.7	1.1 ± 0.1		
AFUA_2G00320	NZR	erg3B	−271.7 ± 31.7	−1.6 ± 0.2		IZR
AFUA_2G01260	DZR	srbA	−15.8 ± 2.1	2.1 ± 0.2	Repression	
AFUA_2G03010	NZR	-	−1.7 ± 0.2	−1.0 ± 0.1		
AFUA_2G03700	NZR	hmg1	−2.6 ± 0.2	−1.1 ± 0.1		
AFUA_2G03860	DZR	zrfB	33.0 ± 2.7	−79.3 ± 7.7	Induction	
AFUA_2G07680	IZR	sidA	−27.3 ± 3.3	−6.9 ± 0.6		
AFUA_2G08740	DZR	zrfF	24.8 ± 1.8	−16.0 ± 2.0	Induction	
AFUA_2G11120	DZR	gtmA	5.7 ± 0.7	−20.3 ± 1.8	Induction	
AFUA_2G14570	NZR	zrcC	−2.1 ± 0.2	−1.1 ± 0.1		
AFUA_2G15010	NZR	-	−1.5 ± 0.2	−1.2 ± 0.1		
AFUA_2G15290	DZR	-	−26.0 ± 3.6	3.2 ± 0.3	Repression	
AFUA_2G17550	IZR	ayg1	−2.7 ± 0.4	−103.5 ± 10.8		
AFUA_3G02270	DZR	cat1	18.9 ± 1.8	−2.0± 0.2	Induction	
AFUA_3G03640	IZR	mirB	−3225.8 ± 304.5	−10.0 ± 1.2		
AFUA_3G14440	NZR	-	−3.5 ± 0.5	−1.3 ± 0.1		
AFUA_4G03410	NZR	fhpA	−3.1 ± 0.3	−1.4 ± 0.2		
AFUA_4G03460	IZR	srbB	−50.4 ± 5.9	−1.1 ± 0.1		NZR
AFUA_4G03930	DZR	cysX	12786.9 ± 1492.2	−7751.9 ± 938.0	Induction	
AFUA_4G06530	NZR	metR	−1.6 ± 0.2	1.2 ± 0.1		
AFUA_4G09560	DZR	zrfC	1270.1 ± 108.0	−5377.1 ± 424.8	Induction	
AFUA_4G10730	Control	rvb1	1.1 ± 0.2	−1.2 ± 0.1		
AFUA_4G12840	IZR	-	−1.6 ± 0.2	−1.7 ± 0.1		
AFUA_5G01030	DZR	gpdB	−1.2 ± 0.1	77.7 ± 7.5		SSA
AFUA_5G01970	NZR	gpdA	1.8 ± 0.2	1.0 ± 0.1		
AFUA_5G02180	NZR	cysB	−2.6 ± 0.2	1.1 ± 0.1		
AFUA_5G03800	NZR	ftrA	−31.0 ± 2.2	1.4 ± 0.2		
AFUA_5G03920	NZR	hapX	−12.2 ± 1.5	−1.2 ± 0.1		
AFUA_5G04130	Control	phoA	1.1 ± 0.1	−1.3 ± 0.2		
AFUA_5G06240	DZR	alcC	−26.8 ± 1.9	3.8 ± 0.5	Repression	
AFUA_5G06270	NZR	hemA	1.8 ± 0.2	−1.0 ± 0.1		
AFUA_5G08090	DZR	pyroA	2.7 ± 0.3	−1.8 ± 0.2	Induction	
AFUA_5G09240	DZR	sodA/sod1	−4.7 ± 0.7	7.3 ± 0.8	Repression	
AFUA_5G09360	Control	calA	1.1 ± 0.1	1.1 ± 0.1		
AFUA_5G09680	NZR	carC/sdh3	3.9 ± 0.4	−1.0 ± 0.1		
AFUA_5G10560	Control	cox5A	−1.2 ± 0.2	1.0 ± 0.1		
AFUA_6G00690	DZR	-	−2.8 ± 0.3	5.2 ± 0.6	Repression	
AFUA_6G04430	NZR	-	2.0 ± 0.2	1.3 ± 0.2		
AFUA_6G04740	Control	actA	−1.0 ± 0.1	1.2 ± 0.1		
AFUA_6G05160	DZR	azf1	1.9 ± 0.2	−23.9 ± 2.9	Induction	
AFUA_6G07720	IZR	acuF	−2.2 ± 0.2	−1.7 ± 0.2		
AFUA_6G09630	DZR	gliZ	45.9 ± 6.0	−67.2 ± 5.3	Induction	
AFUA_6G09710	DZR	gliA	3548.5 ± 87.4	−3039.5 ± 294.8	Induction	
AFUA_6G09740	DZR	gliT	53.6 ± 3.9	−477.3 ± 59.7	Induction	
AFUA_7G00250	IZR	tubB2	−7.0 ± 0.5	−10.0 ± 1.2		
AFUA_7G02560	Control	dld1	−1.1 ± 0.1	1.0 ± 0.1		
AFUA_7G04730	IZR	enb1	−48.3 ± 5.8	−2.4 ± 0.2		
AFUA_7G05070	NZR	frdA	−56.5 ± 7.4	1.9 ± 0.1	Repression	DZR
AFUA_7G06570	DZR	zrcA	−5.9 ± 0.7	16.6 ± 1.5	Repression	
AFUA_7G06790	DZR	yct1	6039.9 ± 362.4	−5555.6 ± 577.8	Induction	
AFUA_8G02450	DZR	gzbA	84.8 ± 8.0	−275.5 ± 32.2	Induction	
AFUA_8G02620	DZR	mchC	1155.6 ± 134.9	−524.3 ± 63.4	Induction	

SSA: strain-specific adapted (genes); DZR: directly ZafA regulated; IZR: indirectly ZafA regulated ; NZR: non-ZafA regulated.

3.2. Zinc Starvation Influences the Expression of Genes Related to Many Different Biological Processes

A functional categorization of the 512 down-regulated and 282 up-regulated genes in a wild-type strain grown under zinc-limiting conditions was performed on the *Aspergillus fumigatus* AF293 (AspGD) genome using the gene ontology (GO) tool integrated in FungiFun2, which is an updated and easy-to-use online resource for functional enrichment analysis of fungal genes [26]. The functional categorization of the down-regulated genes revealed that 26.0% of the input genes were significantly categorized (i.e., 133/512). These genes were related to ergosterol biosynthesis ($p = 1.32 \times 10^{-9}$), cellular response to iron starvation ($p = 1.49 \times 10^{-9}$), fatty acid biosynthesis ($p = 3.65 \times 10^{-5}$), $\alpha(1,3)$-glucan biosynthesis ($p = 0.00013$), calcium transport ($p = 0.00013$) and cellular response to hypoxia ($p = 0.0005$). The functional categorization of the up-regulated genes revealed that 18.8% of the genes were significantly categorized (i.e., 53/282). These genes were related to ascospore formation ($p = 1.74 \times 10^{-5}$), asexual development (conidia formation) ($p = 2.08 \times 10^{-4}$), mitochondrial respiratory chain complex I ($p = 8.5 \times 10^{-5}$), zinc transport ($p = 8.5 \times 10^{-5}$), cellular hyperosmotic response ($p = 0.00021$), and G1 cell cycle arrest in response to nitrogen starvation ($p = 0.00078$). Hence, by using FungiFun2 the functional categorization of 186 genes could be described. In a more careful manual revision we were able to functionally allocate 291 genes into 28 different biological processes and outline the major homeostatic and adaptive responses displayed by the fungus to deal with zinc starvation (Table S6) (see discussion). 247 genes encoded uncharacterized proteins that either were similar to characterized proteins in other organisms or harbored at least one catalytic domain with a predicted function, regardless of its putative subcellular location. Finally, there were also 256 genes that encoded either uncharacterized protein orthologues in other organisms or hypothetical proteins.

Thus, it became apparent that many genes were down regulated under zinc-limiting conditions, mainly those whose expression was induced under iron starvation or influenced by the intracellular calcium levels, those required for fatty acid and phospholipid biosynthesis, and those encoding enzymes involved in ergosterol biosynthesis and whose expression was also strongly co-regulated by iron and hypoxia. Other down-regulated genes encoded proteins related to ethanol metabolism, disulfide bond formation in the endoplasmatic reticulum (ER), enzymes of the citric acid cycle and glycolytic pathway, and components of the mitochondrial electron transport chain (ETC) (Table S6). On the other hand, it was overwhelming the predominance, under zinc-limiting conditions, of up-regulated genes involved in sexual and asexual development, hyphal growth and autophagy, ribosomal assembly and/or translation, response to hyperosmotic stress and regulating and maintaining zinc homeostasis (Table S6). Zinc starvation also influenced, either positively or negatively, the expression of genes that were typically regulated in response to hypoxia and oxidative stress. In addition, the expression of genes involved in biosynthesis of the cell wall, growth factors (riboflavin, thiamine and pyridoxin), heme group, detoxification of D-2-hydroxyglutarate coupled to D-lactate formation, acetate utilization, cytoskeleton and chromatin remodeling and regulation of the cell cycle were also influenced by zinc starvation. Finally, zinc starvation also noticeably influenced the expression levels of genes involved in NAD, sulfur, spermidine and nitrogen metabolism, and others involved in biosynthesis of some secondary metabolites (Table S6). In this regard, it was remarkable that the expression of genes related to gliotoxin biosynthesis was strongly induced under zinc-limiting conditions.

The analysis of the distribution of the 291 genes, categorized functionally based on the most likely role of ZafA in regulating their expression, as defined by their assignation to either the DZR/IZR or NZR group of genes (Table S6), indicated that ZafA influenced either directly or indirectly the expression of ≥60% of the genes involved in the homeostatic response to zinc starvation, fungal development, response to oxidative stress, nitrate assimilation, acetate utilization and in the functioning of the tricarboxylic acid cycle and mitochondrial electron transport chain.

3.3. The ZafA Transcription Factor Binds to Promoter Regions of Homeostatic Genes Whose Transcription Is Induced under Zinc-Limiting Conditions

We assumed that the DZR genes should be the ones most directly targeted by ZafA, such that in their promoter regions there would be at least a copy of a zinc responsive (ZR) consensus sequence to which ZafA might bind. Hence, we used the MEME algorithm as a first approach to identify potential regulatory elements and/or repeated patterns shared among the promoters of 67 genes that could be the most direct ZafA target genes [27]. These genes were chosen on the basis of their fold changes in Exp#1, such that the 54 genes more up-regulated and the 13 genes more down-regulated were selected (Table S5). Thus, we found two motifs (M1 and M2) that were found significantly distributed among the analyzed promoters (E-value for M1 = 3.6×10^{-37}; E-value for M2 = 1.7×10^{-17}) (Figure 3). The M1 motif showed two short CT-rich sequences flanking the 5′-CAAGGT-3′ core sequence and was present in a number of 1 to 5 copies in the promoter regions of 28 genes (i.e., 42.4% of the analyzed genes). The M2 motif corresponded to a relatively homogeneous long AG rich sequence and was present in a number of 1 to 6 copies in the promoters of 36 genes (i.e., 54.5% of the analyzed genes). However, a control MEME search for repeated patterns in the promoter regions of 71 genes (38 SSA genes plus 33 genes whose expression was not influenced by zinc starvation that were selected at random) (Table S7), also produced a M2-like AG-rich motif in 52.1% of the genes (E-value for the M2-like motif = 3.9×10^{-59}). Besides, the M2 motif was not detected in the promoter region of the genes *zrfA* and *zrfC*, whose transcription had been shown to be strongly and directly up-regulated by ZafA under zinc-limiting conditions [11,14]. In contrast, the M1 motif was found in a number of 3 to 5 copies in the promoter regions of the *zafA*, *zrfA*, *zrfB*, *zrfC* and *zrfF* genes, that encoded the major components of the zinc homeostatic system in *A. fumigatus*. Moreover, by using the RSAT algorithm [28], we found that one motif nearly identical to M1 was widely distributed among promoters of all DZR genes, whereas no motif detected among IZR genes showed any similarity with M1. Taken together, these findings suggested that the M1 motif could be a genuine DNA regulatory sequence for binding ZafA.

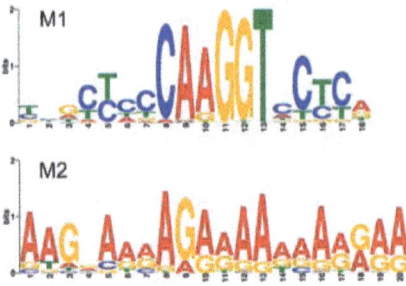

Figure 3. Overrepresented motifs in the promoter regions of the most DZR genes. These motifs were found using the MEME algorithm after analyzing DNA sequences expanding from −20 to −1000 bp from the predicted translation start codons of the ORFs of 67 DZR genes (for numbering purposes, we assigned position −1 to the nucleotide preceding the A of the ATG start codon).

To show whether ZafA was able to bind to the promoter regions of the homeostatic genes *zafA*, *zrfA*, *zrfB* and *zrfC*, we analyzed by EMSA the ability of ZafA to bind to short DNA fragments (210–260 bp) of these promoters harboring the previously identified M1 motif. All attempts to express the full-length ZafA protein in *E. coli* were unsuccessful. However, we had success in expressing a recombinant polypeptide corresponding to the C-terminal 140 amino acids of ZafA (residues 434–570 that included the four most C-terminal zinc fingers ZF3-6) fused to the C-terminus of GST. By using this polypeptide we confirmed the binding and observed that the mobility of all DNA-protein complexes in PAGE gels reduced stepwise as the ZafA:DNA ratio increased, which indicated that ZafA^{ZF3-6} was able to bind, with different affinities, to more than one ZafA binding site in all DNA fragment tested

(Figure 4). Particularly, three steps were observed for the P*zrfA* DNA fragment (Figure 4A) and two for the P*zrfB*, P*zrfC* and P*zafA* DNA fragments (Figure 4B and left panels of C and D), which matched the number of M1-like motifs found in these promoters. Besides, given that at a ZafA:DNA molar ratio of 6.0 both the P*zrfC* and P*zafA* fragments were almost completely shifted (Figure 4C,D, left panels), we performed EMSAs assays at lower ZafA:DNA ratios (between 0.4 and 4.0, right panels). As shown, in both cases, a ZafA:DNA ratio of four was sufficient to shift most of the DNA. This suggested that in both fragments P*zafA* and P*zrfC* there were two ZR motifs to which two ZafA molecules would bind with similar affinities but stronger than those used to bind to the ZR motifs in the P*zrfA* and P*zrfB* fragments.

3.4. The ZafA Transcription Factor Binds Specifically to a 15-bp Zinc-Response Consensus Sequence

To identify the DNA sequence to which ZafA binds, we performed a fluorescent DNase I footprinting analysis by using fluorescently 6-FAM-labeled primers in parallel to dideoxynucleotide DNA sequencing. This procedure allows the accurate separation of 5′-end-labeled DNA fragments using a capillary-based automated DNA sequencer and to identify the nucleotide sequence of sites protected against DNase digestion [29]. To identify the DNA binding site for ZafA we used several 6-FAM-5′-labelled DNA probes carrying short fragments of the promoter regions from the homeostatic genes *zrfA*, *zrfB*, *zrfC* and *zafA* (see material and methods section) that harbored 1–2 copies of the M1 motif detected previously by MEME analyses. Thus, in the P*zafA* fragments used as probes, four different regions protected by ZafA from its enzymatic digestion with DNase I were found, that were called ZRR1-4 (for <u>Z</u>inc-<u>R</u>esponse motif located in the promoter region of gene encoding the <u>R</u>egulatory protein ZafA) (Figure 5 and Figure S2). Similarly, in the P*zrfA* and P*zrfC* fragments used as probes, three different ZafA-protected regions in each one were found, that were called ZRA1-3 and ZRC1-3 (for <u>Z</u>inc-<u>R</u>esponse motif located respectively in the promoter region of *zrf<u>A</u>* and *zrf<u>C</u>*). Besides, in the P*zrfB* fragments, two different ZafA-protected regions were found, that were called ZRB1-2 (for <u>Z</u>inc-<u>R</u>esponse motif located in the promoter region of *zrf<u>B</u>*) (Figure 5 and Figure S2). It is interesting to recall that in nearly all cases, highly hypersensitive regions to DNase I digestion were observed, that contrasted clearly with the adjacent ZafA-protected regions (Figure S2). The comparison of all ZafA-protected regions brought the ZR consensus sequence 5′-DYYVYCARGGTVYYY-3′ (D = A or G or T; V = A or G or C; R = A or G) (Figure 5), which strongly resembled the M1 motif detected after MEME analysis (Figure 3).

3.5. The Highly Conserved 5′-CARGGT-3′ Core of the ZR Motif Is Essential for ZafA Binding

The ZR consensus motif showed a highly conserved 5′-CARGGT-3′ hexanucleotide core that could be essential for ZafA to recognize and bind specifically to it, such that the replacement of a purine by a pyrimidine, or vice-versa, could reduce or impair the binding of ZafA. We thought that this hexanucleotide sequence, rather than the short CT-rich flanking sequences, was required for the specific binding of ZafA. To confirm this, we searched in the *Aspergillus* genome database for a DNA fragment corresponding to the promoter region of any gene with a relatively high number of short CT-rich sequences (≥3 bp) and that contained just one ZR consensus motif, to be used as probe for both EMSA and DNase I footprinting assays. Thus, we selected a 191-bp DNA fragment, the *alcA* gene promoter from *A. nidulans* that met our requirements to be used as a probe. We confirmed by a DNase I footprinting assay that ZafA only protected a region of this DNA fragment that contained the ZR consensus motif (Figure 6A). To ascertain whether the 5′-CARGGT-3′ sequence was essential for ZafA binding, we created a mutated version of this probe in which its 5′-CAAGGT-3′ core had been changed to 5′-CTCAGT-3′. The binding of ZafA to the wild-type P*alcA* probe at ZafA:DNA ratios lower than 6.0 resulted in a retarded DNA:protein complex as shown by EMSA (Figure 6B). In contrast, ZafA did not bind to the P*alcA* probe in which the ZR motif had been mutated. These results showed that ZafA was able to bind to the ZR motif and that the 5′-CARGGT-3′ hexanucleotide core sequence was essential for ZafA to recognize such motif.

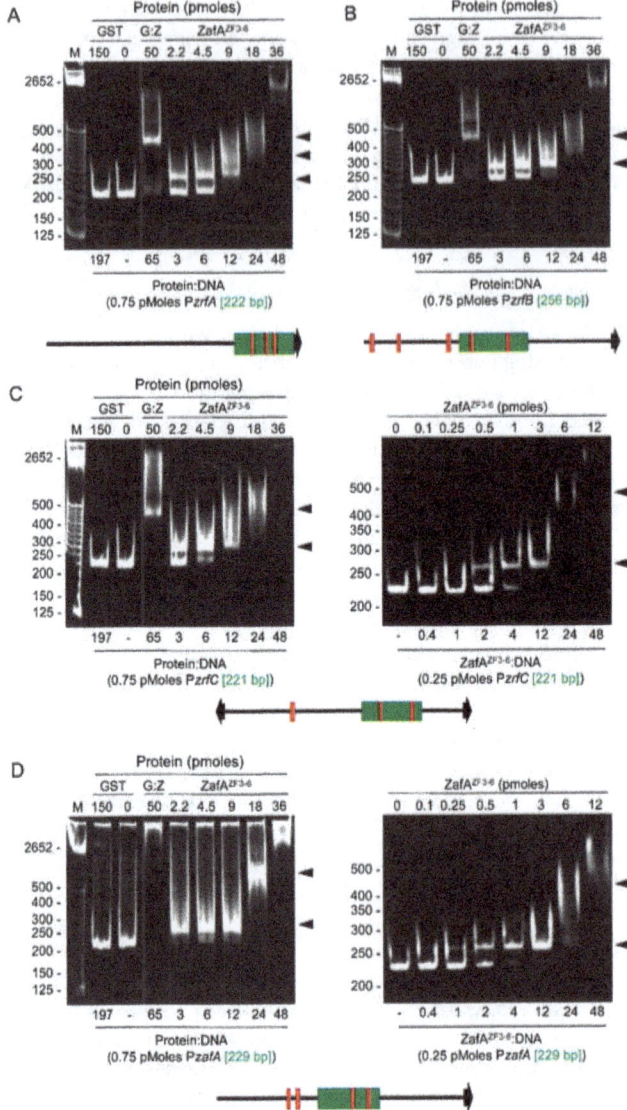

Figure 4. Analyses by Electrophoretic Mobility Shift Assays (EMSA) of the interaction of ZafA^{ZF3-6} with different DNA fragments. (**A**) Interaction of ZafA^{ZF3-6} with a DNA fragment from a promoter region of *zrfA* (green box) that carried three putative ZafA binding motifs (red boxes); (**B**) interaction of ZafA^{ZF3-6} with a DNA fragment from a promoter region of *zrfB* (green box) that carried two putative ZafA binding motifs (red boxes); (**C**) interaction of ZafA^{ZF3-6} with a DNA fragment from a promoter region of *zrfC* (green box) that carried two putative ZafA binding motifs (red boxes); (**D**) interaction of ZafA^{ZF3-6} with a DNA fragment from a promoter region of *zafA* (green box) that carried two putative ZafA binding motifs (red boxes). In the right panel of (**C**) and (**D**), reactions were analyzed with lower ZafA^{ZF3-6}:DNA ratios to gain resolution. All DNA fragments were also incubated without protein and in the presence of both 150 pmoles of purified GST and 50 pmoles of the fusion protein GST-ZafA^{ZF3-6} (as controls) to attain a GST:DNA ratio of 197 and a GST-ZafA^{ZF3-6}:DNA (G:Z) ratio 65. A 25-bp DNA ladder was used as a reference.

ZRA1	-82	tcagtgCGACCTCAAGGTACCCAGCATgccagcgc	-48	
ZRA2'	-84	tcagtctACTGCCAAGGTCCTTCGCCTGGtcggtg	-118	
ZRA3	-159	agaggTAGCTCACAAGGTGTgtctactctcgacca	-130	
ZRB1	-372	aatcacGATCCTCAAGGTCCCTACAGGATTCactt	-338	
ZRB2'	-503	tTGCCTTGCCACCAAGGTCCTCAATTGCGgattca	-469	
ZRC1'	-175	CACACTAACTCCCAAGGTATCTGCGGCCaagagca	-209	
ZRC2'	-292	gcttgttGCTCTCAAGGTCCTCGATCCAAGactct	-326	
ZRC3'	-616	cttgatCGCTACCAAGGTGTCCTTGGCAtcctgta	-650	
ZRR1	-368	gtGGGAAGCTCCCAAGGTCCTCTTGgatttgcaca	-334	
ZRR2	-428	gcccctttCCCTCAAGTTGCtcgcacagatctgga	-394	
ZRR3	-631	ggtgcTTGCTCGCAGGGTCCTCACTCTCCTTCTTG	-597	
ZRR4	-664	cgcttTTTCCCCCAGGGTCCTCATTAAAgaatagg	-630	
Consensus		-------DYYVYCARGGTVYYY--------------		

Figure 5. Alignment of DNA sequences protected against DNase I digestion from the DNA fragments used for EMSA analyses. Protected sequences are in capital letters whereas the surrounding sequences are indicated in gray lower case letters (D = A or G or T; V = A or G or C; R = A or G). ZR motifs labelled with a prime symbol indicate that the DNA sequence shown in the picture corresponded to that in the antiparallel strand.

3.6. ZafA Induces Gene Expression in vivo through Binding to the ZR Motifs

Although mutations within the 5'-CARGGT-3' hexanucleotide core prevented the binding of the recombinant ZafA[ZF3-6] protein to the ZR motifs in vitro, it could be possible that in vivo the wild-type ZafA protein binds to a DNA sequence different from the ZR motif determined in vitro. Hence, to ascertain whether ZafA truly bound to the ZR motifs in the fungus growing in zinc-limiting media, we analyzed in vivo the effect of the inactivation of the three ZR motifs (ZR1, ZR2 and ZR3) present in the divergent promoter that drives the transcription of both *aspf2* and *zrfC* (P*zrfC*) on the transcription of these genes [11]. To test this, we introduced, at the *pyrG* locus of a uridine-uracil-auxotrophic strain (CEA17), the coding sequences of both the luciferase (*luc*) and green fluorescent protein (*gfp*) arranged divergently and separated by a P*zrfC* mutant version (P*zrfC*[ZR123]) whose three ZR motifs had been inactivated by replacing each 5'-CAAGGT-3' hexanucleotide core by the mutated 5'-ACATGT-3' sequence (Figure S1). It must be noted that, in this construction, the ORF of *aspf2* and *zrfC* had been replaced respectively by that of *luc* and *gfp*. Hence, this construction was devised in such a way that the transcription level of *luc* and *gfp* would indicate the functionality of the mutant promoter (P*zrfC*[ZR123]) in a wild-type genetic background, while the expression levels of the endogenous *aspf2* and *zrfC* genes could be used as internal controls of the transcriptional promoter activity of P*zrfC*[wt]. We expected that if these ZR motifs were truly recognized by ZafA in vivo, the mutation of their conserved hexanucleotide cores should prevent the binding of ZafA to P*zrfC*[ZR] and the expression of both *luc* and *gfp*. As shown, the transcription of both reporter genes driven by P*zrfC*[ZR123] in the AFZR1 strain grown under zinc-limiting conditions collapsed as if the fungus was growing in zinc-replete conditions (Figure 7). More precisely, the relative expression ratio (rER) of the *luc* and *gfp* transcripts in AFZR1 was reduced respectively by 523- and 611-fold as measured by RT-qPCR, whereas the relative expression ratios (rERs) of *asfp2* and *zrfC*, which were used as endogenous controls, were identical to that in the control AFZR0 strain that carried the P*zrfC*[wt]. These results undoubtedly showed that ZafA has to bind to the ZR motifs in vivo to induce gene expression under physiological conditions of zinc starvation.

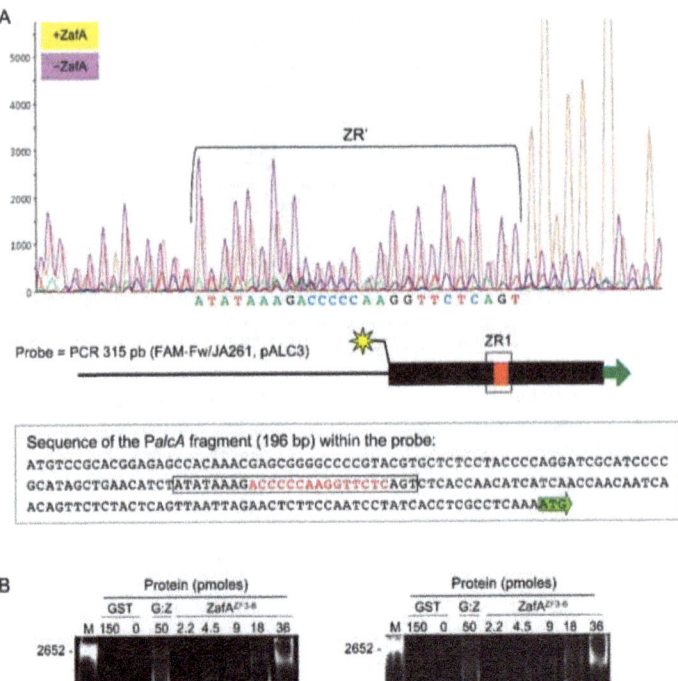

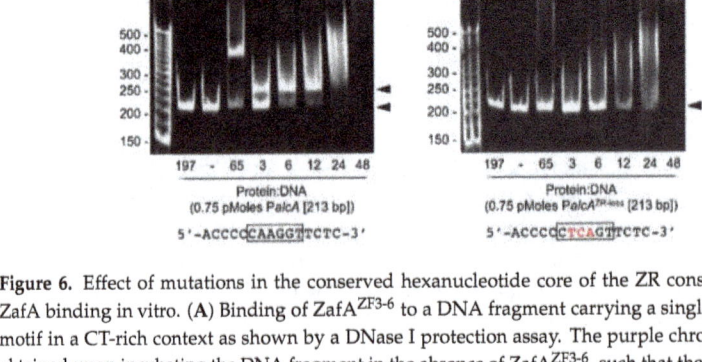

Figure 6. Effect of mutations in the conserved hexanucleotide core of the ZR consensus motif on ZafA binding in vitro. (**A**) Binding of ZafA^{ZF3-6} to a DNA fragment carrying a single ZR consensus motif in a CT-rich context as shown by a DNase I protection assay. The purple chromatogram was obtained upon incubating the DNA fragment in the absence of ZafA^{ZF3-6}, such that the DNA remained naked and, hence, unprotected from DNase I digestion. The yellow chromatogram was obtained upon incubating the DNA in the presence of ZafA^{ZF3-6}. ZafA bound to DNA prevented its digestion with DNase I resulting in a stretch of yellow peaks lower in high than purple peaks. A short stretch of high yellow peaks at the right side of the chromatogram indicated a hypersensitive region to DNase I digestion. The protected region is delimited by a bracket. Below the chromatogram is shown a schematic representation of the whole intergenic region of the *alcA* gene with the beginning of the ORF located at the right side (green arrow). The sequence of the DNA fragment corresponding to the P*alcA* used in the assay is located in the scheme as a black rectangle with the ZR consensus motif in red. The protected region has been squared in the scheme and sequence. The yellow star indicates the end 6-FAM labeled strand detected by the capillary-based automated DNA sequencer; (**B**) Analysis by EMSA of binding of ZafA^{ZF3-6} to a DNA fragment in which the 5′-CAAGGT-3′ hexanucleotide core of its only ZR motif (left panel) had been mutated to 5′-CTCAGT-3′ (right panel). The DNA fragments were also incubated without protein and in the presence of both 150 pmoles of purified GST and 50 pmoles of the fusion protein GST-ZafA^{ZF3-6} (as controls) to attain a GST:DNA ratio of 197 and a GST-ZafA^{ZF3-6}:DNA (G:Z) ratio 65. A 25-bp DNA ladder was used as a reference.

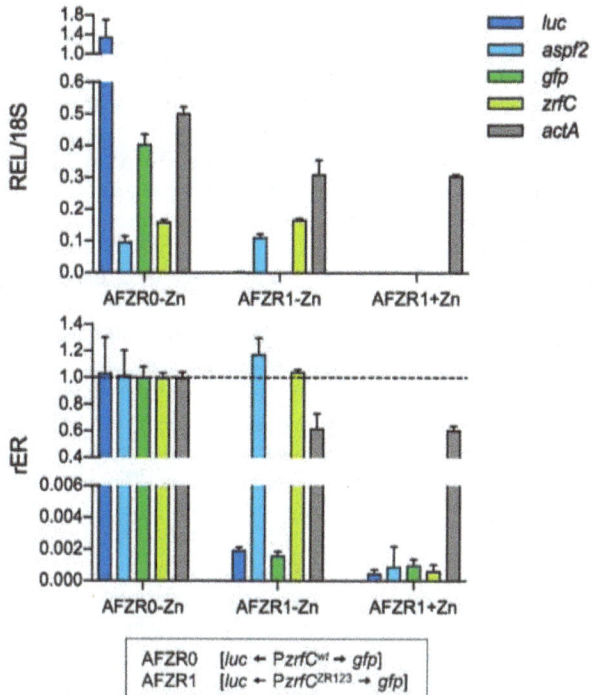

Figure 7. Effect of mutations in the conserved hexanucleotide core of the ZR consensus motif on ZafA binding in vivo. (Upper panel) Relative expression levels (REL) of endogenous *aspf2* and *zrfC* genes and their corresponding reporter genes (*luc* and *gfp*) driven by a wild-type (in the AFZR0 strain) and mutant version (in the AFZR1 strain) or the divergent *aspf2-zrfC* promoter (abbreviated as P*zrfC*). In the mutant version of this promoter the three ZR motifs present in it (P*zrfC*ZR123) had been inactivated by site-directed mutagenesis. Both strains were cultured under zinc-limiting conditions in SDN–Zn for 20 h at 37 °C. The AFZR1 strain was also grown under zinc-replete conditions in SDN + 100 μM Zn to be used as a control of the effect of zinc on the transcriptional activity of ZafA on P*zrfC*wt. (Lower panel) Relative expression ratio (rER) of data represented in the upper panel after their normalization with respect to the expression level in the AFZR0 strain grown under zinc-limiting conditions. The expression level of the actin gene was also measured by RT-qPCR as an additional control. The 18S rRNA was used as an internal reference for all relative quantifications.

3.7. The ZR Motifs Appear to Function as Transcriptional Regulatory Sequences When Located within 1.2 Kb Upstream from the Predicted Translation Start Codon of the ORFs Whose Expression Is Influenced by Zinc

Although the highly conserved 5′-CARGGT-3′ core of the ZR motif was essential for ZafA binding, it exhibited a relatively high degree of variability in the 5′ and 3′-end sequences that flanked the conserved core. This fact could influence the affinity of ZafA binding to the different ZR motifs in the DNA. In addition, it would not be unexpected that ZafA was able to bind also to ZR motifs harboring a certain degree of sequence degeneration. Thus, it could be possible that microarray data under the tested culture conditions had not revealed all ZafA potential targets if the expression of an unknown number of genes were influenced, in addition to zinc starvation, by other environmental signals. Hence, to obtain further evidence on the role of the ZR motif as a regulatory sequence of gene expression under zinc-limiting conditions, we performed a genome-scale search for ZR motifs in the *A. fumigatus* genome by utilizing the RSAT web server [28]. To perform this search, we used as a query pattern the ZR consensus sequence or ZR0 motif, while admitting one substitution out of the conserved hexanucleotide core. Hence, nine additional different degenerated ZR-like motifs (ZR1-9) were possible (Table S8). Since there is one gene every 2938 bp in the genome of *A. fumigatus* and the mean gene length is 1431 bp [30], it can be deducted that the average intergenic sequence is 1482 bp. In order to identify most of the genes harboring ZR-like motifs in their promoter regions, the search for them was extended up to −3000 bp upstream of the predicted translation start codon (TSC) of the open reading frames (ORFs) (i.e., about 2× the size of an average intergenic sequence). By enabling and disabling the overlapping with upstream coding sequences we differentiated two sets of ZR motifs: (1) Non-overlapping ZRs, which were located in the intergenic sequences corresponding to the promoter regions of genes regulated presumably by ZafA; and (2) Overlapping ZRs, which were located either within an upstream ORF or in an intergenic sequence that would correspond to the promoter region of an upstream gene.

A detailed analysis of the non-overlapped ZRs at a genome-scale indicated that there were 545 different ZR motifs distributed among 499 different intergenic sequences harboring the promoter regions that drive the expression of 671 different coding sequences (since 51.3% of such coding sequences were arranged as divergent pairs of genes in the genome). About 91.4% of all genes, including both non-divergent and divergent arranged genes, had in their upstream intergenic regions only one ZR motif, 7.4% had 2 ZRs and 1.2% showed ≥3 ZRs. The most abundant ZR motifs in promoter regions were ZR0, ZR7 and ZR9 (103.3 ± 3.2 per ZR motif). The next most abundant ZR motifs in the promoter regions were ZR2, ZR3, ZR5 and ZR8 (83.5 ± 6.9 per ZR motif). The less abundant ZR motifs were ZR1, ZR4 and ZR6 (31.7 ± 6.7 per ZR motif), as would be expected for the ZR motifs with the lowest variability. Upon analyzing the distribution in the intergenic regions of the most abundant ZRs depending on both the type of ZR and their distance to their correspondent ORFs, it was determined that the overall distribution of the most abundant ZR motifs was influenced very significantly by their distance to the TSC of the ORFs (2-way ANOVA, $p < 0.0001$) but not by the type of ZR motif (2-way ANOVA, $p = 0.322$). Similarly, the overall distribution in the intergenic regions of the less abundant ZR motifs (ZR1, ZR4 and ZR6) was also influenced significantly by their distance to the TSC of the ORFs (2-way ANOVA, $p = 0.0021$) but not by the type of ZR motif (2-way ANOVA, $p = 0.36$). Taken together, these findings suggested that the distance of the ZR motifs to the TSC of the ORFs appeared to be an important factor that determined their distribution in the intergenic promoter regions.

We then analyzed the distribution of all ZR motifs (i.e., non-overlapped plus overlapped ZRs) with respect to their distance to the ORFs. The results revealed that, as expected for a random distribution of ZRs along the DNA, the percentage of ZR motifs within each 200 bp DNA fraction, in which was arbitrarily divided the DNA sequences of 3000 bp located upstream the ORFs, maintained constant between 5.8–7.7% (Figure 8A; blue symbols). Accordingly, the cumulative percent of all ZR motifs (i.e., including both non-overlapped and overlapped ZR motifs) increased directly proportional to the distance to the ORFs ($R^2 = 0.999$), as would be expected for a random distribution (Figure 8B; blue symbols). However, when the distribution of both non-overlapped and overlapped ZR motifs were analyzed separately, they showed opposite distribution patterns. Thus, within each 200 bp DNA fraction, the percent of non-overlapped ZR motifs (Figure 8A; black symbols) reduced exponentially ($R^2 = 0.963$) from 15.3% to 2.2%, whereas that of the overlapped ZR motifs (Figure 8A; red symbols) increased exponentially ($R^2 = 0.955$) as their distance to the ORFs increased from 0.3% to 11.0%. Accordingly, the increase of the cumulative percent of non-overlapped ZR motifs reduced gradually with the distance to the ORFs at a rate that was proportional to the amount of the remaining ZR motifs, i.e., according to an exponential one phase decay relationship ($R^2 = 0.999$) (Figure 8B; black symbols), as would be expected if the ZR motifs were required to be located in a relatively close proximity to the ORFs for them to function as regulatory motifs. In contrast, within the analyzed range (i.e., between −1 to −3000 bp upstream of the ORFs) the increase in the cumulative percent of the overlapped ZR motifs increased with the distance to the ORFs according to a third order polynomial relationship ($R^2 = 0.999$) (Figure 8B; red symbols). Interestingly, the graphical representation of this function showed two phases:

An initial exponential increase of the cumulative percent of the overlapped ZR motifs (until is reached 1200 bp upstream the ORFs) followed by an increase in the cumulative percent of overlapped ZR motifs that was directly proportional to the distance to the ORFs. The exponential part of this representation reflected the distribution of regulatory ZRs located in the intergenic promoter sequences of upstream genes with very short ORFs that were located in the near vicinity of the reference gene; whereas the arithmetic part of the representation reflected the random distribution of the ZR motifs 1200 bp upstream the TSCs. Besides, this distance matched perfectly the average size of the upstream distance to ORFs that was defined by the highest difference between the cumulative percent of overlapped and non-overlapped ZRs located within the 3000 bp upstream the ORFs (Figure 8B).

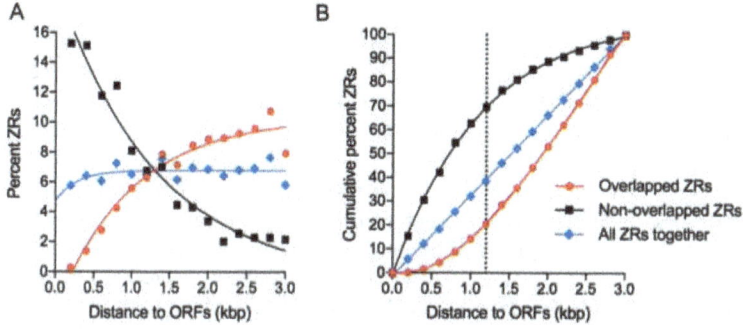

Figure 8. Distribution of ZR motifs at a genome scale in the upstream regions of the predicted open reading frames (ORFs). (**A**) Representation of the percentage of ZR motifs (overlapped, non-overlapped and all together) with respect to their distance to the predicted translation start codons (TSC) of the ORFs; (**B**) representation of the cumulative percent of ZR motifs with respect to their distance to the predicted translation start codons of the ORFs. The dotted line indicates the average size of the DNA sequence that is located upstream of the predicted TSCs containing the ZR motifs with the highest probability of being involved in regulating gene expression (i.e., the functionally relevant ZR motifs). In both graphs, the distance of the ZR motifs to the ORFs represented on the x-axes was divided into 15 fractions of 200 bp each.

Interestingly, the same tendency was observed for pairs of genes arranged divergently whose intergenic regions contained ZR motifs. Thus, the gene having most of the ZR motifs located at less than 1200 bp from the predicted TSC had a higher probability of being regulated by ZafA than the gene that had the ZR motifs at a distance >1200 bp of its TSC. For instance, in the intergenic region of the divergent pair of genes AFUA_1G01560/AFUA_1G01550 there were three motifs located respectively at −75, −105 and −157 bp from the TSC of AFUA_1G01550 (*zrfA*) and at −1784, −1754 and −1702 bp from the TSC of AFUA_1G01560. However, only the AFUA_1G01550 gene (*zrfA*) was induced by ZafA. In contrast, those genes arranged divergently and separated by intergenic sequences < 1200 bp are co-regulated simultaneously by ZafA (e.g., gene pairs AFUA_4G09560/AFUA_4G09580, AFUA_4G03920/AFUA_4G03930 or AFUA_8G02450/AFUA_8G02460). These results suggest that the ZR motifs located within 1200 bp upstream of the TSC of an ORF appear to be functionally relevant in the ZafA-mediated regulation of gene expression under zinc-limiting conditions.

Finally, a comparison of genes detected by microarrays in a wild-type strain as differentially regulated under zinc starvation with that found in silico harboring, non-overlapping ZR motifs in their promoter regions revealed that 31 DZR genes (i.e., 20.3% of the 153 DZR genes) and 55 non-DZR genes (i.e., 8.6% of 641 non-DZR genes) had ZR motifs in their promoters. Interestingly, 150 SSA genes (i.e., 7.0% of 2130 SSA genes) also had non-overlapping ZR motifs in their promoter regions. In addition, 30 DZR genes had their ZR motifs located within 1200 bp from their predicted TSC (i.e., 96.8% of all DZR genes with ZRs). In contrast, the weighted average percent of the non-DZR genes with ZR motifs located within 1200 bp from their predicted TSC was 71.2%. Taken together, these results indicated that ZR motifs predicted to be functionally relevant were located in the promoters of 19.6% of the DZR genes but only in 5.4% of genes not regulated directly by ZafA.

3.8. Putative Role of ZafA as a Repressor under Zinc-Limiting Conditions and Identification of the Most Likely Direct Target Genes of ZafA

The data obtained from the microarray experiments showed that, of all genes from a wild-type strain of *A. fumigatus* whose transcription level was influenced under zinc-limiting conditions, 64.5% were down-regulated whereas 35.5% were up-regulated (Figure 2). In contrast, 22.9% of the DZR genes were down-regulated whereas 77.1% were up-regulated. These results suggested that ZafA functioned mainly as a transcriptional activator rather than as a repressor.

In the microarray version used in this study, the AFUA_7G06570 gene (*zrcA*) had not been included. However, this gene encodes an important putative vacuolar zinc transporter of the CDF family likely involved in the homeostatic response to zinc excess [10]. Besides, it harbors one copy of the ZR consensus motif in its promoter region. Thus, we investigated whether the *zrcA* transcription was also influenced by ZafA under zinc-limiting conditions. It was shown by RT-qPCR that the expression level of *zrcA* reduced by about 6-fold in a wild-type strain grown in a zinc-limiting versus a zinc-replete medium, while it increased by 16-fold in a Δ*zafA* mutant grown in a zinc-limiting medium compared to a wild-type strain grown in the same conditions, i.e., ZafA was formally a repressor of *zrcA* under zinc-limiting conditions (Table 2). Hence, it would be theoretically possible that ZafA also down-regulated the expression of other DZR genes. In this regard, we confirmed by RT-qPCR that ZafA could also be formally considered as a repressor of the genes AFUA_2G01260/*srbA* [ZR3], AFUA_2G15290 [ZR5], AFUA_5G06240/*alcC* [ZR8] and AFUA_6G00690 [ZR5] under zinc-limiting conditions (Table 2). Importantly, all of these genes carried ZR motifs with one substitution in their promoter regions.

On the other hand, it was shown previously through a combination of in vitro and in silico analyses that at least 30 DZR genes with ZR motifs located within 1200 bp from their predicted TSC were putative direct ZafA targets. However, we also observed under zinc-limiting conditions, that some induced (e.g., *gtmA*, *pyroA*, *azf1*, *gliA*, *gliT*) and repressed genes (e.g., *sod1*, *erg24A*) (Table 2), which according to their expression profiles should be formally considered as direct ZafA targets, did not have obvious ZR motifs in their promoter sequences. However, after a search for non-overlapping ZR motifs with two substitutions in the promoter regions (until 1200 bp upstream of the predicted TSCs) using the RSAT software and the ZR consensus motif (ZR0) as a query sequence, 18 additional DZR genes harboring ZR degenerated motifs were found. Interestingly, among these genes were AFUA_6G09310 and AFUA_5G09240/*sodA*, which were ZafA-repressed genes, and *gliA*, which was one of the most strongly induced genes under zinc-limiting conditions. More importantly, this research also allowed us to detect additional ZR motifs in the promoter regions of genes encoding both important zinc homeostatic proteins (e.g., one additional site was detected in *zrfA* and *zafA* thus bringing the total number of ZRs in the promoters of these genes to 3 and 4 respectively) and other genes of unknown function but whose expression was also strongly induced by ZafA under zinc-limiting conditions (e.g., one additional ZR motif was detected in the AFUA_7G06810/*sarA* promoter and the AFUA_8G02460-AFUA_8G02450/*gzbA* divergent promoter; two additional ZR motifs were detected in the AFUA_8G02620/*mchC* promoter and the AFUA_4G03920-AFUA_4G03930/*cysX* divergent promoter).

Finally, it is interesting to recall that unlike the expression level of genes encoding zinc-independent enzymes typically involved in defense against reactive oxygen species (ROS), such as *cat1*, *cat2* and *sodC*; increases under zinc-limiting conditions, the expression level of *sodA*, which encodes a cytosolic Zn/Cu-superoxide dismutase, is reduced most likely as a mechanism to reduce the biosynthesis of the enzyme, since its functionality is strongly reduced under zinc-limiting conditions (Figure 9).

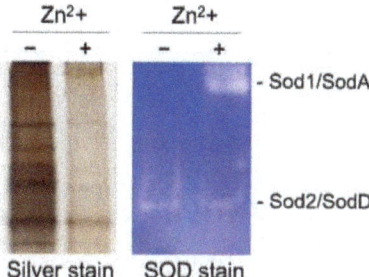

Figure 9. Detection of superoxide activity onto PAGE gels. Total extracts of native proteins were obtained from a wild-type strain of *A. fumigatus* grown under zinc-limiting (SDN–Zn) and zinc-replete conditions (SDN + 100 µM Zn), separated in a PAGE gel (T = 8%) and stained as described in the materials and methods section. Sod1/SodA is a cytosolic Cu/Zn-dependent superoxide dismutase whereas Sod2/SodD is a mitochondrial Mn-dependent superoxide dismutase.

In summary, these results supported the notion that the ZR consensus motif and other ZR motifs with different degeneration degrees out of the 5′-CARGGT-3′ hexanucleotide core, could function as regulatory sequences to which ZafA would bind under zinc-limiting conditions, and be brought to 49, the total number of DZR genes with ZR motifs (i.e., 31.8% of DZR genes) whose expression was regulated directly by ZafA, including 20 confirmed and 29 predicted genes (Table 3).

Table 3. Genes regulated or that were predicted to be regulated by ZafA. The expression of genes in red and blue was shown to be induced and repressed respectively by ZafA, as confirmed by RT-qPCR and/or northern-blot. The expression of genes in **black** and green was predicted to be induced and repressed respectively by ZafA. Start and end positions of the ZR motifs were numbered taking as reference that the base preceding the A nucleotide of the ATG translation start codon is at position −1. Motifs labeled with an asterisk (*) had two substitutions with respect to the ZR consensus motif out of the hexanucleotide core sequence 5′-CARGGC-3′ (ZR0*).

Gene ID	Gene Name	Strand	Start	End	Sequence ZR Motif	Type of ZR Motif
AFUA_1G01550	zrfA	D	−75	−61	GACCTCAAGGTACCC	ZR2
		R	−105	−91	ACTGCCAAGGTCCTT	ZR0
		D	−157	−143	GCTCACAAGGTGTGT	ZR0*
AFUA_1G05900	-	D	−393	−379	TTCGTCAAGGTTGTT	ZR0*
AFUA_1G09810	-	R	−168	−154	ACTCCCAGGGTACTT	ZR0
AFUA_1G10080	zafA	D	−361	−347	GCTCCCAAGGTCCTC	ZR0
		D	−448	−434	CCCCTCAGGGTATTA	ZR0*
		D	−624	−610	GCTCGCAGGGTCCTC	ZR5
		D	−657	−643	TCCCCCAGGGTCCTC	ZR0
AFUA_1G12170	-	R	−364	−350	ATTACCAAGGTGATA	ZR0*
AFUA_1G12850	crnA	R	−500	−486	TGTCCCAGGGTGTAC	ZR0*
AFUA_1G14550	sodC/sod3	R	−123	−109	AGCATCAAGGTCCTC	ZR2
		D	−160	−146	GATCCCAAGGTCCCC	ZR2
AFUA_1G14560	msdS	D	−664	−650	TCTACCAGGGTATAT	ZR8
		D	−928	−914	ATCGCCAGGGTCTTA	ZR9
AFUA_1G14700	-	R	−181	−167	GCACACAAGGTACTC	ZR0*
AFUA_2G01260	srbA	D	−886	−872	TTGGTCAAGGTCCTT	ZR3
AFUA_2G01610	-	D	−841	−827	GTCAGCAGGGTCATC	ZR0*
AFUA_2G02950	-	D	−160	−146	GTCTCCAAGGTCTCC	ZR4
AFUA_2G03860	zrfB	R	−359	−345	ATCCTCAAGGTCCCT	ZR0
		D	−496	−482	GCCACCAAGGTCCTT	ZR0
		D	−794	−780	TCTCCCAAGGTCCCC	ZR0
		D	−893	−879	ACTCCCAAGGTCCTC	ZR0
AFUA_2G06140	-	R	−461	−447	GTTCCCAAGGTCTCC	ZR0

Table 3. *Cont.*

Gene ID	Gene Name	Strand	Start	End	Sequence ZR Motif	Type of ZR Motif
AFUA_2G07810	-	D	−756	−742	GTAGGCAAGGTGCCT	ZR0*
AFUA_2G08280	maeA	R	−910	−896	TTCACCAGGGTAGAC	ZR0*
AFUA_2G08740	zrfF	R	−329	−315	TCTCCCAAGGTCCGC	ZR8
		R	−616	−602	ATAGCCAAGGTAGCT	ZR0*
AFUA_2G15290	-	R	−68	−54	GTCAACAGGGTGCTC	ZR5
AFUA_3G02270	cat1	D	−1052	−1038	TTGACCAGGGTCTCT	ZR3
AFUA_3G10680	-	R	−179	−165	GATCCCAAGGTCCTT	ZR2
		D	−233	−219	TCCCCCAGGGTCCCC	ZR0
AFUA_3G13100	-	D	−581	−567	TTCGTCAAGGTTTCA	ZR0*
AFUA_3G13940	-	D	−1073	−1059	TTTTCCAAGGTGTTG	ZR0 *
AFUA_4G03920	-	R	−307	−293	CTCACCAAGGTCCCC	ZR1
		D	−438	−424	ACCTCCAAGGTCCTG	ZR0*
		R	−576	−562	TCTATCAAGGTAATT	ZR7
		R	−605	−591	GCCTCCAAGGTAGTC	ZR0*
AFUA_4G03930	cysX	D	−272	−258	GCCTCCAAGGTAGTC	ZR0*
		D	−301	−287	TCTATCAAGGTAATT	ZR1
		R	−439	−425	ACCTCCAAGGTCCTG	ZR0*
		D	−570	−556	CTCACCAAGGTCCCC	ZR7
AFUA_4G09560	zrfC	R	−196	−182	ACTCCCAAGGTATCT	ZR0
		R	−313	−299	GCTCTCAAGGTCCTC	ZR0
		R	−637	−623	GCTACCAAGGTGTCC	ZR0
		D	−749	−735	GCCATCAGGGTAGAC	ZR0*
AFUA_4G09580	aspf2	R	−149	−135	GCCATCAGGGTAGAC	ZR0*
		D	−261	−247	GCTACCAAGGTGTCC	ZR0
		D	−585	−571	GCTCTCAAGGTCCTC	ZR0
		D	−702	−688	ACTCCCAAGGTATCT	ZR0
AFUA_4G10460	hcsA	R	−538	−524	ATCGGCAAGGTACAT	ZR0*
AFUA_5G02010	-	D	−122	−108	TTCTCCAGGGTCTTA	ZR0*
AFUA_5G03060	-	R	−807	−793	GCGCGCAAGGTACTT	ZR0*
AFUA_5G05710	-	R	−100	−86	TTCACCAAGGTTTTG	ZR0*
AFUA_5G06240	alcC	D	−595	−581	TCCCCCAGGGTACAT	ZR8
AFUA_5G09240	sodA/sod1	R	−198	−184	ATTCACAGGGTATTA	ZR0*
AFUA_5G12780	-	D	−99	−85	ACTCCCAAGGTACTC	ZR0
AFUA_5G13940	-	R	−231	−217	TCTGTCAGGGTCTGT	ZR8
AFUA_6G00690	-	D	−77	−63	GTCCACAAGGTCTTC	ZR5
AFUA_6G08580	fkbp4	D	−134	−120	GCCCTCAAGGTTCCT	ZR6
AFUA_6G09310	-	D	−1068	−1054	AGTGACAAGGTATTC	ZR0*
AFUA_6G09630	gliZ	D	−790	−776	TCTAACAAGGTCCTC	ZR5
		D	−836	−822	GCCCCCAAGGTGCCT	ZR0
AFUA_6G09710	gliA	R	−469	−455	GTTTGCAAGGTACTC	ZR0*
		R	−522	−508	TCCCCCAAGGTCACA	ZR0*
AFUA_6G10260	akr1	D	−180	−166	GTCGTCAAGGTTCCC	ZR6
AFUA_7G02360	-	R	−323	−309	TCGATCAAGGTGCTT	ZR3
AFUA_7G03970	-	D	−396	−382	TGTACCAAGGTATGT	ZR0*
AFUA_7G06570	zrcA	D	−915	−901	GTCCCCAAGGTACTC	ZR0
AFUA_7G06790	yct1	D	−224	−210	GTTCACAAGGTTCTT	ZR0*
		D	−305	−291	TCTACCAAGGTCCTT	ZR0
AFUA_7G06810	sarA	R	−187	−173	TCCAACAAGGTACCT	ZR5
		R	−276	−262	CCTCCCAGGGTTCTC	ZR0*
AFUA_8G01930	-	R	−603	−589	GCCATCAAGGTCGAT	ZR0*
AFUA_8G02450	gzbA	R	−165	−151	GTCTTCAAGGTTCTC	ZR0*
		D	−401	−387	GTCCCCAAGGTTCTC	ZR6
AFUA_8G02460	-	R	−617	−603	GTCCCCAAGGTTCTC	ZR6
		D	−853	−839	GTCTTCAAGGTTCTC	ZR0*
AFUA_8G02620	mchC	R	−65	−51	ACCCCCAAGGTTCGC	ZR0*
		R	−172	−158	ACTCCCAAGGTACCT	ZR0
		R	−192	−178	GGGCTCAAGGTCCTC	ZR0*

4. Discussion

The effect of zinc-deficiency on the regulation of gene expression and the homeostatic and adaptive response to zinc starvation has been extensively analyzed at a genome-wide scale in the yeast *Saccharomyces cerevisiae* [16–18]. In this yeast, the regulatory response to zinc starvation at the transcriptional level relies on the Zap1 transcription factor [31]. We described for the first time that regulation of zinc homeostasis in a fungal pathogen is essential for virulence [13,14]. Similar studies, conducted later in different *Candida* and *Cryptococcus* species, *Histoplasma capsulatum* and *Blastomyces dermatitidis*, have also highlighted the importance of maintaining zinc homeostasis for the virulence of these pathogenic yeast [19,20,32–36]. In this paper, we report the transcriptional profiling of *A. fumigatus* under zinc starvation and show that ZafA regulates the expression of its most direct target genes through binding to one or more ZR consensus motifs located in their promoter regions.

In both *S. cerevisiae* and the pathogenic fungi studied thus far, the genes more strongly induced by the ZafA/Zap1 orthologues encode membrane transporters of the ZIP family, involved in zinc uptake from zinc-limiting media. Besides, it has been observed that zinc-deficiency increases ROS production [20,33,37] while it reduces sulfate assimilation as a putative mechanism to alleviate oxidative stress under zinc-limiting conditions [38]. In this regard, the adaptive response to zinc deficiency is well mirrored at the transcriptional level by changes in the expression of genes encoding proteins involved in ROS detoxification and assimilation of inorganic sulfur both in *S. cerevisiae* [18] and in all pathogenic yeasts [19,20]. Similarly, in *A. fumigatus* grown under zinc-limiting conditions, genes encoding zinc-independent proteins involved in ROS detoxification (e.g., *cat1*, *cat2* and *sodC*) are up-regulated, while *metR* [39], which is the major regulator of inorganic sulfur assimilation, is down-regulated (Table 2 and Table S6). There are, however, some remarkable differences between species regarding specific aspects. Among the most noteworthy are those related to iron metabolism, ergosterol biosynthesis and protection against oxidative stress, which appears to be linked to the production of certain secondary metabolites.

The major difference between *A. fumigatus* and the yeasts *S. cerevisiae*, *C. albicans* and *C. gatti* regarding iron metabolism is that only *A. fumigatus* is able to synthesize and secrete siderophores for iron uptake under iron-limiting conditions [40]. The only gene related to iron homeostasis that has been reported to be induced by Zap1 in *S. cerevisiae* under zinc-limiting conditions is *FET4* [16,18], which encodes a low-affinity $Zn^{2+}/Fe^{2+}/Cu^+$ transporter [41]. In *C. albicans* grown under zinc-limiting conditions, the expression of some genes encoding proteins of the reductive pathway (e.g., *FRE2* and *FET34*) is up-regulated, whereas the expression of others is down-regulated (e.g., *FRE5*, *FRE7*, *FRE10*, *FET3* and *FTH1*). In *C. gatti* grown under zinc-limiting conditions, the expression of the genes CNBG_3602 and CNBG_2036 is down-regulated [20]. They encode proteins similar to FtrA and MirB of *A. fumigatus* that function respectively in the reductive and non-reductive/siderophore-mediated iron uptake system. In *A. fumigatus* are down-regulated, under zinc-limiting conditions, several genes encoding proteins that are involved in iron uptake, including the FetC/FrtA oxidase/permease complex of the reductive iron uptake system and several non-reductive siderophore transporters of *A. fumigatus* (Table 2 and Table S6). In addition, genes encoding regulators of iron homeostasis (e.g., *hapX* and *srbA*) and siderophore biosynthesis are also down-regulated in *A. fumigatus* under zinc starvation (Table 2 and Table S6). Therefore, provided that all transcriptional profiling studies performed thus far in fungal pathogens have been done under non-iron-limiting conditions, it is concluded that the expression of genes encoding proteins required for iron uptake appear to be reduced under zinc-limiting conditions, which indicates that a reduction of iron intake from non-iron-limiting media appears to be an adaptive response to zinc deficiency. Whether Zap1 plays a direct or indirect role in regulating the expression of genes involved in iron homeostasis in yeast has not been reported [19]. However, it appears that in *A. fumigatus* growing under zinc-limiting conditions, only the expression of some genes related to iron homeostasis is directly or indirectly regulated by ZafA (e.g., *srbA*, *enb1*, *mirB* and *sidA*), whereas most of them appear to be regulated in a ZafA-independent manner (Table 2 and Table S6). On the other hand, the transcriptional profiling of *A. fumigatus* grown in zinc-replete media following a shift from iron-limiting to iron-replete conditions has revealed that the expression of genes typically involved in the homeostatic response to zinc deficiency (e.g., *zafA*, *zrfA* and *zrfB*) is up-regulated whereas that of genes involved in the homeostatic response to zinc excess is down-regulated (e.g., *zrcA*) [42]. The expression profile of these ZafA-regulated genes in the referred conditions has been interpreted as a homeostatic mechanism to prevent zinc excess and zinc toxicity under iron starvation [43]. In addition, in a wild-type strain growing in zinc-replete media under iron-replete conditions HapX mediates both the up-regulation of genes typically involved in the homeostatic response to zinc deficiency (e.g., *zafA* and *zrfB*) and the down-regulation of genes involved in the homeostatic response to zinc excess (e.g., *zrcA* and *zrcC*) [44]. However, it is known that the expression of *hapX* is reduced dramatically under iron-replete conditions [44], which indicates that the predicted action of HapX on the expression of the aforementioned genes must be exerted by the low basal amount of HapX that is synthesized under iron-replete conditions [45]. It is difficult to understand that the HapX-mediated increase of the transcription level of *zafA*, which encodes the key major regulator of zinc homeostasis, only reflects changes in the expression level of *zrfB* and *zrcA* (of all putative direct targets of ZafA). Nevertheless, this could be explained upon considering that *zrfB* is the zinc homeostatic gene with the highest number of ZR motifs in its promoter, whereas *zrcA* is one of the few genes whose expression is strongly repressed by ZafA that harbors one ZR consensus motif in its promoter region (Table 3). Hence, it could be possible that the low increase in the expression level of *zafA* detected in a wild-type strain versus a ΔhapX strain (about 1.8-fold) [44], allows synthesizing a little (although sufficient) amount of ZafA. This regulator could bind preferentially to the promoters of *zrfB* and *zrcA* to induce the former and repress the later. The *zrcC* gene does not have ZR motifs in its promoter region and most likely its repression under zinc-limiting conditions is not directly mediated by ZafA. The HapX-mediated co-regulation of the *zrfB*, *zrcA* and *zrcC* expression in *A. fumigatus* might have evolved as an adaptive response to iron excess caused by a sudden iron supplement under zinc-replete conditions. This response would prevent the rapid storage into the vacuole (mediated by ZrcA and ZrcC) of the zinc taken up by ZrfB. The most likely reason would be that a concentration of cytosolic zinc higher than usual is required to prevent iron toxicity during a transient iron excess. This adaptive response recalls the proactive mechanism that operates in *S. cerevisiae* to protect cells against a sudden high increase of zinc concentration in the medium or "zinc shock" [46]. This mechanism relies on the function of ZRC1 that is the orthologue of the *zrcA* gene of *A. fumigatus*. ZRC1 is up-regulated by Zap1 to enhance the storage into the vacuole of zinc that is incorporated by Zrt1 under zinc-limiting conditions [46,47]. Provided that these studies have been performed in media containing a relatively low amount of iron such as that present in the yeast nitrogen base (i.e., 0.74 μM) used to make the yeast synthetic minimum medium, it could be possible that zinc storage following a "zinc shock" is intended to prevent zinc toxicity due to iron insufficiency. In summary, a mutual relationship between regulation of homeostasis of zinc and iron appears to exist. This zinc/iron cross-homeostasis could be essential for fungal growth in zinc-replete media. It would not only prevent the excessive zinc uptake and zinc toxicity under iron starvation, as proposed previously [43], but also it might increase the cytosolic concentration of zinc to prevent iron toxicity under iron-replete conditions and/or prevent an excessive iron uptake from non-iron-limiting media and iron toxicity under zinc starvation. Consequently, the zinc/iron cross-homeostasis would allow the fungus to grow well in media containing unbalanced Zn:Fe ratios.

Some NZR genes are down-regulated, both in zinc-limiting media under iron-replete conditions and in zinc-replete media after a shift from iron-limiting to iron-replete conditions, such as *hmg1*, which encodes the enzyme of the committed step in ergosterol biosynthesis, and several genes related to siderophore biosynthesis (e.g., *sidC*, *sidF*, *sidG*, *amcA*) [44]. This finding suggests that the reduction of both iron uptake and ergosterol biosynthesis might also have co-evolved as part of the fungal adaptive response both to zinc starvation under non-iron-limiting conditions, and to iron excess under non-zinc-limiting conditions. In addition, a relatively high number of genes whose expression is regulated by the sterol-response element binding proteins (SREPBs), SrbA and SrbB, are also influenced by zinc starvation [48,49]. For instance, several genes involved in ergosterol biosynthesis (i.e., *erg1*, *erg3A/B*, *erg5*, *erg13B*, *erg24A/B*, *erg25A/B* and *hmg1*), ethanol catabolism (*alcC*), heme biosynthesis (*hem13* and *hem14*) and nitrate assimilation (*niiA* and *niaD*), are up-regulated by SREPBs in hypoxia and down-regulated under (normoxic) zinc-limiting conditions (Table 2). Similarly, the own expression of *srbA* and *srbB* is strongly reduced under (normoxic) zinc-limiting conditions (Table 2). On the other hand, it is known that the transcription of *srbB* is autoinduced by SrbB and starts earlier (and is stronger) than that of *srbA*, which is autoinduced by SrbA [49]. Besides, the RNAseq data for the wild-type, Δ*srbA* and Δ*srbB* strains, after 2 h in hypoxia, revealed that the expression level of *srbB* in the Δ*srbA* mutant is lower than in the wild-type strain; whereas the expression level of *srbA* in the Δ*srbB* mutant is higher than in the wild-type strain, i.e., SrbA is formally an inductor of *srbB* whereas SrbB is a repressor of *srbA* [49]. Since *srbA* can be formally considered as a direct ZafA target (Table 2), it could be possible that *srbA* repression initiated by ZafA, under normoxic zinc-limiting conditions, triggers a cascade of transcriptional events that ultimately results in the strong down-regulation of *srbA* and *srbB* (and that of their target genes) under sustained zinc-limiting conditions. However, the same RNAseq experiment also indicated that SrbA and SrbB are formally positive regulators of *zrfA*, *zrfB* and *zrfC* but negative regulators of *zafA* [49]. In addition, SrbA is a positive regulator of *zrcA* and *zrcC* whereas SrbB is a negative regulator of these genes [49]. Hence, provided that these studies have been performed in zinc-replete media, the positive regulation of *zrfA*, *zrfB* and *zrfC* by SrbA and SrbB, along with the early negative regulation of *zrcA* and *zrcC* by SrbB, this suggests that a cytoplasmic zinc concentration higher than usual may be beneficial for fungal growth in hypoxia. Of note, this regulatory mechanism for the adaptive response to hypoxia resembles the one involved in preventing iron toxicity during iron excess under zinc-replete conditions. In addition, it has been reported that the fungal zinc content is about 3-fold higher in mycelium grown in hypoxia than in normoxia [50]. However, the SREPB-mediated down-regulation of *zafA* by SrbA and SrbB under hypoxic zinc-replete conditions suggests that the expression *zrfA*, *zrfB* and *zrfC* should be induced directly by these factors instead of being induced by ZafA.

S. cerevisiae and *C. albicans* lack SREPBs orthologues and appear to have evolved different regulatory mechanism [51]. In either case, under hypoxic (zinc-replete) conditions the expression of the *ERG* genes is down-regulated in *S. cerevisiae* [52], but up-regulated in *C. albicans* [53] and *A. fumigatus* [48]. This suggests that the sterol content in *S. cerevisiae* under hypoxic conditions is largely sufficient to sustain growth whereas in *C. albicans* and *A. fumigatus* it is largely insufficient, thus requiring the increase of the expression of the ergosterol biosynthetic genes encoding enzymes that catalyze oxygen-dependent reactions (Erg1, Erg11, Erg25 and Erg3). However, it has been reported that the expression of the *ERG* genes is down-regulated in a *S. cerevisiae* Δ*zap1* mutant [16], which indicates the ScZap1 is formally a positive regulator of the *ERG* genes. In contrast, in a *C. albicans* Δ*zap1* mutant the expression of the *ERG* genes is up-regulated [19], which indicates the CaZap1 is formally a negative regulator of the *ERG* genes. However, the *ERG* genes in *S. cerevisiae* and *C. albicans* lack Zap1 binding motifs in their promoter regions, which suggests that both ScZap1 and CaZap1 regulate indirectly the expression of these genes [19]. Similarly, the *erg* genes of *A. fumigatus* lack ZR motifs in their promoters and, under zinc-limiting conditions, only a few *erg* genes (e.g., *erg5*, *erg13B*, *erg24A*, *erg24B* and *erg3B*) are likely repressed either directly or indirectly by ZafA, whereas it is likely that the expression of other genes (e.g., *hmg1*, *erg1*, *erg3A*, *erg25A* and *erg25B*) are down-regulated in a ZafA-independent manner (Table 2 and Table S6). In either case, one possibility is that the repression of the *erg* genes under normoxic zinc-limiting conditions is an indirect consequence of *srbA* repression by ZafA.

Zinc starvation increases the expression level of genes encoding components of the electron acceptor complexes I-III of the mitochondrial electron transport chain (ETC) (e.g., *sdh3* and *sdh4* that encode the inner-membrane-anchored subunits of the succinate deshydrogenase), while it reduces the encoding of the mitochondrial fumarate reductase (FrdA) and some genes encoding components of complex IV of the ETC (e.g., AFUA_2G03010 and AFUA_3G14440). The *frdA* gene, which is one of the most strongly induced genes during hypoxia in *A. fumigatus* [54], encodes a FAD-dependent enzyme similar to the yeast Frd1 enzyme that couples the reduction [fumarate → succinate] to the oxidation [ubiquinol → ubiquinone] (i.e., the reverse reaction that is catalyzed by the succinate deshydrogenase) [55]. The complexes I and III are considered as major sites of ROS generation in mitochondria [56]. Thus, it is tempting to speculate that the imbalance between the expression of genes encoding components of the electron acceptor complexes I/II and electron donor complex IV under zinc-limiting conditions may disturb the normal electron flux through the ETC in normoxia. In this scenario, not all electrons that enter the ETC are finally used for O_2 reduction, at the same time that the intracellular partial pressure of O_2 in the fungus growing under zinc-limiting conditions becomes relatively higher than in the fungus growing in zinc-replete media. As a consequence, the intracellular O_2 may reach a slightly hyperbaric concentration while it increases the probability of adventitious reactions of reduced flavins with O_2 leading to the formation of (ROS) [57]. It is known that enzymes of the succinate/fumarate oxidoreductase family, which include the succinate dehydrogenases and fumarate reductases, are prone to react adventitiously with O_2 and autoxidation resulting in the formation of superoxide anion and/or H_2O_2 (because they contain flavins that appear to be the primary site of electron transfer to O_2) [57,58]. For instance, the *E. coli* NadB flavoprotein, which is structurally and evolutionary related to succinate/fumarate oxidoreductases, accounts for about 25% of all H_2O_2 generated endogenously during aerobic growth [59]. Thus, the reduction of the expression level of *frdA* could be interpreted as a mechanism to prevent an excess of ROS formation under zinc-limiting conditions.

Interestingly, the adaptive response against oxidative stress displayed by *A. fumigatus* appears to differ significantly from that displayed by the yeast *S. cerevisiae*, in which only two gene-encoding enzymes for protection against ROS, *SOD1* and *TSA1*, appear to be critical for growth in low zinc [37,60]. The expression level of *SOD1* in *S. cerevisiae*, which encodes the orthologue to the *A. fumigatus* SodA Cu/Zn-superoxide dismutase, is similar under both zinc-replete and zinc-limiting conditions [60]. In contrast, in *A. fumigatus* the expression level of *sodA* reduced significantly under zinc-limiting conditions, most likely as a mechanism to reduce the biosynthesis of the Cu/Zn-SOD whose functionality is strongly reduced under zinc-limiting conditions (Figure 9). Besides, *A. fumigatus* lacks an obvious orthologous of *TSA1*, which encodes a protein with a dual peroxiredoxin-chaperone function that enables yeast cells to adapt both oxidative stress and deficient folding caused by zinc starvation [37,61]. In summary, the adaptive response against oxidative stress in *A. fumigatus* under zinc-limiting conditions might rely on a dual regulatory mechanism that involves both the down-regulation of genes encoding ROS generating enzymes and the up-regulation of genes encoding ROS detoxifying, zinc-independent enzymes.

Unlike pathogenic yeast, *A. fumigatus* has at least 26 biosynthetic gene clusters (named as BGC1-26) that enables it to produce a great diversity of secondary metabolites and mycotoxins [62]. Interestingly, the expression level of 17 genes belonging to 11 different BGCs are differentially regulated under zinc-limiting conditions (Table S6). The BGC-20 dedicated to gliotoxin (GT) biosynthesis appear to be the most largely influenced by zinc starvation, as revealed by the high expression level of *gliZ*, *gliT*, *gliA* and *gtmA*. GliT catalyzes disulfide formation during GT biosynthesis, whereas GliA is involved in GT efflux and both proteins provide self-protection against exogenous and endogenous GT [63–65]. The enzyme GtmA catalyzes the conversion of both the endogenous dithiol-GT (GT-[H_2S]) and the exogenous GT into bisdethiobis(metylthio)gliotoxin (BmGT), which is a negative modulator of GT biosynthesis, using S-adenosylmethionine (SAM) as a methyl donor [66]. Hence, the transcriptional profile of *gliZ*, *gliT* and *gliA* strongly suggests that either a much higher amount of GT is synthesized in zinc-limiting media than in zinc-replete conditions, or GT is far more noxious for fungal growth in zinc-limiting than in zinc-replete media. Besides, the GtmA-mediated bis-thiomethylation of GT is a SAM-dependent reaction that produces as a by-product S-adenosyl-homocysteine (SAH). SAH is then presumably hydrolyzed to homocysteine (Hcy) by the putative S-adenosyl-homocysteinase encoded by AFUA_1G10130/*sahA*, whose expression is also slightly up-regulated under zinc-limiting conditions (Table 2). It has been proposed that GliT, GliA and GtmA work in concert in *A. fumigatus* to resist the effect of exogenous GT in zinc-replete conditions [67], and that GliT and GtmA are up-synthesized in the presence of both GT and GT plus H_2O_2 under zinc-replete conditions [68]. In addition, GT inhibits

H$_2$O$_2$-induced oxidative stress [68,69] because most likely GT is able to replace the function of 2-Cys peroxiredoxins as an electron acceptor in the thioredoxin-peroxiredoxin system [70], which is consistent with the lack of a Tsa1-like protein in *A. fumigatus*. Hence, it is very likely that the concerted action of the GliT, GliA, GtmA and SahA proteins has evolved in *A. fumigatus* as part of the adaptive response against oxidative stress caused by zinc starvation. However, the success of the GT-based strategy to reinforce the ROS detoxification capability of *A. fumigatus* relies on an efficient self-protection mechanism against the toxic effect of an excessive amount of GT. At the same time, a relatively high SAM:SAH ratio must be kept to prevent the competitive inhibition of specific methyltransferases by SAH.

The comparison of genes whose expression levels change in a wild-type strain in zinc-limiting versus zinc-replete conditions with those whose expression levels change in a Δ*zafA* mutant versus a wild-type strain when both are grown under in zinc-limiting conditions, produced a group of genes that was expected to be enriched in the most direct ZafA targets. Thus, we identified 153 putative ZafA target genes (35 down-regulated and 118 up-regulated) that represented about 1.56% of the 9824 predicted ORFs in *A. fumigatus*. Following a similar approach for the characterization of the Zap1 zinc-responsive regulon in *S. cerevisiae*, 111 genes were found to be putatively up-regulated by Zap1 [16], which represent about 1.82% of the 6091 biologically significant ORFs of *S. cerevisiae* [71]. In addition, by using a combination of in vitro and in silico approaches, it was determined that most of the 111 putative Zap1 target genes carried, at their promoter regions, at least one zinc response (ZR) consensus sequence 5'-ACCYYNAAGGT-3' (or a degenerated version of it) to which Zap1 may bind [18,72]. Thus, it was concluded that a total of 81 genes in the yeast genome were direct targets of Zap1 [18]. Using a similar experimental approach, we have found 49 direct targets of ZafA, including 29 potential and 20 confirmed ZafA targets (Table 3). However, several DZR genes (*erg24A*, *gtmA*, *pyroA*, *azf1*, *gliT* and *frdA*) (Table 2), formally considered as direct targets of ZafA, lack obvious ZR motifs. This suggests that ZafA could be able to bind to ZR motifs with a relatively high degeneration degree. In this regard, it must be aware of that, unlike Zap1 of *S. cerevisiae*, which binds to a 11-nucleotide sequence with a relatively low degeneration degree (5'-ACCYYNAAGGT-3'; with 16 [2^2 × 4] different possible ZR sequences), the ZafA transcription factor of *A. fumigatus* binds to a 15-nucleotide zinc response sequence with a higher degeneration degree (5'-DYYVYCARGGTVYYY-3'; with 3456 [3^3 × 2^7] different possible ZR sequences), even without considering that additional substitutions are possible out of the conserved core (5'-CARGGT-3') of the ZR motif. Unlike this hexanucleotide sequence that is essential for ZafA to bind specifically to the ZR motifs, the high variability of both ends of the ZR motif could be intended to determine the affinity of ZafA binding to the ZRs. In either case, it appears that the diversity of ZR sequences to which ZafA can bind in *A. fumigatus* is noticeably higher than those recognized by Zap1 in *S. cerevisiae*. This fact complicates the identification of functionally relevant ZR motifs in many genes of *A. fumigatus*, including some of the most strongly expressed DZR genes, such as *gtmA* and *gliT*.

It is clear that ZafA plays a role in *A. fumigatus* similar to that of Zap1 in *S. cerevisiae* and in the pathogenic yeast *C. albicans* and *C. gattii*, although the only significant identity among these proteins is restricted to the C-terminal DNA binding domain. However, unlike ScZap1 that has four zinc-fingers for DNA binding (one of the CWCH2-type plus three of the C2H2-type), CgZap1 and CaZap1, similar to ZafA, have five (one of the CWCH2-type plus four of the C2H2-type) (Figure 1). Interestingly, given that a zinc-finger binds typically to a 3-nucleotide sequence [73], it may explain that ZafA binds to a 15-nucleotide ZR motif whereas ScZap1 binds to a 11-nucleotide ZR motif. It is interesting as well to note that the high hypersensitivity of the DNA fragments from the *zafA*, *zrfA*, *zrfB* and *zrfC* promoters to DNase I digestion, observed after performing the in vitro footprinting assays, could be an indication that binding of ZafA to the referred promoters causes DNA bending in vivo as part of the transcriptional regulatory mechanism mediated by ZafA, as reported for other regulators [74].

Finally, it is noticeable that most of the DZR genes that harbored ZR motifs in their promoters are up-regulated, which strongly suggests that ZafA mainly functions as an activator rather than as a repressor. Nevertheless, there are also some down-regulated genes that appear to be directly repressed by ZafA (e.g., *zrcA*, *srbA*, *alcC*, *sodA* and others without obvious ZR motifs in their promoters, such as *frdA* and *erg24A*) (Table 3). Similarly, Zap1 of *S. cerevisiae* appears to function nearly exclusively as a transcriptional activator since all 81 proposed Zap1 target genes appear to be induced by Zap1 [16,18], although it has been reported that Zap1 also functions paradoxically as a repressor of *ZRT2* in low zinc [75]. In addition, Zap1 under zinc deficiency represses the expression of the genes *ADH1* and *ADH3*, which encode zinc-dependent alcohol dehydrogenases, most likely as a mechanism to prevent their synthesis under zinc-limiting conditions [76]. Interestingly, a similar rationale could also be applied to explain the repression by ZafA of zinc-dependent enzymes of *A. fumigatus* such as *alcC* and *sodA*.

5. Conclusions

The analysis of the transcriptional profile of the ZafA regulon of *A. fumigatus*, under zinc-limiting conditions, confirmed that ZafA is the major regulator of the zinc homeostatic response to zinc starvation. It also indicated that ZafA, along with other regulators, was involved in a zinc/iron cross-homeostatic controlling network that allows the fungus to grow in media containing unbalanced Zn:Fe ratios. This study suggested as well that reduction of iron uptake and ergosterol biosynthesis might have co-evolved as part of the fungal adaptive response to iron excess under non-zinc-limiting conditions and to zinc starvation under non-iron-limiting conditions, even though this regulation appeared to be exerted only in part and indirectly by ZafA. In agreement with previous reports by other investigators, these results also suggested an increase in ROS production under zinc starvation that *A. fumigatus* counteracted by reducing the expression of genes encoding enzymes involved in ROS production and increasing the expression of genes encoding ROS-detoxifying enzymes and gliotoxin production, along with a self-protection mechanism against endogenous gliotoxin. Finally, it was shown that although ZafA appeared to function mainly as an inductor of gene expression under zinc-limiting conditions, it could also function as a repressor of a limited number of genes. Importantly, ZafA would exert its function as a transcriptional activator through binding to a well-defined zinc response consensus motif (5′-DYYVYCARGGTVYYY-3′) present in the promoters of its most direct target genes. However, ZafA could also bind to ZR motifs with a relatively high degeneration degree in nucleotides out of the conserved hexanucleotide core (5′-CARGGT-3′) and that could be present in the promoter regions of secondary target genes of ZafA.

Supplementary Materials: The following are available online at http://www.mdpi.com/2073-4425/9/7/318/s1, Table S1. Oligonucleotides used for RT-qPCR. Table S2. Oligonucleotides used in this study for construction of plasmids and validation of fungal strains. Table S3. Genes differentially expressed in a wild-type strain grown in zinc-limiting (SDN–Zn) versus zinc-replete conditions (SDN + 100 µM Zn) (Exp#1; FC > 1.5; P < 0.05). Table S4. Genes differentially expressed in a Δ*zafA* strain grown under zinc-limiting conditions with respect to a wild-type strain grown in the same culture conditions (Exp#2; FC > 1.5; P < 0.05). Table S5. Genes up-regulated in Exp#1 but down-regulated in Exp#2 and vice versa (i.e., DZR genes). Table S6. Functional categorization of genes differentially expressed in a wild-type strain grown in zinc-limiting (SDN–Zn) versus zinc-replete conditions (Exp#1). Table S7. Control genes used for MEME analysis. Table S8. All possible ZR motifs harboring one substitution out of the conserved hexanucleotide core (5′-CARGGT-3′) with respect to the ZR consensus motif (ZR0). Figure S1. Construction of strains used to show the function of ZafA in regulating gene expression through binding to the ZR motifs in vivo. Figure S2. Analysis of ZafA binding to DNA fragments from the promoter regions of genes *zafA*, *zrfA*, *zrfB* and *zrfC* by DNase I footprinting assays.

Author Contributions: J.A.C. conceived and designed the experiments; R.V., J.A., L.M., and C.I.S. performed the experiments; J.A.C., R.V. and F.L. analyzed the data; J.A.C. wrote the paper.

Fundings: This work was supported by the "Ministerio de Economía y Competitividad" (Spain) and "Junta de Castilla y León" (Spain) through grants SAF2013-48382-R and SA020G18 respectively to J.A.C.

Acknowledgments: R.V. was supported by a contract through grant SAF2013-48382-R to J.A.C.; C.I.S. was supported by a fellowship from the Universidad Industrial de Santander (Bucaramanga, Colombia) through resolution n° 2604 (18th December 2014).

Conflicts of Interest: The authors declare no conflict of interest.

References

1. Andreini, C.; Bertini, I.; Cavallaro, G.; Holliday, G.L.; Thornton, J.M. Metal ions in biological catalysis: From enzyme databases to general principles. *J. Biol. Inorg. Chem.* **2008**, *13*, 1205–1218. [CrossRef] [PubMed]
2. Auld, D.S. The ins and outs of biological zinc sites. *Biometals* **2009**, *22*, 141–148. [CrossRef] [PubMed]
3. Maret, W. Zinc and sulfur: a critical biological partnership. *Biochemistry* **2004**, *43*, 3301–3309. [CrossRef] [PubMed]
4. Eide, D.J. Zinc transporters and the cellular trafficking of zinc. *Biochim. Biophys. Acta* **2006**, *1763*, 711–722. [CrossRef] [PubMed]
5. Palmiter, R.D.; Findley, S.D. Cloning and functional characterization of a mammalian zinc transporter that confers resistance to zinc. *EMBO J.* **1995**, *14*, 639–649. [PubMed]
6. Maret, W. Zinc biochemistry: From a single zinc enzyme to a key element of life. *Adv. Nutr.* **2013**, *4*, 82–91. [CrossRef] [PubMed]
7. Simons, T.J. Intracellular free zinc and zinc buffering in human red blood cells. *J. Membr. Biol.* **1991**, *123*, 63–71. [CrossRef] [PubMed]
8. Bessman, J.D.; Johnson, R.K. Erythrocyte volume distribution in normal and abnormal subjects. *Blood* **1975**, *46*, 369–379. [PubMed]
9. Karthaus, M.; Buchheidt, D. Invasive aspergillosis: New insights into disease, diagnostic and treatment. *Curr. Pharm. Des.* **2013**, *19*, 3569–3594. [CrossRef] [PubMed]
10. Amich, J.; Calera, J.A. Zinc acquisition: A key aspect in *Aspergillus fumigatus* virulence. *Mycopathologia* **2014**, *178*, 379–385. [CrossRef] [PubMed]
11. Amich, J.; Vicentefranqueira, R.; Leal, F.; Calera, J.A. *Aspergillus fumigatus* survival in alkaline and extreme zinc-limiting environments relies on the induction of a zinc homeostasis system encoded by the *zrfC* and *aspf2* genes. *Eukaryot. Cell* **2010**, *9*, 424–437. [CrossRef] [PubMed]
12. Vicentefranqueira, R.; Moreno, M.A.; Leal, F.; Calera, J.A. The *zrfA* and *zrfB* genes of *Aspergillus fumigatus* encode the zinc transporter proteins of a zinc uptake system induced in an acid, zinc-depleted environment. *Eukaryot. Cell* **2005**, *4*, 837–848. [CrossRef] [PubMed]
13. Amich, J.; Vicentefranqueira, R.; Mellado, E.; Ruiz-Carmuega, A.; Leal, F.; Calera, J.A. The ZrfC alkaline zinc transporter is required for *Aspergillus fumigatus* virulence and its growth in the presence of the Zn/Mn-chelating protein calprotectin. *Cell Microbiol.* **2014**, *16*, 548–564. [CrossRef] [PubMed]
14. Moreno, M.A.; Ibrahim-Granet, O.; Vicentefranqueira, R.; Amich, J.; Ave, P.; Leal, F.; Latge, J.P.; Calera, J.A. The regulation of zinc homeostasis by the ZafA transcriptional activator is essential for *Aspergillus fumigatus* virulence. *Mol. Microbiol.* **2007**, *64*, 1182–1197. [CrossRef] [PubMed]
15. Eide, D.J. Homeostatic and adaptive responses to zinc deficiency in *Saccharomyces cerevisiae*. *J. Biol. Chem.* **2009**, *284*, 18565–18569. [CrossRef] [PubMed]
16. Lyons, T.J.; Gasch, A.P.; Gaither, L.A.; Botstein, D.; Brown, P.O.; Eide, D.J. Genome-wide characterization of the Zap1p zinc-responsive regulon in yeast. *Proc. Natl. Acad. Sci. USA* **2000**, *97*, 7957–7962. [CrossRef] [PubMed]
17. North, M.; Steffen, J.; Loguinov, A.V.; Zimmerman, G.R.; Vulpe, C.D.; Eide, D.J. Genome-wide functional profiling identifies genes and processes important for zinc-limited growth of *Saccharomyces cerevisiae*. *PLoS Genet.* **2012**, *8*, e1002699. [CrossRef] [PubMed]
18. Wu, C.Y.; Bird, A.J.; Chung, L.M.; Newton, M.A.; Winge, D.R.; Eide, D.J. Differential control of Zap1-regulated genes in response to zinc deficiency in *Saccharomyces cerevisiae*. *BMC Genom.* **2008**, *9*, 370. [CrossRef] [PubMed]
19. Nobile, C.J.; Nett, J.E.; Hernday, A.D.; Homann, O.R.; Deneault, J.S.; Nantel, A.; Andes, D.R.; Johnson, A.D.; Mitchell, A.P. Biofilm matrix regulation by *Candida albicans* Zap1. *PLoS Biol.* **2009**, *7*, e1000133. [CrossRef] [PubMed]
20. Schneider, R.O.; Fogaca, N.S.; Kmetzsch, L.; Schrank, A.; Vainstein, M.H.; Staats, C.C. Zap1 regulates zinc homeostasis and modulates virulence in *Cryptococcus gattii*. *PLoS ONE* **2012**, *7*, e43773. [CrossRef] [PubMed]
21. Hatayama, M.; Aruga, J. Gli protein nuclear localization signal. *Vitam. Horm.* **2012**, *88*, 73–89. [PubMed]
22. Espeso, E.A.; Tilburn, J.; Sanchez-Pulido, L.; Brown, C.V.; Valencia, A.; Arst, H.N., Jr.; Peñalva, M.A. Specific DNA recognition by the *Aspergillus nidulans* three zinc finger transcription factor PacC. *J. Mol. Biol.* **1997**, *274*, 466–480. [CrossRef] [PubMed]

23. Evans-Galea, M.V.; Blankman, E.; Myszka, D.G.; Bird, A.J.; Eide, D.J.; Winge, D.R. Two of the five zinc fingers in the Zap1 transcription factor DNA binding domain dominate site-specific DNA binding. *Biochemistry* **2003**, *42*, 1053–1061. [CrossRef] [PubMed]

24. Pavletich, N.P.; Pabo, C.O. Crystal structure of a five-finger GLI-DNA complex: New perspectives on zinc fingers. *Science* **1993**, *261*, 1701–1707. [CrossRef] [PubMed]

25. Calera, J.A.; Ovejero, M.C.; López-Medrano, R.; Segurado, M.; Puente, P.; Leal, F. Characterization of the *Aspergillus nidulans aspnd1* gene demonstrates that the ASPND1 antigen, which it encodes, and several *Aspergillus fumigatus* immunodominant antigens belong to the same family. *Infect. Immun.* **1997**, *65*, 1335–1344. [PubMed]

26. Priebe, S.; Kreisel, C.; Horn, F.; Guthke, R.; Linde, J. FungiFun2: A comprehensive online resource for systematic analysis of gene lists from fungal species. *Bioinformatics* **2015**, *31*, 445–446. [CrossRef] [PubMed]

27. Bailey, T.L.; Boden, M.; Buske, F.A.; Frith, M.; Grant, C.E.; Clementi, L.; Ren, J.; Li, W.W.; Noble, W.S. MEME SUITE: Tools for motif discovery and searching. *Nucleic Acids Res.* **2009**, *37*, W202–W208. [CrossRef] [PubMed]

28. Medina-Rivera, A.; Defrance, M.; Sand, O.; Herrmann, C.; Castro-Mondragon, J.A.; Delerce, J.; Jaeger, S.; Blanchet, C.; Vincens, P.; Caron, C.; et al. RSAT 2015: Regulatory Sequence Analysis Tools. *Nucleic Acids Res.* **2015**, *43*, W50–W56. [CrossRef] [PubMed]

29. Zianni, M.; Tessanne, K.; Merighi, M.; Laguna, R.; Tabita, F.R. Identification of the DNA bases of a DNase I footprint by the use of dye primer sequencing on an automated capillary DNA analysis instrument. *J. Biomol. Tech.* **2006**, *17*, 103–113. [PubMed]

30. Nierman, W.C.; Pain, A.; Anderson, M.J.; Wortman, J.R.; Kim, H.S.; Arroyo, J.; Berriman, M.; Abe, K.; Archer, D.B.; Bermejo, C.; et al. Genomic sequence of the pathogenic and allergenic filamentous fungus *Aspergillus fumigatus*. *Nature* **2005**, *438*, 1151–1156. [CrossRef] [PubMed]

31. Zhao, H.; Eide, D.J. Zap1p, a metalloregulatory protein involved in zinc-responsive transcriptional regulation in *Saccharomyces cerevisiae*. *Mol. Cell Biol.* **1997**, *17*, 5044–5052. [CrossRef] [PubMed]

32. Do, E.; Hu, G.; Caza, M.; Kronstad, J.W.; Jung, W.H. The ZIP family zinc transporters support the virulence of *Cryptococcus neoformans*. *Med. Mycol.* **2016**, *54*, 605–615. [CrossRef] [PubMed]

33. Schneider Rde, O.; Diehl, C.; dos Santos, F.M.; Piffer, A.C.; Garcia, A.W.; Kulmann, M.I.; Schrank, A.; Kmetzsch, L.; Vainstein, M.H.; Staats, C.C. Effects of zinc transporters on *Cryptococcus gattii* virulence. *Sci. Rep.* **2015**, *5*, 10104. [CrossRef] [PubMed]

34. Xu, W.; Solis, N.V.; Ehrlich, R.L.; Woolford, C.A.; Filler, S.G.; Mitchell, A.P. Activation and alliance of regulatory pathways in *C. albicans* during mammalian infection. *PLoS Biol.* **2015**, *13*, e1002076. [CrossRef] [PubMed]

35. Bottcher, B.; Palige, K.; Jacobsen, I.D.; Hube, B.; Brunke, S. Csr1/Zap1 maintains zinc homeostasis and influences virulence in *Candida dubliniensis* but is not coupled to morphogenesis. *Eukaryot. Cell* **2015**, *14*, 661–670. [CrossRef] [PubMed]

36. Kujoth, G.C.; Sullivan, T.D.; Merkhofer, R.; Lee, T.J.; Wang, H.; Brandhorst, T.; Wuthrich, M.; Klein, B.S. CRISPR/Cas9-mediated gene disruption reveals the importance of zinc metabolism for fitness of the dimorphic fungal pathogen *Blastomyces dermatitidis*. *MBio* **2018**, *9*, e00412-18. [CrossRef] [PubMed]

37. Wu, C.Y.; Bird, A.J.; Winge, D.R.; Eide, D.J. Regulation of the yeast TSA1 peroxiredoxin by ZAP1 is an adaptive response to the oxidative stress of zinc deficiency. *J. Biol. Chem.* **2007**, *282*, 2184–2195. [CrossRef] [PubMed]

38. Wu, C.Y.; Roje, S.; Sandoval, F.J.; Bird, A.J.; Winge, D.R.; Eide, D.J. Repression of sulfate assimilation is an adaptive response of yeast to the oxidative stress of zinc deficiency. *J. Biol. Chem.* **2009**, *284*, 27544–27556. [CrossRef] [PubMed]

39. Amich, J.; Schafferer, L.; Haas, H.; Krappmann, S. Regulation of sulphur assimilation is essential for virulence and affects iron homeostasis of the human-pathogenic mould *Aspergillus fumigatus*. *PLoS Pathog.* **2013**, *9*, e1003573. [CrossRef] [PubMed]

40. Haas, H. Fungal siderophore metabolism with a focus on *Aspergillus fumigatus*. *Nat. Prod. Rep.* **2014**, *31*, 1266–1276. [CrossRef] [PubMed]

41. Dix, D.R.; Bridgham, J.T.; Broderius, M.A.; Byersdorfer, C.A.; Eide, D.J. The *FET4* gene encodes the low affinity Fe(II) transport protein of *Saccharomyces cerevisiae*. *J. Biol. Chem.* **1994**, *269*, 26092–26099. [PubMed]

42. Schrettl, M.; Kim, H.S.; Eisendle, M.; Kragl, C.; Nierman, W.C.; Heinekamp, T.; Werner, E.R.; Jacobsen, I.; Illmer, P.; Yi, H.; et al. SreA-mediated iron regulation in *Aspergillus fumigatus*. *Mol. Microbiol.* **2008**, *70*, 27–43. [CrossRef] [PubMed]

43. Yasmin, S.; Abt, B.; Schrettl, M.; Moussa, T.A.; Werner, E.R.; Haas, H. The interplay between iron and zinc metabolism in *Aspergillus fumigatus*. *Fungal. Genet. Biol.* **2009**, *46*, 707–713. [CrossRef] [PubMed]

44. Schrettl, M.; Beckmann, N.; Varga, J.; Heinekamp, T.; Jacobsen, I.D.; Jochl, C.; Moussa, T.A.; Wang, S.; Gsaller, F.; Blatzer, M.; et al. HapX-mediated adaption to iron starvation is crucial for virulence of *Aspergillus fumigatus*. *PLoS Pathog.* **2010**, *6*, e1001124. [CrossRef] [PubMed]

45. Gsaller, F.; Hortschansky, P.; Beattie, S.R.; Klammer, V.; Tuppatsch, K.; Lechner, B.E.; Rietzschel, N.; Werner, E.R.; Vogan, A.A.; Chung, D.; et al. The Janus transcription factor HapX controls fungal adaptation to both iron starvation and iron excess. *EMBO J.* **2014**, *33*, 2261–2276. [CrossRef] [PubMed]

46. MacDiarmid, C.W.; Milanick, M.A.; Eide, D.J. Induction of the ZRC1 metal tolerance gene in zinc-limited yeast confers resistance to zinc shock. *J. Biol. Chem.* **2003**, *278*, 15065–15072. [CrossRef] [PubMed]

47. Miyabe, S.; Izawa, S.; Inoue, Y. Expression of *ZRC1* coding for suppressor of zinc toxicity is induced by zinc-starvation stress in Zap1-dependent fashion in *Saccharomyces cerevisiae*. *Biochem. Biophys. Res. Commun.* **2000**, *276*, 879–884. [CrossRef] [PubMed]

48. Blatzer, M.; Barker, B.M.; Willger, S.D.; Beckmann, N.; Blosser, S.J.; Cornish, E.J.; Mazurie, A.; Grahl, N.; Haas, H.; Cramer, R.A. SREBP coordinates iron and ergosterol homeostasis to mediate triazole drug and hypoxia responses in the human fungal pathogen *Aspergillus fumigatus*. *PLoS Genet.* **2011**, *7*, e1002374. [CrossRef] [PubMed]

49. Chung, D.; Barker, B.M.; Carey, C.C.; Merriman, B.; Werner, E.R.; Lechner, B.E.; Dhingra, S.; Cheng, C.; Xu, W.; Blosser, S.J.; et al. ChIP-seq and in vivo transcriptome analyses of the *Aspergillus fumigatus* SREBP SrbA reveals a new regulator of the fungal hypoxia response and virulence. *PLoS Pathog.* **2014**, *10*, e1004487. [CrossRef] [PubMed]

50. Vödisch, M.; Scherlach, K.; Winkler, R.; Hertweck, C.; Braun, H.P.; Roth, M.; Haas, H.; Werner, E.R.; Brakhage, A.A.; Kniemeyer, O. Analysis of the *Aspergillus fumigatus* proteome reveals metabolic changes and the activation of the pseurotin A biosynthesis gene cluster in response to hypoxia. *J. Proteome Res.* **2011**, *10*, 2508–2524. [CrossRef] [PubMed]

51. Maguire, S.L.; Wang, C.; Holland, L.M.; Brunel, F.; Neuveglise, C.; Nicaud, J.M.; Zavrel, M.; White, T.C.; Wolfe, K.H.; Butler, G. Zinc finger transcription factors displaced SREBP proteins as the major Sterol regulators during Saccharomycotina evolution. *PLoS Genet.* **2014**, *10*, e1004076. [CrossRef] [PubMed]

52. Baumann, K.; Dato, L.; Graf, A.B.; Frascotti, G.; Dragosits, M.; Porro, D.; Mattanovich, D.; Ferrer, P.; Branduardi, P. The impact of oxygen on the transcriptome of recombinant *S. cerevisiae* and *P. pastoris*—A comparative analysis. *BMC Genom.* **2012**, *12*, 218. [CrossRef] [PubMed]

53. Synnott, J.M.; Guida, A.; Mulhern-Haughey, S.; Higgins, D.G.; Butler, G. Regulation of the hypoxic response in *Candida albicans*. *Eukaryot. Cell* **2010**, *9*, 1734–1746. [CrossRef] [PubMed]

54. Kroll, K.; Pahtz, V.; Hillmann, F.; Vaknin, Y.; Schmidt-Heck, W.; Roth, M.; Jacobsen, I.D.; Osherov, N.; Brakhage, A.A.; Kniemeyer, O. Identification of hypoxia-inducible target genes of *Aspergillus fumigatus* by transcriptome analysis reveals cellular respiration as an important contributor to hypoxic survival. *Eukaryot. Cell* **2014**, *13*, 1241–1253. [CrossRef] [PubMed]

55. Camarasa, C.; Faucet, V.; Dequin, S. Role in anaerobiosis of the isoenzymes for *Saccharomyces cerevisiae* fumarate reductase encoded by OSM1 and FRDS1. *Yeast* **2007**, *24*, 391–401. [CrossRef] [PubMed]

56. Holmstrom, K.M.; Finkel, T. Cellular mechanisms and physiological consequences of redox-dependent signalling. *Nat. Rev. Mol. Cell Biol.* **2014**, *15*, 411–421. [CrossRef] [PubMed]

57. Messner, K.R.; Imlay, J.A. Mechanism of superoxide and hydrogen peroxide formation by fumarate reductase, succinate dehydrogenase, and aspartate oxidase. *J. Biol. Chem.* **2002**, *277*, 42563–42571. [CrossRef] [PubMed]

58. Chaiyen, P.; Fraaije, M.W.; Mattevi, A. The enigmatic reaction of flavins with oxygen. *Trends Biochem. Sci.* **2012**, *37*, 373–380. [CrossRef] [PubMed]

59. Korshunov, S.; Imlay, J.A. Two sources of endogenous hydrogen peroxide in *Escherichia coli*. *Mol. Microbiol.* **2010**, *75*, 1389–1401. [CrossRef] [PubMed]

60. Wu, C.Y.; Steffen, J.; Eide, D.J. Cytosolic superoxide dismutase (SOD1) is critical for tolerating the oxidative stress of zinc deficiency in yeast. *PLoS ONE* **2009**, *4*, e7061. [CrossRef] [PubMed]

61. MacDiarmid, C.W.; Taggart, J.; Kerdsomboon, K.; Kubisiak, M.; Panascharoen, S.; Schelble, K.; Eide, D.J. Peroxiredoxin chaperone activity is critical for protein homeostasis in zinc-deficient yeast. *J. Biol. Chem.* **2013**, *288*, 31313–31327. [CrossRef] [PubMed]

62. Bignell, E.; Cairns, T.C.; Throckmorton, K.; Nierman, W.C.; Keller, N.P. Secondary metabolite arsenal of an opportunistic pathogenic fungus. *Philos. Trans. R Soc. Lond. B Biol. Sci.* **2016**, *371*, 20160023. [CrossRef] [PubMed]

63. Scharf, D.H.; Remme, N.; Heinekamp, T.; Hortschansky, P.; Brakhage, A.A.; Hertweck, C. Transannular disulfide formation in gliotoxin biosynthesis and its role in self-resistance of the human pathogen *Aspergillus fumigatus*. *J. Am. Chem. Soc.* **2010**, *132*, 10136–10141. [CrossRef] [PubMed]

64. Schrettl, M.; Carberry, S.; Kavanagh, K.; Haas, H.; Jones, G.W.; O'Brien, J.; Nolan, A.; Stephens, J.; Fenelon, O.; Doyle, S. Self-protection against gliotoxin—A component of the gliotoxin biosynthetic cluster, GliT, completely protects *Aspergillus fumigatus* against exogenous gliotoxin. *PLoS Pathog.* **2010**, *6*, e1000952. [CrossRef] [PubMed]

65. Wang, D.N.; Toyotome, T.; Muraosa, Y.; Watanabe, A.; Wuren, T.; Bunsupa, S.; Aoyagi, K.; Yamazaki, M.; Takino, M.; Kamei, K. GliA in *Aspergillus fumigatus* is required for its tolerance to gliotoxin and affects the amount of extracellular and intracellular gliotoxin. *Med. Mycol.* **2014**, *52*, 506–518. [CrossRef] [PubMed]

66. Dolan, S.K.; Owens, R.A.; O'Keeffe, G.; Hammel, S.; Fitzpatrick, D.A.; Jones, G.W.; Doyle, S. Regulation of nonribosomal peptide synthesis: Bis-thiomethylation attenuates gliotoxin biosynthesis in *Aspergillus fumigatus*. *Chem. Biol.* **2014**, *21*, 999–1012. [CrossRef] [PubMed]

67. Owens, R.A.; O'Keeffe, G.; Smith, E.B.; Dolan, S.K.; Hammel, S.; Sheridan, K.J.; Fitzpatrick, D.A.; Keane, T.M.; Jones, G.W.; Doyle, S. Interplay between gliotoxin resistance, secretion, and the methyl/methionine cycle in *Aspergillus fumigatus*. *Eukaryot. Cell* **2015**, *14*, 941–957. [CrossRef] [PubMed]

68. Owens, R.A.; Hammel, S.; Sheridan, K.J.; Jones, G.W.; Doyle, S. A proteomic approach to investigating gene cluster expression and secondary metabolite functionality in *Aspergillus fumigatus*. *PLoS ONE* **2014**, *9*, e106942. [CrossRef] [PubMed]

69. Gallagher, L.; Owens, R.A.; Dolan, S.K.; O'Keeffe, G.; Schrettl, M.; Kavanagh, K.; Jones, G.W.; Doyle, S. The *Aspergillus fumigatus* protein GliK protects against oxidative stress and is essential for gliotoxin biosynthesis. *Eukaryot. Cell* **2012**, *11*, 1226–1238. [CrossRef] [PubMed]

70. Choi, H.S.; Shim, J.S.; Kim, J.A.; Kang, S.W.; Kwon, H.J. Discovery of gliotoxin as a new small molecule targeting thioredoxin redox system. *Biochem. Biophys. Res. Commun.* **2007**, *359*, 523–528. [CrossRef] [PubMed]

71. Lin, D.; Yin, X.; Wang, X.; Zhou, P.; Guo, F.B. Re-annotation of protein-coding genes in the genome of *Saccharomyces cerevisiae* based on support vector machines. *PLoS ONE* **2013**, *8*, e64477. [CrossRef] [PubMed]

72. Zhao, H.; Butler, E.; Rodgers, J.; Spizzo, T.; Duesterhoeft, S.; Eide, D. Regulation of zinc homeostasis in yeast by binding of the *ZAP1* transcriptional activator to zinc-responsive promoter elements. *J. Biol. Chem.* **1998**, *273*, 28713–28720. [CrossRef] [PubMed]

73. Pavletich, N.P.; Pabo, C.O. Zinc finger-DNA recognition: Crystal structure of a Zif268-DNA complex at 2.1 A. *Science* **1991**, *252*, 809–817. [CrossRef] [PubMed]

74. Nagaich, A.K.; Appella, E.; Harrington, R.E. DNA bending is essential for the site-specific recognition of DNA response elements by the DNA binding domain of the tumor suppressor protein p53. *J. Biol. Chem.* **1997**, *272*, 14842–14849. [CrossRef] [PubMed]

75. Bird, A.J.; Blankman, E.; Stillman, D.J.; Eide, D.J.; Winge, D.R. The Zap1 transcriptional activator also acts as a repressor by binding downstream of the TATA box in ZRT2. *EMBO J.* **2004**, *23*, 1123–1132. [CrossRef] [PubMed]

76. Bird, A.J.; Gordon, M.; Eide, D.J.; Winge, D.R. Repression of *ADH1* and *ADH3* during zinc deficiency by Zap1-induced intergenic RNA transcripts. *EMBO J.* **2006**, *25*, 5726–5734. [CrossRef] [PubMed]

Review

Host-Pathogen Interactions Mediated by MDR Transporters in Fungi: As Pleiotropic as it Gets!

Mafalda Cavalheiro [1,2], Pedro Pais [1,2], Mónica Galocha [1,2] and Miguel C. Teixeira [1,2,]*

[1] Department of Bioengineering, Instituto Superior Técnico, Universidade de Lisboa, 1049-001 Lisbon, Portugal; mafalda.cavalheiro@tecnico.ulisboa.pt (M.C.); pedrohpais@tecnico.ulisboa.pt (P.P.); monicagalocha@tecnico.ulisboa.pt (M.G.)

[2] Biological Sciences Research Group, iBB—Institute for Bioengineering and Biosciences, Instituto Superior Técnico, Universidade de Lisboa, 1049-001 Lisboa, Portugal

* Correspondence: mnpct@tecnico.ulisboa.pt; Tel.: +351-21-841-7772; Fax: +351-21-841-9199

Received: 18 May 2018; Accepted: 27 June 2018; Published: 2 July 2018

Abstract: Fungal infections caused by *Candida*, *Aspergillus*, and *Cryptococcus* species are an increasing problem worldwide, associated with very high mortality rates. The successful prevalence of these human pathogens is due to their ability to thrive in stressful host niche colonization sites, to tolerate host immune system-induced stress, and to resist antifungal drugs. This review focuses on the key role played by multidrug resistance (MDR) transporters, belonging to the ATP-binding cassette (ABC), and the major facilitator superfamilies (MFS), in mediating fungal resistance to pathogenesis-related stresses. These clearly include the extrusion of antifungal drugs, with *C. albicans CDR1* and *MDR1* genes, and corresponding homologs in other fungal pathogens, playing a key role in this phenomenon. More recently, however, clues on the transcriptional regulation and physiological roles of MDR transporters, including the transport of lipids, ions, and small metabolites, have emerged, linking these transporters to important pathogenesis features, such as resistance to host niche environments, biofilm formation, immune system evasion, and virulence. The wider view of the activity of MDR transporters provided in this review highlights their relevance beyond drug resistance and the need to develop therapeutic strategies that successfully face the challenges posed by the pleiotropic nature of these transporters.

Keywords: fungal pathogens; multidrug transporters; host-pathogen interaction; virulence

1. Introduction

Species belonging to the *Aspergillus*, *Cryptococcus*, and *Candida* genera constitute the most relevant human fungal pathogens. Infections caused by these pathogens are especially severe in immunocompromised patients, particularly HIV-infected patients, cancer patients, and transplant recipients [1–3]. *Candida albicans* and *Candida glabrata* are the most prevalent of the pathogenic *Candida* species, being responsible for more than 400,000 life-threatening infections worldwide every year [4], as well as persistent mucosal infections [5,6]. *Aspergillus fumigatus* is the most frequent pathogenic species of the *Aspergillus* genus found to cause life-threatening pulmonary disease [7]. Central nervous system manifestations of meningitis or meningoencephalitis are recurrent manifestations associated with *Cryptococcus neoformans* infections [8].

One very resourceful feature of pathogenic fungi is the expression of multidrug resistance (MDR) transporters, which allow the development of antifungal drug resistance, being responsible for many cases of therapeutic failure [9,10]. MDR transporters belong mainly to two superfamilies, the ATP-binding cassette (ABC) and the major facilitator superfamilies (MFS). The ABC transporters have two main domains, each including a transmembrane domain (TMD) with six trans-membrane segments and a nucleotide-binding domain (NBD). This structure usually allows the transport of different molecules against an electrochemical gradient at the direct expense of ATP hydrolysis [11–13]. On the other hand, the transport performed by MFS transporters is driven by a proton-motive force. The MFS-MDR transporters are clustered into two families in fungi, the drug:H⁺ antiporter 1 (DHA1), including 12 transmembrane segment proteins, and the drug:H⁺ antiporter 2 (DHA2), including 14 transmembrane segment proteins [14]. Numerous molecules are proposed to be transported by MDR transporters of both superfamilies, including steroids, lipids, anti-cancer molecules, antifungals, herbicides, antibiotics, fluorescent dyes, carbohydrates, metabolites, neurotransmitters, nucleosides, amino acids, peptides, organic and inorganic anions, cations, and various Kreb's cycle intermediates [15,16], strongly suggesting either a promiscuous nature for MDR transporters or that they may indirectly affect the accumulation of these very diverse compounds, as a consequence of the transport of their physiological substrates.

The best-studied families of fungal drug efflux pumps are those from the model yeast *Saccharomyces cerevisiae* [17], in part, because it was the first eukaryote to have its genome sequenced [18]. In this model eukaryote, the ABC transporters have been classified into three main subfamilies, namely, the pleiotropic drug resistance (PDR), MDR, and multidrug resistance-associated protein (MRP), composing a total of 21 ABC drug resistance-related transporters [17,19]. Additionally, a total of 22 MFS drug transporters were reported in the genome of *S. cerevisiae* [20].

Genome sequencing has revealed that there is a wide repertoire of predicted ABC drug efflux pumps among pathogenic fungal species. However, only a few of these proteins have been functionally characterised, including those reported to be involved in multidrug resistance in human pathogens, such as *Candida albicans* [21–23], *Candida glabrata* [24–27], *A. fumigatus* [28,29], and *C. neoformans* [30,31]. Gaur et al. [32] constructed a complete inventory of ABC proteins in the genome of *C. albicans*, based on sequence similarities with ABC systems in other living organisms, and they found that *C. albicans* possesses 28 putative ABC proteins. The genome of the second most prevalent *Candida* species, *C. glabrata*, is predicted to have approximately two-thirds the number of ABC transporters predicted for *C. albicans* (18) [33]. Otherwise, much larger numbers of ABC proteins are predicted in the genomes of *A. fumigatus* and *C. neoformans*. The *A. fumigatus* genome is predicted to encode 49 ABC transporters, 35 of which are predicted to be multidrug efflux pumps [7,34]. *C. neoformans* is predicted to encode 54 ABC transporters [35].

The MFS transporters constitute the largest group of secondary active transporters, functioning as uniporters, symporters, or antiporters [16]. A subset of these transporters is involved in drug efflux [36]. Costa et al. [20] listed the DHA-MFS transporters found to occur, based on phylogenetic analysis with the model yeast *S. cerevisiae*, in the pathogenic yeasts of the *Candida* genus and also in *C. neoformans*, and *A. fumigatus*. Within the DHA1-MFS family, it was found 18, 10, 9, and 54 predicted transporters in the genome of *C. albicans*, *C. glabrata*, *C. neoformans*, and *A. fumigatus*, respectively. Regarding the DHA2-MFS family, 8, 5, 7, and 32 were found as predicted transporters in the genome of *C. albicans*, *C. glabrata*, *C. neoformans*, and *A. fumigatus*, respectively.

Although the main studied function of MDR transporters in fungi is related to the efflux of antifungal drugs, whose importance in azole resistance is reviewed herein, these transporters have different physiological roles that are also determinant for the survival of fungal pathogens in the human host. In this review, a collection of the identified physiological functions of the MDR transporters is compiled, highlighting how certain roles are necessary for the adaptation of fungi to the host niches, as well as in the fight against stresses imposed by the immune response of the host. Clues on the function of MDR transporters are also extracted from the gathered information on their transcription regulators.

2. MDR Transporters in Fungal Pathogens: Mediators of Azole Drug Resistance

Efflux-mediated drug resistance appears to be the most widespread mechanism of azole drug resistance among pathogenic fungal species [37–39]. The ABC drug efflux pumps have, in general, been linked to greater clinical significance [40]. However, and unlike what has been observed for ABC drug efflux pumps, which are widespread from bacteria to man, the MFS-MDR family appears to be strictly conserved within bacteria and fungi, turning these proteins into interesting candidates for targets for the development of new antifungal drugs [20].

The ABC transporters implicated in azole drug resistance in *C. albicans* are described as *Candida* Drug Resistance *(CDR)* genes , with *CaCDR1* encoding the best characterised multidrug transporter [41]. Together with *CaCDR1*, *CaCDR2* also contributes to azole resistance mediated by increased drug efflux [23,42]. In the non-*albicans Candida* species, the *CDR* genes have also been shown to play a role in azole resistance. In *Candida dubliniensis*, *CdCDR1* is overexpressed in the fluconazole-resistant strains [43]. In addition, *CDR1* is also induced in azole resistant *Candida parapsilosis* and *Candida tropicalis* clinical isolates [44–46]. In turn, the expression of the ABC transporters CgCdr1 and CgCdr2 in *C. glabrata* is related with the frequent azole resistance observed in this species [47]. Additionally, the ABC transporters CgSnq2 and CgYor1 have also been implicated in azole stress response in *C. glabrata* [26,48]. *Candida krusei*, which displays intrinsic azole resistance, harbors two ABC drug efflux pumps (CkAbc1 and CkAbc2) that are upregulated during azole stress [49]. *Candida krusei* CkAbc1 is involved in intrinsic fluconazole resistance [50] and confers fluconazole resistance through drug efflux upon overexpression in *S. cerevisiae* [51]. *CkABC2* expression was correlated with itraconazole resistance [52].

In *A. fumigatus*, the mechanisms of antifungal resistance have deserved more intensive investigations in the past ten years. Alteration of the azole target *cyp51A* is a major mechanism in clinical and environmental isolates. However, there are now also several non-*cyp51A*-mediated azole-resistant isolates in which the underlying mechanisms remain partially unsolved [53]. The reduced intracellular accumulation of drugs has also been correlated with the overexpression of MDR efflux transporter genes in this pathogen [38]. In contrast to the extensive number of genes predicted to encode transporters in *A. fumigatus* [54], there are very few studies characterizing these transporters and their relationship with multidrug resistance. However, decreased intracellular accumulation of itraconazole was verified in *A. fumigatus*, hinting for a possible participation of efflux pumps in azole resistance in this fungus. Slaven et al. [29] and Nascimento et al. [55] have characterised AfuAtrF and AfuMdr4, respectively, as ABC transporters, and correlated them with itraconazole resistance. Additionally, genes encoding the set of transporters AfuAbcA-E, AtrF, and AtrI are induced during voriconazole stress [53,56].

In *C. neoformans*, the reduced intracellular accumulation of drugs has been correlated with the overexpression of MDR efflux transporter genes [39]. The genome of this pathogen is predicted to encode nearly 86 DHA-MFS transporters and 54 ABC transporters. Nevertheless, only three efflux pumps have been related with drug extrusion, CnAfr1, CnAfr2, and CnMdr1 [30,57,58]. A recent study by Chang et al. [57] demonstrated that CnAfr1 is a crucial for azole efflux and is important for handling other xenobiotics, including cycloheximide, nocodazole, and trichostatin A. The overexpression of the *C. neoformans* ABC transporter CnAfr1 is known to underlie clinical fluconazole resistance [31], while CnMdr1 confers itraconazole resistance upon overexpression in *S. cerevisiae* [51]. However, these three efflux pumps appeared to play no clear role in susceptibility towards amphotericin B and 5-fluorocytosine.

Although transporter-mediated azole resistance has been initially associated with ABC drug efflux pumps, further insights have shown that MFS-MDR transporters also play a relevant role in this phenomenon. In fact, several transporters from the MFS-MDR family are also relevant players in clinical azole resistance phenotypes, as is the case with the *C. albicans* CaMdr1. Considered a major mediator of azole resistance, it is overexpressed in some resistant clinical isolates, underlying fluconazole resistance [59]. A similar case was also identified in *Candida dubliniensis*, with increased *CdMDR1* transcript levels associated with clinical resistance phenotypes to fluconazole, but not to ketoconazole [43,60]. Moreover, *C. parapsilosis* azole resistant clinical isolates overexpress CpMdr1 [44,45], and such a response is also observed during the in vitro induction of azole resistance [61]. Resistant clinical isolates of *C. tropicalis* were found to overexpress the CtMdr1 transporter [46,62]. The *C. albicans* MFS transporter CaFlu1 confers fluconazole resistance in *S. cerevisiae*, but it plays a secondary role in *C. albicans* [63,64].

A relevant role of MFS efflux pumps in azole response is also observed in *C. glabrata*, translated by the increased expression of CgAqr1, CgQdr2, CgTpo1_1, and CgTpo3, found to mediate clotrimazole resistance in clinical isolates [10]. These transporters were previously found to mediate azole resistance in laboratory strains as well [24,25,27,65]. Interestingly, the level of correlation between the increased expression of these transporters and the azole drug resistance was similar to that observed for the *CgCDR2* gene, thus highlighting the importance of MFS-MDR transporters in the clinical setting [10]. More recently, the MFS-MDR transporters CgTpo1_2, CgFlr1, and CgFlr2, were also found to confer azole resistance, by mediating the decreased intracellular accumulation of these drugs [65,66]. In *A. fumigatus*, the MFS transporters can potentially have a relevant role in azole resistance, since AfuMdr3 displays an increased expression in itraconazole-resistant mutants [55]. In addition, three transporters (AfuMfsA–C) are highly expressed during voriconazole stress [56].

Within the MFS-MDR transporters, only eight (CaMdr1, CaFlu1, CaNag3, CaNag4, CaNag6, CaQdr1, CaQdr2, and CaQdr3), seven (CgAqr1, CgFlr1, CgFlr2, CgTpo1_1, CgTpo1_2, CgTpo3, CgQdr2), three (CNA07070, CNC03290, and CND00440/aflT), and six (gliA, mfs56, Mdr3, and MfsA–C), were already characterised in *C. albicans*, *C. glabrata*, *C. neoformans*, and *A. fumigatus*, respectively [20].

Not only is there a wide variety of MDR transporters involved in azole resistance in several fungal pathogens, but the regulation of their expression is also diverse. The most prominent mechanism of azole resistance acquisition involves increased drug efflux pumps gene expression, which is mediated by master regulators of azole resistance. They include Tac1 and Mrr1 in *C. albicans* and related species; and CgPdr1 in *C. glabrata*. These transcription factors are subjected to gain-of-function (GOF) mutations in their protein sequence, which result in hyperactive forms, responsible for constitutive increased transcription of their target genes [44,61,67–72]. Additionally, each transcription factor regulates its own expression, increasing their own transcript levels in response to azole stress [67,73–75]. ABC transporters in particular have been first described as targets of CaTac1 and CgPdr1 [26,48,76–78], but several MFS transporters are also activated by CaMrr1 and CgPdr1 [65,66,71,72], thus reinforcing the role of multiple transporters in azole resistance pathways.

Other than mutations in transcriptional regulators, another regulatory mechanism mediating azole resistance is chromosomal abnormalities. These can include loss of heterozygosity (LOH), chromosomal aneuploidies, or increased gene copy number. Development of azole resistance by LOH events is known to occur in *C. albicans*, namely in the genomic regions containing *CaTAC1* or *CaMRR1*. Mutations in *MRR1* followed by the loss of heterozygosity, contribute to the overexpression of this gene [72]. This phenomenon also occurs to give rise to *CaTAC1* homozygous mutations because of the loss of heterozygosity in chromosome 5 [70]. Chromosomal rearrangements also contribute to the amplification of the *CaTAC1* gene [79]. In *C. glabrata*, the existence of differential chromosome configurations and segmental aneuploidies was observed in azole-resistant strains [80]. Alterations in chromosome copy number is related with *C. neoformans* resistance to azoles. In particular, CnAfr1 overexpression due to chromosome 1 duplication results in resistant populations in response to selection [81].

Although the transcriptional regulation of MDR transporters in *A. fumigatus* and *C. neoformans* have not yet been deeply studied, Hagiwara and colleagues [82] reported a novel Zn_2–Cys_6 transcription factor, AfuAtrR, as playing a key role in an azole resistance mechanism of *A. fumigatus*, by regulating the drug target AfuCyp51A (14-α sterol demethylase) and the putative drug efflux pump AfuCdr1B (AfuAbcC) expression. Additionally, through the screening of a library of transcription factor mutants, different transcription factors regulating the expression of efflux pumps genes in response to fluconazole exposure were identified in *C. neoformans*. These include *CnCRZ1* and *CnYAP1*, which had been previously characterised as transcription factors functioning in response to environmental stress, including thermal, hypoxic, and fluconazole, as well as oxidative stress in *C. neoformans* [83,84]. Chang and co-workers [57] determined that CnAfr1 expression was positively regulated by CnCrz1 and CnYap1 in response to fluconazole.

3. MDR Transporters in Fungal Pathogens: Mediating the Transport of Physiological Substrates as a Way to Adapt to Host Niches

Fungal pathogens have different strategies to colonize, survive, and infect the human host. Each species has preferential niches, developing specific mechanisms that allow adaptation to each microenvironment and evasion from the immune system. For instance, *C. albicans* exhibits different morphologies to better adapt to each niche, for example switching from the hyphal morphology used to penetrate host epithelial cells to the yeast shape to circulate in the bloodstream [85]. The preferential niches for *C. albicans* and other *Candida* species include the oral cavity, the skin, the urogenital, and gastrointestinal tracts, whose infection might lead to disseminated disease [85,86]. In the case of the *Cryptococcus* species, the preferable niches of the human host include the pulmonary tissue and the central nervous system. The establishment of a *Cryptococcus* invasive disease often starts with infection of the lung tissue, followed by invasion of the bloodstream, and subsequent dissemination of *Cryptococcus* cells to the central nervous system [87,88]. Likewise, *A. fumigatus* infects primarily the lungs, disseminating from there to any possible organ [89].

In each specific niche, different stresses like heat shock, osmotic, oxidative and nitrosative stresses, pH variations, and hypoxia might be encountered, forcing the development of stress resistance mechanisms by fungal pathogens [90]. Given the different stresses and different nutrient availability, fungal pathogens must have increased flexibility to adapt and respond to the environment. This wide capacity of metabolic adaptation and resistance to stress also promotes virulence in fungal pathogens, as it lessens their vulnerability to the surrounding environment [86].

In the oral cavity, different factors protect the host from the presence of fungal pathogens. First of all, different cell types are ready to identify them, including polymorphonuclear leukocytes, monocytes/macrophages, non-major histocompatibility complex (MHC)-restricted CD8+ T cells, and oral epithelial cells [91]. Upon the presence of *Candida* cells, the epithelial cells from the oral mucosa induce expression of a nitric oxide synthase, which leads to increased levels of nitric oxide believed to have a candidacidal activity, helping in the resistance to mucosal candidiasis [91]. These cells are also able to secrete different antimicrobial peptides, including members of the β-defensin family [92]. Additionally, the saliva has powerful antimicrobial peptides, such as histatins. These are known to promote cell cycle disturbance, increased intracellular reactive oxygen species (ROS), and loss of cell volume in *C. albicans*, ultimately leading to the apoptosis of yeast cells. Histatin-1, -3, and -5 are produced by acinar cells present in human salivary glands [93].

The lungs constitute the first site for *Cryptococcus* infection. This niche also has its strategies of protection against invasion. As a first defense mechanism of the innate immune system is the presence of bronchial and alveolar Type I and Type II epithelial cells that, besides forming a physical barrier, also release several cytokines and chemokines to activate an immunologic response [94]. Also important for this response is the presence in the lung tissue of inflammatory monocytes, alveolar macrophages, dendritic cells, and neutrophils [95,96]. M1 macrophages produce ROS and reactive nitrogen species (RNS), including nitric oxide, known to have anti-cryptococcal properties [96]. *Aspergillus fumigatus* also enters the host by inhalation, facing the action of alveolar macrophages, neutrophils, and dendritic cells [97,98].

In turn, the vaginal mucosa is a low pH environment with the presence of weak acids, like acetic and lactic acids [99,100], protected by *Candida*-specific cell-mediated immunity. This immunity is mediated by T helper (Th) 1-type responses [101]. Additionally, the established microbiome present in this niche is also responsible for its protection. A good example is the role of *Lactobacillus crispatus* in this context, found to mediate the immune response of vaginal epithelial cells deployed in the presence of *C. albicans* [102].

Although generally associated to multidrug resistance, some MDR transporters have been found to contribute to the adaptation of fungal pathogens to human host niches. This role of MDR transporters seems to be linked to a general function of cellular detoxification, given their ability to catalyse the efflux of antifungal drugs, but also other toxic metabolites present mainly in the stationary phase of cellular growth [103]. Interestingly, the ability to control the concentration of some of these physiological substrates is likely to confer a selective advantage when growing, both as commensals and pathogens in human host niches.

For example, the *C. albicans* ABC transporter Cdr1 was found to transport steroids. Accumulation assays with radiolabelled β-estradiol and corticosterone have demonstrated the ability of CaCdr1, expressed in the AD1234568 (AD)-*CDR1 S. cerevisiae* strain, to enhance the efflux of these human steroids, that appear to compete with drugs such as cycloheximide, O-phenanthroline, chloramphenicol, fluconazole, or rhodamine 123 as Cdr1 substrates [104]. However, the transport of [^{3}H]-β-estradiol was not affected by the truncation of domain 12 of CaCdr1, a mutation that affects drug transport, suggesting that Cdr1 displays different binding sites for these different substrates [105]. Interestingly, increased incidence of vulvovaginal candidiasis has been linked to the presence of elevated estrogen levels during pregnancy or to the presence of exogenous estrogens from oral contraceptives [106–109]. The effect of β-estradiol or ethynyl estradiol, but not α-estradiol or estriol, on the formation of germ tubes has been reported, revealing the specificity of such an induction [110,111]. *Candida albicans* strain SC5314 and the vaginal isolate GC15 in Roswell Park Memorial Institute (RPMI)-free supplemented with supraphysiological (10^{-5} M) or physiological (10^{-10} M) concentrations of 17-β-estradiol and estradiol, respectively, during different time points, exhibit the upregulation of *CaCDR1* and *CaCDR2*. When both genes were disrupted, the *C. albicans* cells exposed to 17-β-estradiol had a decreased germ tube formation [111]. These evidences point out the importance that MDR efflux pump-mediated resistance to chemical compounds present in human body niches have in the success of fungal infections.

In *S. cerevisiae*, sterol uptake requires the involvement of the ABC transporters Pdr11 and Aus1. These are necessary for the nonvesicular movement, from the plasma membrane to the endoplasmatic reticulum, a movement needed for the sterol ester synthesis required for sterol uptake [112]. The *S. cerevisiae* Upc2 transcription factor is known to induce the *AUS1* and *PDR11* expression under anaerobiosis conditions, when the sterol uptake is essential. The absence of these two genes leads to defects in sterol uptake [113]. Although, in *S. cerevisiae*, no connection to MDR has been observed for these transporters; in *Cryptococcus gattii*, Pdr11 was found to be necessary for fluconazole resistance in the VGII clinical strain [114]; and *C. glabrata* Aus1 has been linked to sterol transport associated with the presence of azole drugs, being upregulated in anaerobic conditions [115]. In fact, sterol uptake was found to be used by *C. glabrata* strains defective in ergosterol [116] and in *C. glabrata* clinical isolates [117]. In fact, upon the use of azoles, which affect the production of ergosterol, the sterol uptake response is activated as a resistance mechanism that allows *C. glabrata* cells to survive ergosterol defects with host sterols, like cholesterol [118]. Sterol uptake might also be necessary for *C. glabrata* virulence, as Aus1 is necessary in a mice model of disseminated infection, being significantly upregulated in such conditions [115]. Therefore, sterol uptake is an important strategy developed by *C. glabrata* to survive in the human host, ensuring its virulence and antifungal resistance in mucosal niches.

Major facilitator superfamilies multidrug transporters have also been found to contribute to lipid homeostasis in *C. albicans* and *C. glabrata*. The MFS-MDR transporters Qdr1, Qdr2, and Qdr3 of *C. albicans* were found to be localized in the plasma membrane, with preferential localization of Qdr1 and Qdr2 in lipid rafts. These transporters do not participate in the resistance to any type of known antifungals [119], although their homologs in *C. glabrata* and in *S. cerevisiae* are recognized drug transporters [25,120–122]. Nevertheless, they seem to have other important functions for the survival of *C. albicans*. When absent, *CaQDR1*, *CaQDR2*, or *CaQDR3* lead to the accumulation of phosphatidylinositol and phosphatidylserine, as well as of sphingolipids, thus indicating a role of these transporters in lipid homeostasis [119]. On the other hand, the *C. glabrata* Tpo1_2 MFS-MDR transporter has also been linked to fatty acid and sterol homeostasis, especially during biofilm formation. The absence of *CgTPO1_2* upon biofilm formation leads to a decrease of very long fatty acids. Moreover, ergosterol content in the Δ*cgtpo1_2* deletion mutant was found to be 30% higher than in its parental strain [123]. This physiological function of lipid transportation in *C. glabrata* seems to be important for to the adaptation to low pH environments with organic acids, as alterations in the membrane lipid composition have been described as a mechanism to surpass the stress on such environment [124].

The CgTpo1_2 homolog, CgTpo1_1, is also an MFS drug transporter with an interesting proposed role in the survival of *C. glabrata* in the oral cavity. One of the mechanisms used by the host to resist candidiasis is the release of the antimicrobial peptide histatin-5 in the saliva [93], as described above. Interestingly, the CgTpo1_1 transporter was found to be necessary for the resistance of *C. glabrata* to this antimicrobial peptide [123]. Another similar player in histatin-5 resistance is the *C. albicans* Flu1 MFS transporter involved in the resistance towards fluconazole, ketoconazole, and itraconazole [63]. In fact, the absence of *CaFLU1* significantly defects cellular growth and biofilm formation upon the exposure to this antimicrobial peptide [125]. Both studies give evidence of the role of MDR transporters' action against antimicrobial peptides, a usually effective weapon of the human immune system against microbial infections.

The roles of the Ste6 ABC transporter of *S. cerevisiae* are also different and interesting. This transporter was found to be responsible for the secretion of a lipopeptide mating pheromone, designated a-factor, but also for the resistance of this species to valinomycin, an antibiotic peptide [126]. Interestingly, *C. albicans* homolog of Ste6, designated Hst6, was found to be able to complement the role of *S. cerevisiae* Ste6 in a Δscste6 deletion mutant [127]. Nevertheless, its specific function in *C. albicans* remains to be discovered, as this species is not known to secrete mating pheromones. On the other hand, the Ste6 homolog in *C. neoformans* has been shown to be involved in the mating process of this pathogen [128]. In turn, *A. fumigatus* Afu3g03430, which is also similar to *S. cerevisiae* Ste6, has been indicated as being involved in iron uptake [129–131]. Iron uptake is an important function for the survival of any fungal pathogen, as this cofactor is necessary for several enzymatic reactions and is withheld in the human host as a protective mechanism against microbial infections [132].

Another example of the importance of MDR transporters in the adaptation to human niches is their role in polyamine resistance. Polyamines like spermine, spermidine, and putrescine are essential organic polycations, which are usually involved in the regulation of nucleic acid and protein synthesis, as well as cell growth [14]. The plasma membrane MFS-MDR transporters, Tpo1–4 and Qdr3, were the first identified as important players in the resistance to polyamines in *S. cerevisiae* [133–135]. In *C. glabrata*, MFS-MDR transporters have also been found to contribute to the resistance to polyamines, which may become toxic above certain concentrations, such as those found in the human urogenital tract [136]. The prevalence of *Candida* species in this niche is likely facilitated by two homologs of ScTpo1 in *C. glabrata*, CgTpo1_1 and CgTpo1_2, which have been identified as determinants in spermine resistance [65]. In turn, the CgTpo3 transporter, known to be involved in *C. glabrata* azole resistance, presents the same physiological role as its *S. cerevisiae* homolog, Tpo3, in polyamine resistance [24].

Since the vaginal tract is characterised by an acidic pH of about 4, as well as the presence of weak acids, like acetic and lactic acids [99,100], the ability to tolerate the stress induced by weak acids is crucial for *Candida* colonization and successful infection. Interestingly, *S. cerevisiae* MFS-MDR transporters, like Aqr1 and Qdr1, are important determinants in short-chain monocarboxylic acid resistance [120,137]. *S. cerevisiae* Aqr1 has also been implicated in the excretion of homoserine, threonine, alanine, glutamate, and aspartate, probably through vesicles, releasing the amino acids to the extracellular environment through exocytosis [138]. The homolog of *S. cerevisiae* Aqr1 in *C. glabrata*, CgAqr1, has been linked to flucytosine and clotrimazole resistance, as well as resistance against acetic acid [27]. Such a role in the resistance to acetic acid is also performed by the MFS-MDR CgDtr1 and CgTpo3 transporters [139,140], being the later controlled by the CgHaa1 transcription factor, known to regulate the response of *C. glabrata* to acetic acid [140]. These different roles of MFS transporters [14,20] point out their importance for the survival of such pathogens in acidic human microenvironments.

4. MDR Transporters in Fungal Pathogens: Playing a Role in Virulence, Biofilm Formation, and Phagocytosis Evasion

To persist in the human host, fungal pathogens need to survive in the first site of infection, and disseminate to other niches of the host. To achieve this goal, several barriers imposed by the niches need to be surpassed, as well as the ones imposed by the immune response (Figure 1). To face and resist such offenses, fungal pathogens develop strategies of survival, including biofilm formation, macrophage evasion mechanisms, and virulence factors, where the role of given MDR transporters becomes important.

For example, similarly to the *S. cerevisiae* Pdr5 and Yor1 ABC transporters [17], *C. albicans* CaCdr1, CaCdr2, and CaCdr3 have been shown to play a role in phospholipid translocation when expressed in *S. cerevisiae*. CaCdr1 and CaCdr2 conduct an in-to-out movement of phospholipids through the plasma membrane bilayer in an energy-dependent manner, while CaCdr3, which is not involved in antifungal drug resistance, produces the opposite movement, from the outer to the inner leaflet of the plasma membrane [141]. The transport of phospholipids by Cdr1 may be extremely important for *C. albicans* evasion to the host immune system, as some phospholipids have been identified as lipid antigens and precursors of lysophospholipids, also known to be lipid antigens recognized by human natural killer T lymphocytes [142,143]. For instance, to avoid identification by the host immune response, *C. albicans* relies on CaQdr2 and CaQdr3 transporters, given that they are necessary for the maintenance of phosphatidylinositol, which is a lipid antigen recognized by the CD1 antigen-presenting cells from the host [144]. Interestingly, the absence of *CaQDR2* and *CaQDR3* genes have shown to attenuated *C. albicans* virulence in a murine model of hematogenously disseminated candidiasis [119].

When fungal evasion is not effective, pathogens are phagocyted and a new environment is set upon them, to which they have to adapt to survive. Different host cells are able to perform phagocytosis, including monocytes/macrophages, neutrophils, and dendritic cells [145,146]. All these lead to the formation of phagolysosomes that attempt to kill pathogens by producing ROS and RNS, activating the action of proteases and decreasing the pH, resorting to K^+ fluxes [146]. *Candida* species display an activation of oxidative stress response in this environment, as well as other survival mechanisms [147]. In particular, *C. glabrata* cells are known to be able to resist and proliferate inside mammalian macrophages for a very long time [148]. Together with other more specific mechanisms, *C. glabrata* is known to count with two transporters from the MFS-MDR family, CgAqr1 and CgDtr1, which are responsible for the resistance to weak acids and oxidative stress [27,139]. CgAqr1 was found to confer resistance to acetic acid [27], like its homolog in *S. cerevisiae* [137], and CgDtr1 was identified as an acetic acid exporter, allowing the resistance to this acid and oxidative stress [139]. It is likely that the role of CgDtr1 in weak acid and oxidative stress resistance underlies its contribution to the survival of *C. glabrata* in the macrophages and, indirectly, to increase virulence. Indeed, the *CgDTR1* expression is upregulated in cells phagocytozed by *Galleria mellonella*, hemocytes and the overexpression of this gene leads to an increase in the survival of *C. glabrata* cells in *G. mellonella* hemocytes. Furthermore, CgDtr1 was found to be necessary for the full virulence of *C. glabrata* in the infection model *G. mellonella* [139].

Cryptococcus neoformans survival in phagocytic cells also depends on its reaction to the new environment with low pH, ROS, and nitric oxide [96]. Upon *C. neoformans* phagocytosis by macrophages in vitro, the overexpression of the *CnAFR1* gene takes place [39]. The Afr1 ABC transporter is known for the in vitro and in vivo resistance to fluconazole [31,39,58], but has also been linked to the virulence of *C. neoformans*, in intravenous and in inhalation mouse models [39]. Although not helping to avoid phagocytosis, CnAfr1 delays the maturation of the phagolysosome, which exhibits less acid vacuoles. CnAfr1 seems to interfere with the pathway of phagolysosome maturation, as the early endosome marker Rab5 and late marker Rab7 are less detected upon the overexpression of the *CnAFR1* gene [149]. Moreover, Goulart and colleagues have reported the upregulation of other ABC transporters in *C. neoformans* cells phagocyted by peritoneal macrophages [150], highlighting the importance these transporters within the study of the *Cryptococcus* species survival in the host.

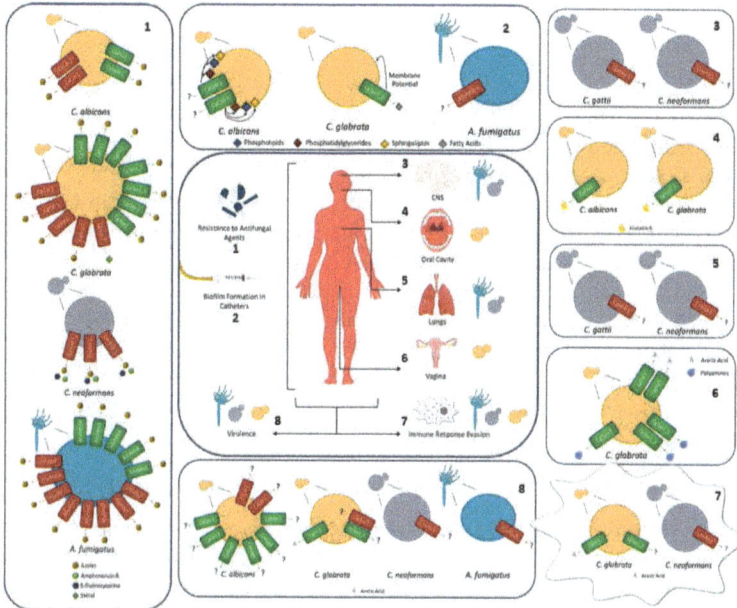

Figure 1. The role of multidrug resistance (MDR) transporters in the survival of *Candida*, *Cryptococcus*, and *Aspergillus* species upon infection in the human host. (1) MDR transporters involved in multiple azole resistance; (2) MDR transporters necessary for biofilm formation; (3) MDR transporters involved in *Cryptococcus* species virulence in the central nervous system (CNS); (4) MDR transporters involved in the efflux of histatin-5 in the oral cavity; (5) MDR transporters involved in *Cryptococcus* species virulence in the lungs; (6) *Candida glabrata* MDR transporters involved in polyamine or acetic acid export for the survival in the vaginal tract; (7) MDR transporters involved in the fight against the immune response of the host; and (8) MDR transporters known to be essential for normal virulence of fungal pathogens. The central picture summarizes the niches of infection in which the referred fungal pathogens are found. ATP-binding cassette (ABC) transporters are highlighted in red and major facilitator superfamilies (MFS) transporters are highlighted in green.

Cryptococcus gattii seems to count with drug transporters for its full virulence in the brain and the pulmonary tissue, like the MFS Mdr1 and the ABC Pdr11 transporters, determinants of *C. gattii* MDR [58,114]. The *MDR1* gene was found to be upregulated in cells recovered from the brain and lungs of infected mice, while *PDR11* was overexpressed in cells recovered from the lungs of infected mice [151]. Both Mdr1 and Pdr11 were proposed to play an additional role *C. gattii* virulence and adaptation to the host environment, although the exact mechanism underlying this observation is unclear.

Likewise, *A. fumigatus* infects primarily the lungs disseminating to any possible organ [89]. Such infections are believed to make use of the 49 ABC-like genes and 278 MFS-like genes present in this species [54], which, given their elevated number, indicate a major role in *A. fumigatus*. From all these, some have been identified as antifungal resistant players [28,29,55,56], but having additional roles in expelling toxic molecules [54]. abcA and abcB are two *A. fumigatus* transporters, with a high similarity to the *S. cerevisiae* Pdr5 drug efflux pump. abcB is necessary for full virulence of *A. fumigatus* in the *G. mellonella* model of infection and the overexpression of the *abcA* gene leads to an augmentation of virulence in the presence of voriconazole, in the same infection model [152].

The *C. albicans* Mdr1 MFS transporter is well known for its role in MDR [9,153,154]. Nevertheless, this transporter has been identified as necessary for full virulence in *C. albicans* in immunocompetent and immunocompromised mice models [155], although its specific functions in virulence have not yet been assessed. Moreover, the CaNag3, CaNag4, and CaNag6 MFS transporters are necessary for the full virulence of *C. albicans* in a mouse model [156].

Several MDR transporters have also been shown to have roles in biofilm formation, an important feature that allows the persistence of fungal pathogens in the host [157]. One specific case is the activity displayed by CaQdr2 and CaQdr3 in the architecture of *C. albicans* biofilm, which, although not contributing for the total biomass produced, help to shape the structure of the biofilm [119]. In *C. glabrata*, the CgTpo1_2 has been highlighted as an important player in biofilm formation, being necessary for the normal expression of *CgALS1*, *CgEAP1*, and *CgEPA1* genes encoding adhesins. Its role in biofilm formation seems to be linked to its importance in fatty acids homeostasis, as it affects the incorporation of very long chain fatty acids, which are usually found in *C. glabrata* biofilms [123]. In turn, *A. fumigatus* Mdr4 transporter has been identified as upregulated upon voriconazole exposure and biofilm formation conditions [129,158], indicating a possible role in the development of biofilm. Interestingly, GOF mutations in the multidrug resistance transcription factor Pdr1 were also found to be required for increased virulence and biofilm formation, by controlling the expression of the adhesion encoding gene *EPA1* [159,160].

5. Hints on the Function of MDR Transporters Based on Transcription Regulation

Recently, clues on the physiological roles of MDR transporters, including the transport of lipids, ions, and small metabolites, have emerged from transcriptional regulation data, linking these transporters to important pathogenesis features such as resistance to host niche environments, biofilm formation, immune system evasion, and virulence. In this section, all available information on the transcriptional regulation of ABC/MFS-MDR transporter genes in the human pathogens *C. albicans* and *C. glabrata* is reviewed. Although a similar analysis would be very interesting for *A. fumigatus* and *C. neoformans*, the available information is very limited and scattered. On the contrary, information on *C. albicans* and *C. glabrata* has been recently gathered in the PathoYeastract database (www.pathoyeastract.org) [161], enabling the analysis of MDR transporter regulation, not only in the multidrug resistance context, but also within the perspective of their physiological roles and their implication in other pathogenesis-related phenotypes.

Besides the expected role of the MDR transcription factors, transcription factors controlling stress response, cell cycle, cell wall dynamics, biofilm formation, lipid and carbohydrate metabolism, and nutrient availability have been found to regulate the expression of the MDR transporter encoding genes in *C. albicans*. In fact, the number of regulatory associations with MDR transporter genes involving transcription factors virtually not related to MDR is much higher than that involving recognized MDR transcription factors. Surprisingly, with regards to multidrug resistance, only 36% (12 out of 33) of the MDR-related transporters in *C. albicans* are reported to be regulated by at least one MDR-related transcription factor, more specifically, 57% (8 out of 14) of the ABC transporters and nearly 21% (4 out of 19) of the MFS-MDR transporters (Figure 2A). Expectably, the transcription factor CaMrr1, a multidrug resistance regulator in *C. albicans*, was found to regulate 11 out of the 12 MDR transporters reported to be regulated by at least one MDR-related transcription factor, including the MFS-MDR transporter, CaMdr1, a key player in the acquisition of azole resistance in clinical isolates [71]. Transporters that have been clearly linked to resistance of a wider range of drugs are indeed those that are regulated by a higher number of MDR transcription factors. This is the case for *CaCDR1* and *CaCDR2*, which are the main ABC transporter genes responsible for azole drug resistance in *C. albicans*. *CaCDR1* is regulated by CaFcr1, CaMrr1, CaMrr2, and CaTac1, whereas *CaCDR2* is reported to be regulated by CaMrr1 and CaTac1. In particular, the *CaCDR1* regulation appears to be highly complex, being controlled by at least 19 transcription factors not only related to MDR, but also with biofilm formation, stress response, cell-wall dynamics, carbohydrate metabolism, cell-cycle, and lipid metabolism. Besides CaMrr1, *C. albicans* carries yet another major regulator of multidrug resistance transporters, CaTac1, which is described as the major factor needed for the regulation of

CaCDR1 and *CaCDR2* [76]. Nevertheless, contrasting with CaMrr1, CaTac1 appears to regulate no other MDR transporter genes besides *CaCDR1* and *CaCDR2* (Figure 2A).

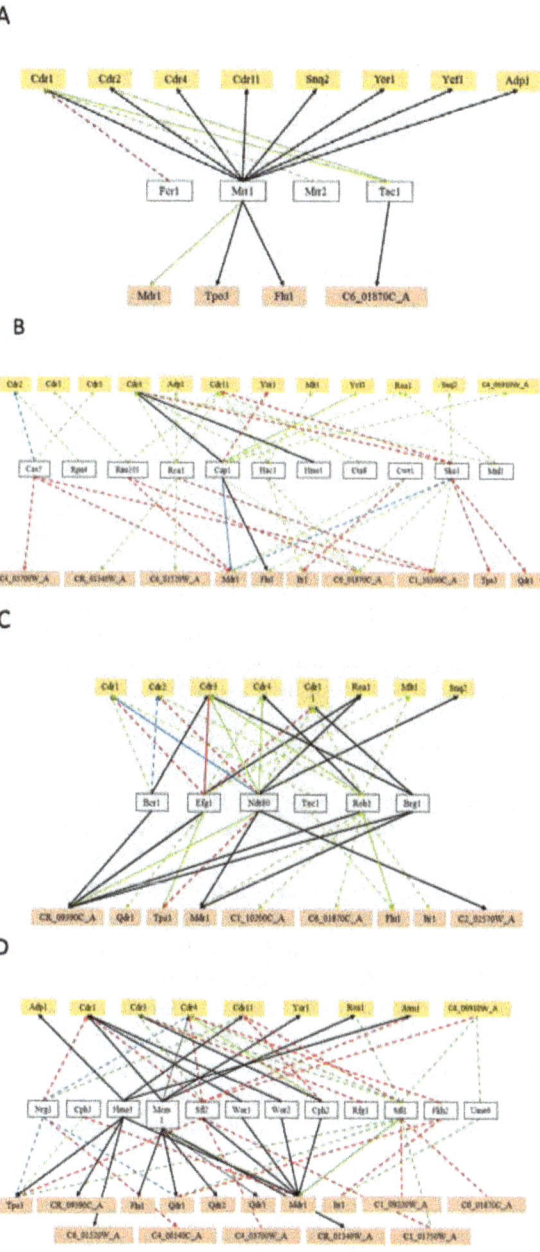

Figure 2. *Cont.*

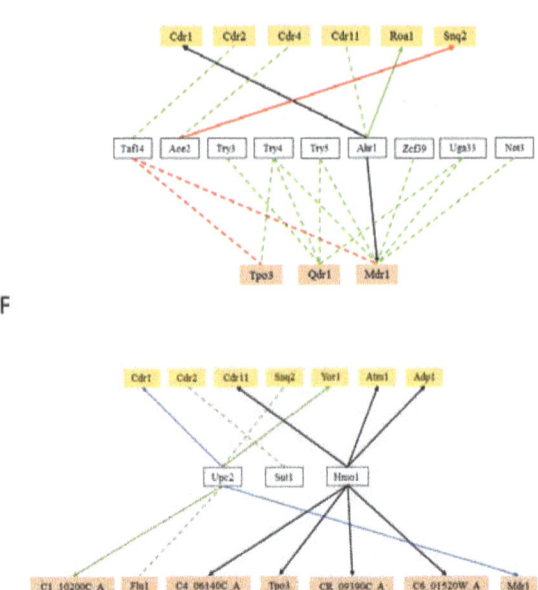

Figure 2. Transcriptional regulatory networks that control the expression of MDR transporter genes in *C. albicans*, considering the subgroups of transcription factors known to be involved in multidrug resistance (**A**); in stress response (**B**); biofilm formation (**C**); cell-cycle/morphology (**D**); cell-wall dynamics (**E**); and lipid (**F**). The displayed regulatory associations are according to the data present in the Pathoyeastract database (http://pathoyeastract.org/) [161]. The ABC transporters are highlighted by the yellow colour and the MFS transporters are highlighted by the orange colour. Arrows indicate the experimental basis of the documented regulatory associations, either expression evidence (dashed line) or DNA-binding evidence (filled line). Green, red, blue, or black arrows indicate a positive, negative, positive and negative, or unspecified association, respectively.

It is surprising the number of stress-related transcription factors reported to regulate MDR transporters in *C. albicans*. Nearly 67% (22 out of 33) of the *C. albicans* MDR transporters have been found to be regulated by at least one stress-related transcription factor, with the transcription factors CaCap1 and CaSko1 regulating the highest number of transporters (eight and nine, respectively) (Figure 2B). These two transcription factors are known to be required for oxidative stress tolerance in *C. albicans* [162,163], which is a vital process for this pathogen to survive in healthy hosts, especially during phagocytosis. Additionally, the transcription factor CaRim101 was found to regulate the expression of a considerable number of MDR transporter genes (7). This regulator plays a crucial role in pH-response [164], and the Rim101 pathway is required for host-pathogen interactions, as it regulates the expression of genes that stimulate host cell damage [165]. Among other stress-related transcription factors that were found to control MDR transporter genes are CaCta8, an essential transcription factor that mediates heat shock response [166]; CaRpn4 and CaHac1, key regulators of unfolded protein response (UPR), which is a crucial phenomenon for cellular protein homeostasis maintenance, which is often lost when cells are under stress such as that induced by antifungals [167]; and Cas5, a zinc finger transcription factor that controls the response to cell wall stress, including those induced by echinocandins [168], but also the response to membrane stress exerted by the azole antifungal drugs [169].

Transcription factors related to biofilm formation were also found to regulate a significant number of MDR transporters in *C. albicans* (nearly 52%, 17 out of 33) (Figure 2C). Nobile et al. [170] described and analysed the transcriptional network controlling the biofilm formation in *C. albicans*, whose biofilms are a major source of medical device-associated infections. They demonstrated that CaBcr1, CaTec1, CaEfg1, CaNdt80, CaRob1, and CaBrg1 are the major players in the transcriptional network controlling the biofilm development in this human pathogen, including many MDR transporters. In fact, genes encoding for drug efflux pumps had been previously been reported in biofilms to be differentially regulated during development, as well as upon exposure to antimicrobial agents, including *CaCDR1*, *CaCDR2*, *CaMDR1*, *CaNAG3*, and *CaNAG4* [171–173]. CaRob1 and CaNdt80 were found to regulate the highest number of MDR transporters (nearly 65%, 11 out of 17 each), whereas CaTec1 was found to regulate a smaller portion of those (approximately 12%, 2 out of 17). These associations corroborate the observation that efflux pump-mediated multidrug resistance is an important trait of biofilm cells [174].

In *C. albicans*, the ability to undergo morphological switching from yeast cells to hyphae, in response to various environmental signals, is an important virulence factor that contributes to biofilm formation, invasion and dissemination of *Candida* in host tissues, and resistance to macrophage and neutrophil engulfment [171,175,176]. Interestingly, cell-cycle/morphology related transcription factors were found to regulate the highest portion of MDR transporters, 72% (24 out of 33), including the *CDR* genes and *MDR1* (Figure 2D). In fact, despite the fact that cell-cycle related transcriptional regulators have never been demonstrated to be crucial for efflux pump expression, a positive correlation between the level of antifungal drug resistance and the ability to form hyphae in the presence of azole drugs has been identified [176]. For instance, Mcm1, which is an essential transcription factor in *C. albicans* crucial for morphogenesis [177], was found to regulate the expression of several MDR transporters (10 out of 24), including *CaCDR1* and *CaMDR1*. CaMcm1 was previously demonstrated to be dispensable for *CaMDR1* upregulation in response to H_2O_2 , but was required for full *CaMDR1* induction by benomyl [177]. The transcription factors CaSfl1 and CaSfl2, two homologous heat shock factor-type transcriptional regulators that antagonistically control morphogenesis in *C. albicans*, while being required for full pathogenesis and virulence [178], were also found to regulate a number of MDR transporters (11 and 9, respectively). Finally, CaWor1 and CaWor2 are transcriptional regulators of *C. albicans* opaque cell formation, and were also found to regulate MDR transporters.

Also, cell-wall dynamics related transcription factors regulate MDR transporters in *C. albicans* (9), including the *CDR* gene family and *CaMDR1* (Figure 2E). The Try transcription factors and Taf14, Ahr1, Uga33, and Zcf39 are all reported regulators of *C. albicans* yeast form adherence, and they were found to regulate at least one MDR transporter.

Additionally, CaUpc2, the master regulator of the ergosterol biosynthesis (*ERG*) genes [179], and CaHmo1, which was also shown to bind promoters of ergosterol metabolism genes [180], regulate the expression of 13 MDR transporters (Figure 2F). The regulation of ABC drug efflux pumps by the lipid metabolism related transcription factor, is likely to be due to the activity of some of these MDR transporters in phospholipid translocation or ergosterol transport. It is not likely that all drug pumps involved in multidrug resistance are phospholipid translocators. For instance, *CaMDR1* shows no detectable phospholipid exchange activity [181]. However, the observation that these transcription factors also control the expression of MDR-MFS transporters, suggest that they may also play a role in this lipid metabolism. This hypothesis is consistent with the observation that the deletion of MFS-MDR genes *CaQDR1*, *CaQDR2*, and *CaQDR3* [119], or *CgTPO1_2* [123], does affect lipid composition.

In the case of *C. glabrata*, only a few regulators of MDR transporter genes have yet been unveiled, reflecting the fact that the study of transcriptional regulation in this yeast is still in its infancy, especially in biological processes beyond drug resistance. The transcription factor CgPdr1 is described as the master regulator of multidrug resistance in this organism, regulating the expression of several MDR transporter genes [48]. CgPdr1 is thought to form a heterodimer with CgStb5, as it happens in the closely related *S. cerevisiae*. The overexpression of *CgSTB5* in *C. glabrata* represses azole resistance, while its deletion produces a shy intensification in resistance. Expression analysis assays established that CgStb5p shares many transcriptional targets with CgPdr1, but, unlike the second, it is a negative regulator of pleiotropic drug resistance [68,182]. These two MDR-related transcription factors were found to regulate a total of 12 MDR transporters (seven ABC and five MFS) (Figure 3A).

Besides the expected role of MDR transcription factors, stress response transcription factors have been found to regulate the MDR transporter encoding genes in *C. glabrata*. Five stress-responsive transcription factors are reported to regulate the expression of at least one MDR transporter gene (Figure 3B). CgYap1 is the major regulator of oxidative stress response genes in *C. glabrata* [183], and it was also demonstrated to induce the expression of multidrug transporters [184]. As antifungals induce the endogenous production of ROS, thus inducing oxidative stress response mediated by CgYap1, it is likely that CgYap1 targets the MDR transporters that play a role in the extrusion of oxidative stress generating molecules from the cells. Otherwise, CgYap7 is a transcriptional repressor of nitric oxide oxidase and also regulates the iron–sulfur cluster biogenesis [185]. The remaining transcription factors present in this group were not yet characterised. However, they are thought to be involved in salt tolerance (CgHal9 and CgYap6) and weak acid response (CgWar1), based on the function of their *S. cerevisiae* homologs. It is interesting to note that the ABC transporter encoding ORF *CAGL0M07293g*, although uncharacterised, is reported to be regulated by CgYap7, CgYap6, CgHal9, and CgWar1 transcription factors. This is in accordance with its predicted function as a weak-acid-inducible multidrug transporter required for weak organic acid resistance, based on the function of its closest *S. cerevisiae* homolog ScPdr12.

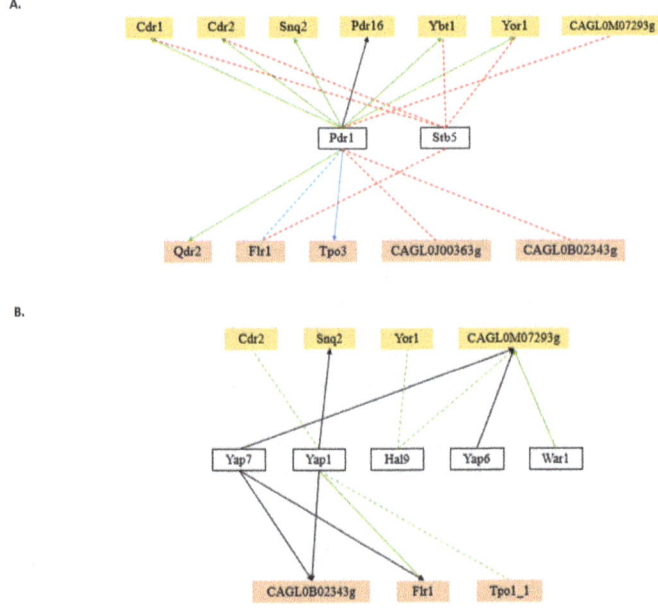

Figure 3. Transcriptional regulatory networks that control the expression of MDR transporter genes in *C. glabrata*, considering the subgroups of transcription factors known to be involved in multidrug resistance (**A**) and in stress response (**B**). The displayed regulatory associations are according to the data present in the Pathoyeastract database (http://pathoyeastract.org/) [161]. ABC transporters are highlighted by the yellow colour and MFS transporters are highlighted by the orange colour. Arrows indicate the experimental basis of the documented regulatory associations, either expression evidence (dashed line) or DNA-binding evidence (filled line). Green, red, blue, or black arrows indicate a positive, negative, positive and negative, or unspecified association, respectively.

6. Conclusions and Perspectives.

MDR transporters are undoubtedly necessary players for the successful survival of fungal pathogens in the human host (Tables 2 and 2). As reviewed herein, their prominent role in MDR has a real impact in the clinical acquisition of drug resistance, allowing these pathogens to persist even upon treatment with different antifungal agents. Indeed, within the characterised MDR transporters, 88.5% and 71.4% of the ABC and MFS, respectively, have been shown to contribute to drug resistance, suggesting that this is their most relevant feature.

This review, however, highlights the observation that both ABC and MFS-MDR transporters contribute to the ability of fungal pathogens to colonize the host and evade host induced defenses, by executing other functions not directly connected with MDR. The study of the transcriptional networks controlling the expression of drug transporters in *C. albicans* and *C. glabrata* points towards possible links between MDR transporters and several cellular processes, including stress response, morphological switching, cell wall, and lipid homeostasis. Indeed, mounting evidence suggests that the function of MDR transporters goes well beyond their traditional role in drug resistance. Clues on what may be the physiological role of these transporters, suggest that their natural activity is not linked to the transport of chemical compounds that are not found in nature, but rather the transport of metabolites that can be found in the natural ecosystems where pathogenic fungi thrive. Among the characterised ABC and MFS transporters, 19.2% and 42.8%, respectively have been associated to the transport of such biomolecules. Significantly, MDR transporter roles associated to phospholipid and ergosterol incorporation, as well as in the excretion of metabolites that reach toxic concentrations in host niches, such as polyamines and weak acids, establish a strong link between the MDR transporters and the survival of fungal pathogens in several human microenvironments. Additionally, a few MDR transporters have been linked to the resistance to stress imposed by the host immune system, including phagocytosis and antimicrobial peptides, such as histatin-5. Not surprisingly, a high proportion of the characterised ABC (43.5%) and MFS (39.3%)-MDR transporters were found to be required for full virulence in infection models, several of which, however, for unknown reasons. It will be interesting to ascertain the precise function of these transporters in the context of virulence.

Despite their crucial role in several aspects of pathogenesis, there is still a striking lack of knowledge about the function of the majority of the MDR transporters. Indeed, among the 241 predicted ABC drug transporters in the fungal pathogens approached in this review, only 10.8%, that is 26, have been characterised. In the case of the predicted MFS-MDR transporters, only 5.8% (28 out of 479), have been functionally analysed. This observation highlights the pressing need to invest in the study of these two families of drug transporters, with special emphasis on less studied organisms, which in this context includes *A. fumigatus*, for which only 2.1% of the MFS-MDR transporters have been characterised, or *C. tropicalis*, for which only 4.5% of the ABC drug efflux pumps have been studied.

Overall, this review gathers evidence of the multitasking capacity of MDR transporters in fungal pathogens, while highlighting their key role in the successful colonization, persistence, and virulence in the human host. The truly pleiotropic activity of the ABC and MFS-MDR transporters underlines their importance in fungal pathogenesis, and highlights them as very promising drug targets for the development of new antifungals.

Table 1. ATP-binding cassette (ABC) transporters of *Candida*, *Aspergillus*, and *Cryptococcus* species focusing their roles in multidrug resistance (MDR) and physiological functions, as well as their pathogenicity and virulence features.

	Species	Total #	Characterised #	Characterised ORFs	Role in MDR	Physiological Role	Pathogenicity and Virulence Features
ABC Proteins	*Candida albicans*	28	4	orf19.6000/CDR1	Multidrug transporter of ABC superfamily	Transport of phospholipids (in-to-out direction), steroids	Induced by β-estradiol, progesterone, corticosteroid, or cholesterol; Spider biofilm induced
				orf19.5958/CDR2	Multidrug transporter of ABC superfamily, overexpressed in azole-resistant isolates	Transports phospholipids (in-to-out direction)	Repressed in young biofilms
				orf19.1313/CDR3	-	Transporter of the Pdr/Cdr family of the ATP-binding cassette superfamily; transports phospholipids (out-to-in direction); expressed in opaque-phase cells	Induced by macrophage interaction; Spider biofilm induced
				orf19.5079/CDR4	-	-	Rat catheter and flow model biofilm induced
	Candida glabrata	18	5	CAGL0M01760g/CDR1	Multidrug transporter of ABC superfamily, involved in resistance to azoles, expression regulated by Pdr1p, increased abundance in azole resistant strains	Expression increased by loss of the mitochondrial genome	-
				CAGL0F02717g/CDR2 (PDH1)	Multidrug transporter, predicted plasma membrane ATP-binding cassette (ABC) transporter; regulated by Pdr1p; involved in fluconazole resistance	-	-
				CAGL0I04862g /SNQ2	Predicted plasma membrane ATP-binding cassette (ABC) transporter, putative transporter involved in multidrug resistance; involved in Pdr1p-mediated azole resistance		
				CAGL0G00242g/YOR1	Putative ABC transporter involved in multidrug efflux; gene is upregulated in azole-resistant strain	-	-
				CAGL0F01419g/AUS1		ATP-binding cassette transporter involved in sterol uptake	Necessary for *C. glabrata* virulence in a mice model of disseminated infection
	Candida tropicalis	22	1	CDR1	Induced in clinical azole resistant isolates	-	-
	Candida parapsilosis	19	1	CPAR2_405290/CDR1	Induced in clinical azole resistant isolates	-	-
	Candida dubliniensis	19	1	Cd36_85210/CDR1	Predicted multidrug transporter of ABC superfamily, overexpressed in fluconazole resistance, overexpressed in fluconazole-resistant derivatives obtained in vitro	-	-
	Candida krusei	9	2	ABC1	Upregulated during azole stress, involved in innate fluconazole resistance, confers	-	-

Table 1. *Cont.*

Organism		Gene	Function		Notes	
Aspergillus fumigatus	49	ABC2	fluconazole resistance through drug efflux upon hyperexpression in *S. cerevisiae*	-	-	
			Upregulated during azole stress, expression correlated with itraconazole resistance	-	-	
		Afu6g04360/ATRF	Putative ABC transporter; drug efflux pump; involved in itraconazole resistance	-	-	
		Afu1g12690/MDR4	ABC multidrug transporter; induced by voriconazole exposure in vitro and in mice; involved in itraconazole resistance	-	Biofilm growth regulated protein	
		Afu1g17440/ABCA	ABC drug exporter; induced during voriconazole stress	-	Overexpression leads to an augmentation of virulence in the presence of voriconazole in the *G. mellonella* model of infection	
		Afu1g10390/ABCB	Putative ABC multidrug transporter; transcript induced by voriconazole	-	Necessary for full virulence of *A. fumigatus* in the *G. mellonella* model of infection	
		Afu1g14330/ABCC	Putative ABC transporter; induced during voriconazole stress; mutation causes increased itraconazole, voriconazole and posaconazole sensitivity	-	-	
	7	Afu6g03470/ABCD	Putative MDR1 family ABC transporter; induced during voriconazole stress	-	Biofilm growth regulated protein	
		Afu7g00480/ABCE	Putative ABC transporter; induced during voriconazole stress	-	Biofilm growth regulated protein	
		Afu3g07300/ATRI	Putative ABC transporter; induced during voriconazole stress mutation causes increased itraconazole and voriconazole sensitivity	-	-	
Cryptococcus neoformans	54	AFR1	Pump required for azole efflux and other xenobiotics, including cycloheximide, nocodazole, and trichostatin A; involved in clinical fluconazole resistance; role in susceptibility towards amphotericin B and 5-fluorocytosine; expression was positively regulated by CnCrz1 and CnYap1 in response to fluconazole	-	Necessary for full virulence of *C. neoformans* in intravenous and in inhalation mouse models; overexpression upon *C. neoformans* phagocytosis; and involved in the resistance against microglia	
	3	AFR2	Role in susceptibility towards amphotericin B and 5-fluorocytosine	-	-	
		MDR1	Confers itraconazole resistance upon hyperexpression in *S. cerevisiae*; role in susceptibility towards amphotericin B and 5-fluorocytosine	-	-	
Cryptococcus gattii	23	MDR1	Confers fluconazole resistance in *S. cerevisiae*	-	Overexpressed in cells recovered from the brain of infected mice	
	2	PDR11	Necessary for fluconazole resistance in the VGII clinical strain	-	Overexpressed in cells recovered from the lungs of infected mice	

Table 2. MFS transporters of *Candida*, *Aspergillus*, and *Cryptococcus* species, focusing their roles in MDR and physiological functions, as well as their pathogenicity and virulence features.

	Species	Total #	Characterised #	Characterised ORFs	Role in MDR	Physiological Role	Pathogenicity and Virulence Features
MFS Proteins	*Candida albicans*	26	8	orf19.5604/MDR1	Major mediator of azole resistance; methotrexate is preferred substrate; overexpression in drug-resistant clinical isolates confers fluconazole resistance; repressed in young biofilm	-	Necessary for full virulence in *C. albicans* in immunocompetent and immunocompromised mice models
				orf19.6577/FLU1	Involved in the resistance towards fluconazole, ketoconazole, and itraconazole; confers fluconazole resistance in *S. cerevisiae*	Involved in histatin-5 efflux	-
				orf19.2160/NAG4	Required for wild-type cycloheximide resistance	-	Required for wild-type mouse virulence
				orf19.2158/NAG3	Required for wild-type cycloheximide resistance	-	Required for wild-type mouse virulence; Spider biofilm repressed
				orf19.2151/NAG6	Required for wild-type cycloheximide resistance	-	Required for wild-type mouse virulence
				orf19.508/QDR1	-	Involved in lipid homeostasis	Involved in biofilm architecture and thickness and virulence
				orf19.6992/QDR2		Involved in lipid homeostasis	Involved in biofilm architecture and thickness and virulence in a murine model of hematogenously disseminated candidiasis
				orf19.136/QDR3		Involved in lipid homeostasis	Involved in biofilm architecture and thickness and virulence in a murine model of hematogenously disseminated candidiasis
	Candida glabrata	15	8	CAGL0J09944g/AQR1	Involved in resistance to flucytosine and imidazoles	Involved in resistance to acetic acid	-
				CAGL0G03927g/TPO1_1	Putative drug:H$^+$ antiporter, involved in efflux of clotrimazole; required for resistance to clotrimazole and other drugs	Involved in the resistance to histatin-5; involved in spermine resistance	Involved in virulence
				CAGL0E03674g/TPO1_2	Putative drug:H$^+$ antiporter, involved in efflux of clotrimazole; required for resistance to clotrimazole and other drugs	Involved in fatty acid and sterol homeostasis upon biofilm formation; involved in spermine resistance	Involved in virulence in the *G. mellonella* model and biofilm formation
				CAGL0I10384g/TPO3	Confers imidazole and triazole drug resistance; activated by CgPdr1	Involved in polyamine homeostasis	-
				CAGL0G08624g/QDR2	Confers imidazole drug resistance, involved in clotrimazole efflux; activated by CgPdr1; upregulated in azole-resistant strain	-	-
				CAGL0H06017g/FLR1	Confers resistance to benomyl; gene is downregulated in azole-resistant strain	-	-
				CAGL0H06039g/FLR2	Multidrug transporter of the major facilitator superfamily involved in 5-flucytosine resistance	-	-
				CAGL0M06281g/DTR1	-	Acetate exporter in the plasma membrane	Required for virulence in *G. mellonella* model
	Candida tropicalis	26	1	MDR1	Overexpression in resistant clinical isolates and upon biofilm formation	-	-

Table 2. *Cont.*

Candida parapsilosis	34	1	CPAR2_301760/MDR1	Member of the MDR family of major facilitator transporter superfamily; putative drug transporter; expression increased in fluconazole and voriconazole resistant strains	–	–
Candida dubliniensis	14	1	Cd36_63890/MDR1	Predicted multidrug transporter of ABC superfamily, involved in multidrug resistance	–	–
			Afu6g09710/GLIA		Predicted major facilitator type glioxin transporter, encoded in the putative gliotoxin biosynthetic gene cluster	
			Afu1g05010/MFS56	Putative MFS transporter; mutation causes increased azole sensitivity	–	–
Aspergillus fumigatus	278	6	Afu3g03500/MDR3	Putative multidrug resistance protein; transcript upregulated in response to amphotericin B; displays itraconazole-increased expression in resistant mutants	–	–
			Afu8g05710/MFSA	Highly expressed during voriconazole stress	Putative major facilitator superfamily (MFS) sugar transporter	Calcium induced; transcript upregulated in conidia exposed to neutrophils
			Afu1g15490/MFSB	Putative major facilitator superfamily (MFS) transporter; highly expressed during voriconazole stress	–	–
			Afu1g03200/MFSC	Putative major facilitator superfamily (MFS) transporter; highly expressed during voriconazole stress	–	–
Cryptococcus neoformans	86	3	CNA07070	–	Dityrosine transporter	–
			CNC03290	–		–
			CND00440/AFLT			–

Funding: Work conducted in this field has been financially supported by "Fundação para a Ciência e a Tecnologia" (FCT) (Contract PTDC/BBB-BIO/4004/2014 and PhD scholarships to MC and PP).

Conflicts of Interest: The authors declare no conflict of interest.

References

1. Sims, C.R.; Ostrosky-Zeichner, L.; Rex, J.H. Invasive candidiasis in immunocompromised hospitalized patients. *Arch. Med. Res.* **2005**, *36*, 660–671. [CrossRef] [PubMed]
2. Chauhan, N.; Latge, J.-P.; Calderone, R. Signalling and oxidant adaptation in *Candida albicans* and *Aspergillus fumigatus*. *Nat. Rev. Microbiol.* **2006**, *4*, 435–444. [CrossRef] [PubMed]
3. Schmalzle, S.A.; Buchwald, U.K.; Gilliam, B.L.; Riedel, D.J. *Cryptococcus neoformans* infection in malignancy. *Mycoses Diagn. Ther. Prophyl. Fungal Dis.* **2016**, *59*, 542–552. [CrossRef]
4. Wisplinghoff, H.; Bischoff, T.; Tallent, S.M.; Seifert, H.; Wenzel, R.P.; Edmond, M.B. Nosocomial bloodstream infections in US hospitals: Analysis of 24,179 cases from a prospective nationwide surveillance study. *Clin. Infect. Dis.* **2004**, *39*, 309–317. [CrossRef] [PubMed]
5. Fidel, P.L.; Vazquez, J.A.; Sobel, J.D. *Candida glabrata*: Review of epidemiology, pathogenesis, and clinical disease with comparison to *C albicans*. *Clin. Microbiol. Rev.* **1999**, *12*, 80–96. [PubMed]
6. Mccullough, M.J.; Ross, B.C.; Reade, P.C. *Candida albicans*: A review of its history, taxonomy, epidemiology, virulence attributes, and methods of strain differentiation. *Oral Maxillofac. Surg.* **1996**, *25*, 136–144. [CrossRef]
7. Da Silva Ferreira, M.E.S.; Colombo, A.L.; Paulsen, I.; Ren, Q.; Wortman, J.; Huang, J.; Goldman, M.H.S.; Goldman, G.H. The ergosterol biosynthesis pathway, transporter genes, and azole resistance in *Aspergillus fumigatus*. *Med. Mycol.* **2005**, *43*, S313–S319. [CrossRef]
8. Dou, H.-T.; Xu, Y.-C.; Wang, H.-Z.; Li, T.-S. Molecular epidemiology of *Cryptococcus neoformans* and *Cryptococcus gattii* in China between 2007 and 2013 using multilocus sequence typing and the DiversiLab system. *Eur. J. Clin. Microbiol. Infect. Dis.* **2015**, *34*, 753–762. [CrossRef] [PubMed]
9. Sanglard, D.; Kuchler, K.; Ischer, F.; Pagani, J.L.; Monod, M.; Bille, J. Mechanisms of resistance to azole antifungal agents in *Candida albicans* isolates from AIDS patients involve specific multidrug transporters. *Antimicrob. Agents Chemother.* **1995**, *39*, 2378–2386. [CrossRef] [PubMed]
10. Costa, C.; Ribeiro, J.; Miranda, I.M.; Silva-dias, A.I.; Cavalheiro, M.; Costa-de-oliveira, S.; Rodrigues, A.G.; Teixeira, M.C. Clotrimazole drug resistance in *Candida glabrata* clinical isolates correlates with increased expression of the drug: H^+ antiporters CgAqr1, CgTpo1_1, CgTpo3 and CgQdr2. *Front. Microbiol.* **2016**. [CrossRef] [PubMed]
11. Prasad, R.; Khandelwal, N.K.; Banerjee, A. Yeast ABC transporters in lipid trafficking. *Fungal Genet. Biol.* **2016**, *93*, 25–34. [CrossRef] [PubMed]
12. Cannon, R.D.; Holmes, A.R. Learning the ABC of oral fungal drug resistance. *Mol. Oral Microbiol.* **2015**, *30*, 425–437. [CrossRef] [PubMed]
13. Del Sorbo, G.; Schoonbeek, H.J.; De Waard, M.A. Fungal transporters involved in efflux of natural toxic compounds and fungicides. *Fungal Genet. Biol.* **2000**, *30*, 1–15. [CrossRef] [PubMed]
14. Sá-Correia, I.; dos Santos, S.C.; Teixeira, M.C.; Cabrito, T.R.; Mira, N.P. Drug: H^+ antiporters in chemical stress response in yeast. *Trends Microbiol.* **2008**, *17*, 22–31. [CrossRef] [PubMed]
15. Prasad, R.; Banerjee, A.; Khandelwal, N.K.; Dhamgaye, S. The ABCs of *Candida albicans* multidrug transporter Cdr1. *Eukaryot. Cell* **2015**, *14*, 1154–1164. [CrossRef] [PubMed]
16. Pao, S.S.; Paulsen, I.T.; Saier, M.H. Major facilitator superfamily. *Microbiol. Mol. Biol. Rev.* **1998**, *62*, 1–34. [PubMed]
17. Decottignies, A.; Grant, A.M.; Nichols, J.W.; de Wet, H.; McIntosh, D.B.; Goffeau, A. ATPase and multidrug transport activities of the overexpressed yeast ABC protein Yor1p. *J. Biol. Chem.* **1998**, *273*, 12612–12622. [CrossRef] [PubMed]
18. Goffeau, A.; Barrell, B.G.; Bussey, H.; Davis, R.W.; Dujon, B.; Feldmann, H.; Galibert, F.; Hoheisel, J.D.; Jacq, C.; Johnston, M.; et al. Life with 6000 genes. *Science* **1996**, *274*, 546–567. [CrossRef] [PubMed]
19. Bauer, B.E.; Wolfger, H.; Kuchler, K. Inventory and function of yeast ABC proteins: About sex, stress, pleiotropic drug and heavy metal resistance. *Biochim. Biophys. Acta* **1999**, *1461*, 217–236. [CrossRef]
20. Costa, C.; Dias, P.J.; Sá-Correia, I.; Teixeira, M.C. MFS multidrug transporters in pathogenic fungi: Do they have real clinical impact? *Front. Physiol.* **2014**, *5*, 197. [CrossRef] [PubMed]

21. Franz, R.; Michel, S.; Morschhäuser, J. A fourth gene from the *Candida albicans CDR* family of ABC transporters. *Gene* **1998**, *220*, 91–98. [CrossRef]

22. Balan, I.; Alarco, A.M.; Raymond, M. The *Candida albicans CDR3* gene codes for an opaque-phase ABC transporter. *J. Bacteriol.* **1997**, *179*, 7210–7218. [CrossRef] [PubMed]

23. Sanglard, D.; Ischer, F.; Monod, M.; Bille, J. Cloning of *Candida albicans* genes conferring resistance to azole antifungal agents: Characterization of *CDR2*, a new multidrug ABC transporter gene. *Microbiology* **1997**, *143*, 405–416. [CrossRef] [PubMed]

24. Costa, C.; Nunes, J.; Henriques, A.; Mira, N.P.; Nakayama, H.; Chibana, H.; Teixeira, M.C. *Candida glabrata* drug: H$^+$ antiporter CgTpo3 (ORF CAGL0I10384g): Role in azole drug resistance and polyamine homeostasis. *J. Antimicrob. Chemother.* **2014**, *69*, 1767–1776. [CrossRef] [PubMed]

25. Costa, C.; Pires, C.; Cabrito, T.R.; Renaudin, A.; Ohno, M.; Chibana, H.; Sá-Correia, I.; Teixeira, M.C. *Candida glabrata* drug: H$^+$ antiporter CgQdr2 confers imidazole drug resistance, being activated by transcription factor CgPdr1. *Antimicrob. Agents Chemother.* **2013**, *57*, 3159–3167. [CrossRef] [PubMed]

26. Torelli, R.; Posteraro, B.; Ferrari, S.; La Sorda, M.; Fadda, G.; Sanglard, D.; Sanguinetti, M. The ATP-binding cassette transporter—Encoding gene CgSNQ2 is contributing to the CgPDR1-dependent azole resistance of *Candida glabrata. Mol. Microbiol.* **2008**, *68*, 186–201. [CrossRef] [PubMed]

27. Costa, C.; Henriques, A.; Pires, C.; Nunes, J.; Ohno, M.; Chibana, H.; Sá-Correia, I.; Teixeira, M.C. The dual role of *Candida glabrata* drug: H$^+$ antiporter CgAqr1 (ORF CAGL0J09944g) in antifungal drug and acetic acid resistance. *Front. Microbiol.* **2013**, *4*, 170. [CrossRef] [PubMed]

28. Tobin, M.B.; Peery, R.B.; Skatrud, P.L. Genes encoding multiple drug resistance-like proteins in *Aspergillus fumigatus* and *Aspergillus flavus. Gene* **1997**, *200*, 11–23. [CrossRef]

29. Slaven, J.W.; Anderson, M.J.; Sanglard, D.; Dixon, G.K.; Bille, J.; Roberts, I.S.; Denning, D.W. Increased expression of a novel *Aspergillus fumigatus* ABC transporter gene, *atrF*, in the presence of itraconazole in an itraconazole resistant clinical isolate. *Fungal Genet. Biol.* **2002**, *36*, 199–206. [CrossRef]

30. Thornewell, S.J.; Peery, R.B.; Skatrud, P.L. Cloning and characterization of CneMDR1: A *Cryptococcus neoformans* gene encoding a protein related to multidrug resistance proteins. *Gene* **1997**, *201*, 21–29. [CrossRef]

31. Posteraro, B.; Sanguinetti, M.; Sanglard, D.; La Sorda, M.; Boccia, S.; Romano, L.; Morace, G.; Fadda, G.; Cassette, B. Identification and characterization of a *Cryptococcus neoformans* ATP binding cassette (ABC) transporter-encoding gene, *CnAFR1*, involved in the resistance to fluconazole. *Mol. Microbiol.* **2003**, *47*, 357–371. [CrossRef] [PubMed]

32. Gaur, M.; Choudhury, D.; Prasad, R. Complete inventory of ABC proteins in human pathogenic yeast, *Candida albicans. J. Mol. Microbiol. Biotechnol.* **2005**, *9*, 3–15. [CrossRef] [PubMed]

33. Dujon, B.; Sherman, D.; Fischer, G.; Durrens, P.; Casaregola, S.; Lafontaine, I.; de Montigny, J.; Marck, C.; Neuvéglise, C.; Talla, E.; et al. Genome evolution in yeasts. *Nature* **2004**, *430*, 35–44. [CrossRef] [PubMed]

34. Nierman, W.C.; Pain, A.; Anderson, M.J.; Wortman, J.R.; Kim, H.S.; Arroyo, J.; Berriman, M.; Abe, K.; Archer, D.B.; Bermejo, C.; et al. Genomic sequence of the pathogenic and allergenic filamentous fungus *Aspergillus fumigatus. Nature* **2005**, *438*, 1151–1156. [CrossRef] [PubMed]

35. Loftus, B.J.; Fung, E.; Roncaglia, P.; Rowley, D.; Amedeo, P.; Vamathevan, J.; Miranda, M.; Anderson, I.J.; Fraser, J.A.; Allen, J.E.; et al. The genome of the Basidiomycetous yeast and human pathogen *Cryptococcus neoformans. Science* **2005**, *307*, 1321–1324. [CrossRef] [PubMed]

36. Holmes, A.R.; Cardno, T.S.; Strouse, J.J.; Ivnitski-Steele, I.; Keniya, M.V.; Lackovic, K.; Monk, B.C.; Sklar, L.A.; Cannon, R.D. Targeting efflux pumps to overcome antifungal drug resistance. *Future Med. Chem.* **2016**, *8*, 1485–1501. [CrossRef] [PubMed]

37. K Redhu, A.; Shah, A.H.; Prasad, R. MFS transporters of *Candida* species and their role in clinical drug resistance. *FEMS Yeast Res.* **2016**, *16*. [CrossRef] [PubMed]

38. Lupetti, A.; Danesi, R.; Campa, M.; Del Tacca, M.; Kelly, S. Molecular basis of resistance to azole antifungals. *Trends Mol. Med.* **2002**, *8*, 76–81. [CrossRef]

39. Sanguinetti, M.; Posteraro, B.; La Sorda, M.; Torelli, R.; Fiori, B.; Santangelo, R.; Delogu, G.; Fadda, G. Role of AFR1, an ABC transporter-encoding gene, in the in vivo response to fluconazole and virulence of *Cryptococcus neoformans. Infect. Immun.* **2006**, *74*, 1352–1359. [CrossRef] [PubMed]

40. Cannon, R.D.; Lamping, E.; Holmes, A.R.; Niimi, K.; Baret, P.V.; Keniya, M.V.; Tanabe, K.; Niimi, M.; Goffeau, A.; Monk, B.C. Efflux-mediated antifungal drug resistance. *Clin. Microbiol. Rev.* **2009**, *22*, 291–321. [CrossRef] [PubMed]

41. Prasad, R.; De Wergifosse, P.; Goffeau, A.; Balzi, E. Molecular cloning and characterization of a novel gene of *Candida albicans, CDR1*, conferring multiple resistance to drugs and antifungals. *Curr. Genet.* **1995**, *27*, 320–329. [CrossRef] [PubMed]

42. Siikala, E.; Rautemaa, R.; Richardson, M.; Saxen, H.; Bowyer, P.; Sangland, D. Persistent *Candida albicans* colonization and molecular mechanisms of azole resistance in autoimmune polyendocrinopathy-candidiasis-ectodermal dystrophy (APECED) patients. *J. Antimicrob. Chemother.* **2010**, *65*, 2505–2513. [CrossRef] [PubMed]

43. Moran, G.P.; Sangland, D.; Donnelly, S.M.; Shanley, D.B.; Sullivan, D.J.; Coleman, D.C. Identification and expression of multidrug transporters responsible for fluconazole resistance in *Candida dubliniensis. Antimicrob. Agents Chemother.* **1998**, *42*, 1819–1830. [PubMed]

44. Berkow, E.L.; Manigaba, K.; Parker, J.E.; Barker, K.S.; Kelly, S.L.; Rogers, P.D. Multidrug transporters and alterations in sterol biosynthesis contribute to azole antifungal resistance in *Candida parapsilosis. Antimicrob. Agents Chemother.* **2015**, *59*, 5942–5950. [CrossRef] [PubMed]

45. Souza, A.C.R.; Fuchs, B.B.; Pinhati, H.M.S.; Siqueira, R.A.; Hagen, F.; Meis, J.F.; Mylonakis, E.; Colombo, A.L. *Candida parapsilosis* resistance to fluconazole: Molecular mechanisms and in vivo impact in infected *Galleria mellonella* larvae. *Antimicrob. Agents Chemother.* **2015**, *59*, 6581–6587. [CrossRef] [PubMed]

46. Choi, M.J.; Won, E.J.; Shin, J.H.; Kim, S.H.; Lee, W.G.; Kim, M.N.; Lee, K.; Shin, M.G.; Suh, S.P.; Ryang, D.W.; et al. Resistance mechanisms and clinical features of fluconazole-nonsusceptible *Candida tropicalis* isolates compared with fluconazole-less-susceptible isolates. *Antimicrob. Agents Chemother.* **2016**, *60*, 3653–3661. [CrossRef] [PubMed]

47. Sangland, D.; Ischer, F.; Bille, J. Role of ATP-binding-cassette transporter genes in high-frequency acquisition of resistance to azole antifungals in *Candida glabrata. Antimicrob. Agents Chemother.* **2001**, *45*, 1174–1183. [CrossRef] [PubMed]

48. Vermitsky, J.-P.; Earhart, K.D.; Smith, W.L.; Homayouni, R.; Edlind, T.D.; Rogers, P.D. Pdr1 regulates multidrug resistance in *Candida glabrata*: Gene disruption and genome-wide expression studies. *Mol. Microbiol.* **2006**, *61*, 704–722. [CrossRef] [PubMed]

49. Katiyar, S.K.; Edlind, T.D. Identification and expression of multidrug resistance-related ABC transporter genes in *Candida krusei. Med. Mycol.* **2001**, *39*, 109–116. [CrossRef] [PubMed]

50. Lamping, E.; Ranchod, A.; Nakamura, K.; Tyndall, J.D.A.; Niimi, K.; Holmes, A.R.; Niimi, M.; Cannon, R.D. Abc1p is a multidrug efflux transporter that tips the balance in favor of innate azole resistance in *Candida krusei. Antimicrob. Agents Chemother.* **2009**, *53*, 354–369. [CrossRef] [PubMed]

51. Lamping, E.; Monk, B.C.; Niimi, K.; Holmes, A.R.; Tsao, S.; Tanabe, K.; Niimi, M.; Uehara, Y.; Cannon, R.D. Characterization of three classes of membrane proteins involved in fungal azole resistance by functional hyperexpression in *Saccharomyces cerevisiae. Eukaryot. Cell* **2007**, *6*, 1150–1165. [CrossRef] [PubMed]

52. He, X.; Zhao, M.; Chen, J.; Wu, R.; Zhang, J.; Cui, R.; Jiang, Y.; Chen, J.; Cao, X.; Xing, Y.; et al. Overexpression of both *ERG11* and *ABC2* genes might be responsible for itraconazole resistance in clinical isolates of *Candida krusei. PLoS ONE* **2015**, *10*, e0136185. [CrossRef] [PubMed]

53. Meneau, I.; Coste, A.T.; Sangland, D. Identification of *Aspergillus fumigatus* multidrug transporter genes and their potential involvement in antifungal resistance. *Med. Mycol.* **2016**, *54*, 616–627. [CrossRef] [PubMed]

54. Abad, A.; Victoria Fernández-Molina, J.; Bikandi, J.; Ramírez, A.; Margareto, J.; Sendino, J.; Luis Hernando, F.; Pontón, J.; Garaizar, J.; Rementeria, A. What makes *Aspergillus fumigatus* a successful pathogen? Genes and molecules involved in invasive aspergillosis. *Rev. Iberoam. Micol.* **2010**, *27*, 155–182. [CrossRef] [PubMed]

55. Nascimento, A.M.; Goldman, G.H.; Park, S.; Marras, S.A.E.; Delmas, G.; Oza, U.; Lolans, K.; Dudley, M.N.; Mann, P.A.; Perlin, D.S. Multiple resistance mechanisms among *Aspergillus fumigatus* mutants with high-level resistance to itraconazole. *Antimicrob. Agents Chemother.* **2003**, *47*, 1719–1726. [CrossRef] [PubMed]

56. Da Silva Ferreira, M.E.; Malavazi, I.; Savoldi, M.; Brakhage, A.A.; Goldman, M.H.S.; Kim, H.S.; Nierman, W.C.; Goldman, G.H. Transcriptome analysis of *Aspergillus fumigatus* exposed to voriconazole. *Curr. Genet.* **2006**, *50*, 32–44. [CrossRef] [PubMed]

57. Chang, M.; Sionov, E.; Khanal Lamichhane, A.; Kwon-Chung, K.J.; Chang, Y.C. Roles of three *Cryptococcus neoformans* and *Cryptococcus gattii* efflux pump-coding genes in response to drug treatment. *Antimicrob. Agents Chemother.* **2018**, *62*. [CrossRef] [PubMed]

58. Basso, L.R.; Gast, C.E.; Bruzual, I.; Wong, B. Identification and properties of plasma membrane azole efflux pumps from the pathogenic fungi *Cryptococcus gattii* and *Cryptococcus neoformans*. *J. Antimicrob. Chemother.* **2015**, *70*, 1396–1407. [CrossRef] [PubMed]

59. Wirsching, S.; Michel, S.; Morschhäuser, J. Targeted gene disruption in *Candida albicans* wild-type strains: The role of the *MDR1* gene in fluconazole resistance of clinical *Candida albicans* isolates. *Mol. Microbiol.* **2000**, *36*, 856–865. [CrossRef] [PubMed]

60. Wirsching, S.; Moran, G.P.; Sullivan, D.J.; Coleman, D.C.; Morschhäuser, J. *MDR1*-mediated drug resistance in *Candida dubliniensis*. *Antimicrob. Agents Chemother.* **2001**, *45*, 3416–3421. [CrossRef] [PubMed]

61. Silva, A.P.; Miranda, I.M.; Guida, A.; Synnott, J.; Rocha, R.; Silva, R.; Amorim, A.; Pina-Vaz, C.; Butler, G.; Rodrigues, A.G. Transcriptional profiling of azole-resistant *Candida parapsilosis* strains. *Antimicrob. Agents Chemother.* **2011**, *55*, 3546–3556. [CrossRef] [PubMed]

62. Bizerra, F.C.; Nakamura, C.V.; De Poersch, C.; Estivalet Svidzinski, T.I.; Borsato Quesada, R.M.; Goldenberg, S.; Krieger, M.A.; Yamada-Ogatta, S.F. Characteristics of biofilm formation by *Candida tropicalis* and antifungal resistance. *FEMS Yeast Res.* **2008**, *8*, 442–450. [CrossRef] [PubMed]

63. Calabrese, D.; Bille, J.; Sanglard, D. A novel multidrug efflux transporter gene of the major facilitator superfamily from *Candida albicans (FLU1)* conferring resistance to fluconazole. *Microbiology* **2000**, *146*, 2743–2754. [CrossRef] [PubMed]

64. Park, S.; Perlin, D.S. Establishing surrogate markers for fluconazole resistance in *Candida albicans*. *Microb. Drug Resist.* **2005**, *11*, 232–238. [CrossRef] [PubMed]

65. Pais, P.; Costa, C.; Pires, C.; Shimizu, K.; Chibana, H.; Teixeira, M.C. Membrane proteome-wide response to the antifungal drug clotrimazole in *Candida glabrata*: role of the transcription factor CgPdr1 and the drug: H$^+$ antiporters CgTpo1_1 and CgTpo1_2. *Mol. Cell. Proteom.* **2016**, *15*, 57–72. [CrossRef] [PubMed]

66. Pais, P.; Pires, C.; Costa, C.; Okamoto, M.; Chibana, H.; Teixeira, M.C. Membrane proteomics analysis of the *Candida glabrata* response to 5-flucytosine: Unveiling the role and regulation of the drug efflux transporters CgFlr1 and CgFlr2. *Front. Microbiol.* **2016**, *7*, 2045. [CrossRef] [PubMed]

67. Tsai, H.-F.; Krol, A.A.; Sarti, K.E.; Bennett, J.E. *Candida glabrata PDR1*, a transcriptional regulator of a pleiotropic drug resistance network, mediates azole resistance in clinical isolates and petite mutants. *Antimicrob. Agents Chemother.* **2006**, *50*, 1384–1392. [CrossRef] [PubMed]

68. Paul, S.; Schmidt, J.A.; Moye-Rowley, W.S. Regulation of the CgPdr1 transcription factor from the pathogen *Candida glabrata*. *Eukaryot. Cell* **2011**, *10*, 187–197. [CrossRef] [PubMed]

69. Coste, A.; Turner, V.; Ischer, F.; Morschhäuser, J.; Forche, A.; Selmecki, A.; Berman, J.; Bille, J.; Sanglard, D. A mutation in Tac1p, a transcription factor regulating *CDR1* and *CDR2*, is coupled with loss of heterozygosity at chromosome 5 to mediate antifungal resistance in *Candida albicans*. *Genetics* **2006**, *172*, 2139–2156. [CrossRef] [PubMed]

70. Coste, A.; Selmecki, A.; Forche, A.; Diogo, D.; Bougnoux, M.E.; D'Enfert, C.; Berman, J.; Sanglard, D. Genotypic evolution of azole resistance mechanisms in sequential *Candida albicans* isolates. *Eukaryot. Cell* **2007**, *6*, 1889–1904. [CrossRef] [PubMed]

71. Morschhäuser, J.; Barker, K.S.; Liu, T.T.; BlaB-Warmuth, J.; Homayouni, R.; Rogers, P.D. The transcription factor Mrr1p controls expression of the *MDR1* efflux pump and mediates multidrug resistance in *Candida albicans*. *PLoS Pathog.* **2007**, *3*, e164. [CrossRef] [PubMed]

72. Dunkel, N.; Blaß, J.; Rogers, P.D.; Morschhäuser, J. Mutations in the multi-drug resistance regulator *MRR1*, followed by loss of heterozygosity, are the main cause of *MDR1* overexpression in fluconazole-resistant *Candida albicans* strains. *Mol. Microbiol.* **2008**, *69*, 827–840. [CrossRef] [PubMed]

73. Liu, T.T.; Znaidi, S.; Barker, K.S.; Xu, L.; Homayouni, R.; Saidane, S.; Morschhäuser, J.; Nantel, A.; Raymond, M.; Rogers, P.D. Genome-wide expression and location analyses of the *Candida albicans* Tac1p regulon. *Eukaryot. Cell* **2007**, *6*, 2122–2138. [CrossRef] [PubMed]

74. Znaidi, S.; De Deken, X.; Weber, S.; Rigby, T.; Nantel, A.; Raymond, M. The zinc cluster transcription factor Tac1p regulates *PDR16* expression in *Candida albicans*. *Mol. Microbiol.* **2007**, *66*, 440–452. [CrossRef] [PubMed]

75. Schubert, S.; Barker, K.S.; Znaidi, S.; Schneider, S.; Dierolf, F.; Dunkel, N.; Aïd, M.; Boucher, G.; Rogers, P.D.; Raymond, M.; et al. Regulation of efflux pump expression and drug resistance by the transcription factors Mrr1, Upc2, and Cap1 in *Candida albicans*. *Antimicrob. Agents Chemother.* **2011**, *55*, 2212–2223. [CrossRef] [PubMed]

76. Coste, A.T.; Karababa, M.; Ischer, F.; Bille, J.; Sanglard, D. *TAC1*, transcriptional activator of *CDR* genes, is a new transcription factor involved in the regulation of *Candida albicans* ABC transporters *CDR1* and *CDR2*. *Eukaryot. Cell* **2004**, *3*, 1639–1652. [CrossRef] [PubMed]

77. Caudle, K.E.; Barker, K.S.; Wiederhold, N.P.; Xu, L.; Homayouni, R.; Rogers, P.D. Genomewide expression profile analysis of the *Candida glabrata* Pdr1 regulon. *Eukaryot. Cell* **2011**, *10*, 373–383. [CrossRef] [PubMed]

78. Tsai, H.-F.; Sammons, L.R.; Zhang, X.; Suffis, S.D.; Su, Q.; Myers, T.G.; Marr, K.A.; Bennett, J.E. Microarray and molecular analyses of the azole resistance mechanism in *Candida glabrata* oropharyngeal isolates. *Antimicrob. Agents Chemother.* **2010**, *54*, 3308–3317. [CrossRef] [PubMed]

79. Selmecki, A.; Forche, A.; Berman, J. Genomic plasticity of the human fungal pathogen *Candida albicans*. *Eukaryot. Cell* **2010**, *9*, 991–1008. [CrossRef] [PubMed]

80. Poláková, S.; Blume, C.; Zárate, J.A.; Mentel, M.; Jørck-Ramberg, D.; Stenderup, J.; Piskur, J. Formation of new chromosomes as a virulence mechanism in yeast *Candida glabrata*. *Proc. Natl. Acad. Sci. USA* **2009**, *106*, 2688–2693. [CrossRef] [PubMed]

81. Sionov, E.; Lee, H.; Chang, Y.C.; Kwon-Chung, K.J. *Cryptococcus neoformans* overcomes stress of azole drugs by formation of disomy in specific multiple chromosomes. *PLoS Pathog.* **2010**, *6*, e1000848. [CrossRef] [PubMed]

82. Hagiwara, D.; Miura, D.; Shimizu, K.; Paul, S.; Ohba, A.; Gonoi, T.; Watanabe, A.; Kamei, K.; Shintani, T.; Moye-Rowley, W.S.; et al. A novel Zn_2-Cys_6 transcription factor AtrR plays a key role in an azole resistance mechanism of *Aspergillus fumigatus* by Co-regulating *cyp51A* and *cdr1B* expressions. *PLoS Pathog.* **2017**, *13*. [CrossRef] [PubMed]

83. Paul, S.; Doering, T.L.; Moye-Rowley, W.S. *Cryptococcus neoformans* Yap1 is required for normal fluconazole and oxidative stress resistance. *Fungal Genet. Biol.* **2015**, *74*, 1–9. [CrossRef] [PubMed]

84. Lev, S.; Desmarini, D.; Chayakulkeeree, M.; Sorrell, T.C.; Djordjevic, J.T. The Crz1/Sp1 transcription factor of *Cryptococcus neoformans* is activated by calcineurin and regulates cell wall integrity. *PLoS ONE* **2012**, *7*, e51403. [CrossRef] [PubMed]

85. Noble, S.M.; Gianetti, B.A.; Witchley, J.N. *Candida albicans* cell-type switching and functional plasticity in the mammalian host. *Nat. Rev. Microbiol.* **2017**, *15*, 96–108. [CrossRef] [PubMed]

86. Brown, A.J.P.; Brown, G.D.; Netea, M.G.; Gow, N.A.R. Metabolism impacts upon *Candida* immunogenicity and pathogenicity at multiple levels. *Trends Microbiol.* **2014**, *22*, 614–622. [CrossRef] [PubMed]

87. Levitz, S.M. The Ecology of *Cryptococcus neoformans* and the epidemiology of cryptococcosis. *Rev. Infect. Dis.* **1991**, *13*, 1163–1169. [CrossRef] [PubMed]

88. Chen, S.C.-A.; Meyer, W.; Sorrell, T.C. *Cryptococcus gattii* Infections. *Clin. Microbiol. Rev.* **2014**, *27*, 980–1024. [CrossRef] [PubMed]

89. Latgé, J.-P. *Aspergillus fumigatus* and Aspergillosis. *Clin. Microbiol. Rev.* **1999**, *12*, 310–350. [CrossRef] [PubMed]

90. Polke, M.; Hube, B.; Jacobsen, I.D. Candida survival strategies. In *Advances in Applied Microbiology*; Elsevier Ltd.: New York, NY, USA, 2015; Volume 91, pp. 139–235, ISBN 9780128022504.

91. Villar, C.C.; Dongari-Bagtzoglou, A. Immune defence mechanisms and immunoenhancement strategies in oropharyngeal candidiasis. *Expert Rev. Mol. Med.* **2008**, *10*. [CrossRef] [PubMed]

92. Mathews, M.; Jia, H.P.; Guthmiller, J.M.; Losh, G.; Graham, S.; Johnson, G.K.; Tack, B.F.; McCray, P.B. Production of β-defensin antimicrobial peptides by the oral mucosa and salivary glands. *Infect. Immun.* **1999**, *67*, 2740–2745. [PubMed]

93. Wunder, D.; Dong, J.; Baev, D.; Edgerton, M. Human Salivary Histatin 5 Fungicidal action does not induce programmed cell death pathways in *Candida albicans*. *Antimicrob. Agents Chemother.* **2004**, *48*, 110–115. [CrossRef] [PubMed]

94. Taylor-Smith, L.M. *Cryptococcus*—Epithelial interactions. *J. Fungi* **2017**, *3*, 53. [CrossRef] [PubMed]

95. Heung, L.J. Innate immune responses to *Cryptococcus*. *J. Fungi* **2017**, *3*, 35. [CrossRef] [PubMed]

96. Leopold Wager, C.M.; Hole, C.R.; Wozniak, K.L.; Wormley, F.L. *Cryptococcus* and phagocytes: Complex interactions that influence disease outcome. *Front. Microbiol.* **2016**, *7*, 105. [CrossRef] [PubMed]

97. Serrano-Gomez, D.; Dominguez-Soto, A.; Ancochea, J.; Jimenez-Heffernan, J.A.; Leal, J.A.; Corbi, A.L. Dendritic cell-specific intercellular adhesion molecule 3-grabbing nonintegrin mediates binding and internalization of *Aspergillus fumigatus* conidia by dendritic cells and macrophages. *J. Immunol.* **2018**, *173*, 5635–5643. [CrossRef]

98. Ellett, F.; Jorgensen, J.; Frydman, G.H.; Jones, C.N.; Irimia, D. Neutrophil interactions stimulate evasive hyphal branching by *Aspergillus fumigatus*. *PLoS Pathog.* **2017**, *13*, e1006154. [CrossRef] [PubMed]

99. O'Hanlon, D.E.; Moench, T.R.; Cone, R.A. Vaginal pH and microbicidal lactic acid when lactobacilli dominate the microbiota. *PLoS ONE* **2013**, *8*, e80074. [CrossRef] [PubMed]

100. Chaudry, A.N.; Travers, P.J.; Yuenger, J.; Colletta, L.; Evans, P.; Zenilman, J.M.; Tummon, A. Analysis of vaginal acetic acid in patients undergoing treatment for bacterial vaginosis. *J. Clin. Microbiol.* **2004**, *42*, 5170–5175. [CrossRef] [PubMed]

101. Fidel, P.L., Jr. Immunity in vaginal candidiasis. *Curr. Opin. Infect. Dis.* **2005**, *18*, 107–111. [CrossRef] [PubMed]

102. Niu, X.X.; Li, T.; Zhang, X.; Wang, S.X.; Liu, Z.H. *Lactobacillus crispatus* modulates vaginal epithelial cell innate response to *Candida albicans*. *Chin. Med. J.* **2017**, *130*, 273–279. [CrossRef] [PubMed]

103. Sipos, G.; Kuchler, K. Fungal ATP-binding cassette (ABC) transporters in drug resistance & detoxification. *Curr. Drug Targets* **2006**, *7*, 471–481. [PubMed]

104. Krishnamurthy, S.; Gupta, V.; Snehlata, P.; Prasad, R. Characterisation of human steroid hormone transport mediated by Cdr1p, a multidrug transporter of *Candida albicans*, belonging to the ATP binding cassette super family. *FEMS Microbiol. Lett.* **1998**, *158*, 69–74. [CrossRef] [PubMed]

105. Krishnamurthy, S.; Chatterjee, U.; Gupta, V.; Prasad, R.; Das, P.; Snehlata, P.; Hasnain, S.E.; Prasad, R. Deletion of transmembrane domain 12 of *CDR1*, a multidrug transporter from *Candida albicans*, leads to altered drug specificity: Expression of a yeast multidrug transporter in baculovirus expression system. *Yeast* **1998**, *14*, 535–550. [CrossRef]

106. Bauters, T.G.M.; Dhont, M.A.; Temmerman, M.I.L.; Nelis, H.J. Prevalence of vulvovaginal candidiasis and susceptibility to fluconazole in women. *Am. J. Obstet. Gynecol.* **2002**, *187*, 569–574. [CrossRef] [PubMed]

107. Spinillo, A.; Bernuzzi, A.M.; Cevini, C.; Gulminetti, R.; Luzi, S.; De Santolo, A. The relationship of bacterial vaginosis, *Candida* and *Trichomonas* infection to symptomatic vaginitis in postmenopausal women attending a vaginitis clinic. *Maturitas* **1997**, *27*, 253–260. [CrossRef]

108. Spinillo, A.; Capuzzo, E.; Nicola, S.; Baltaro, F.; Ferrari, A.; Monaco, A. The impact of oral contraception on vulvovaginal candidiasis. *Contraception* **1995**, *51*, 293–297. [CrossRef]

109. Fidel, P.L.; Cutright, J.; Steele, C. Effects of reproductive hormones on experimental vaginal candidiasis. *Infect. Immun.* **2000**, *68*, 651–657. [CrossRef] [PubMed]

110. White, S.; Larsen, B. *Candida albicans* morphogenesis is influenced by estogen. *Cell. Mol. Life Sci.* **1997**, *53*, 744–749. [CrossRef] [PubMed]

111. Cheng, G.; Yeater, K.M.; Hoyer, L.L. Cellular and molecular biology of *Candida albicans* estrogen response. *Eukaryot. Cell* **2006**, *5*, 180–191. [CrossRef] [PubMed]

112. Li, Y.; Prinz, W.A. ATP-binding Cassette (ABC) transporters mediate nonvesicular, raft-modulated sterol movement from the plasma membrane to the endoplasmic reticulum. *J. Biol. Chem.* **2004**, *279*, 45226–45234. [CrossRef] [PubMed]

113. Wilcox, L.J.; Balderes, D.A.; Wharton, B.; Tinkelenberg, A.H.; Rao, G.; Sturley, S.L. Transcriptional profiling identifies two members of the ATP-binding cassette transporter superfamily required for sterol uptake in yeast. *J. Biol. Chem.* **2002**, *277*, 32466–32472. [CrossRef] [PubMed]

114. Yang, M.L.; Uhrig, J.; Vu, K.; Singapuri, A.; Dennis, M.; Gelli, A.; Thompson, G.R. Fluconazole susceptibility in *Cryptococcus gattii* is dependent on the ABC transporter Pdr11. *Antimicrob. Agents Chemother.* **2016**, *60*, 1202–1207. [CrossRef] [PubMed]

115. Nagi, M.; Tanabe, K.; Ueno, K.; Nakayama, H.; Aoyama, T.; Chibana, H.; Yamagoe, S.; Umeyama, T.; Oura, T.; Ohno, H.; et al. The *Candida glabrata* sterol scavenging mechanism, mediated by the ATP-binding cassette transporter Aus1p, is regulated by iron limitation. *Mol. Microbiol.* **2013**, *88*, 371–381. [CrossRef] [PubMed]

116. Bard, M.; Sturm, A.M.; Pierson, C.A.; Brown, S.; Rogers, K.M.; Nabinger, S.; Eckstein, J.; Barbuch, R.; Lees, N.D.; Howell, S.A.; et al. Sterol uptake in *Candida glabrata*: Rescue of sterol auxotrophic strains. *Diagn. Microbiol. Infect. Dis.* **2005**, *52*, 285–293. [CrossRef] [PubMed]

117. Hazen, K.C.; Stei, J.; Darracott, C.; Breathnach, A.; May, J.; Howell, S.A. Isolation of cholesterol-dependent *Candida glabrata* from clinical specimens. *Diagn. Microbiol. Infect. Dis.* **2005**, *52*, 35–37. [CrossRef] [PubMed]

118. Li, Q.Q.; Tsai, H.F.; Mandal, A.; Walker, B.A.; Noble, J.A.; Fukuda, Y.; Bennett, J.E. Sterol uptake and sterol biosynthesis act coordinately to mediate antifungal resistance in *Candida glabrata* under azole and hypoxic stress. *Mol. Med. Rep.* **2018**, *17*, 6585–6597. [CrossRef] [PubMed]

119. Shah, A.H.; Singh, A.; Dhamgaye, S.; Chauhan, N.; Vandeputte, P.; Suneetha, K.J.; Kaur, R.; Mukherjee, P.K.; Chandra, J.; Ghannoum, M.A.; et al. Novel role of a family of major facilitator transporters in biofilm development and virulence of *Candida albicans*. *Biochem. J.* **2014**, *460*, 223–235. [CrossRef] [PubMed]

120. Nunes, P.A.; Tenreiro, S.; Sá-Correia, I. Resistance and adaptation to quinidine in *Saccharomyces cerevisiae*: role of *QDR1* (YIL120w), encoding a plasma membrane transporter of the major facilitator superfamily required for multidrug resistance. *Antimicrob. Agents Chemother.* **2001**, *45*, 1528–1534. [CrossRef] [PubMed]

121. Vargas, R.C.; Tenreiro, S.; Teixeira, M.C.; Fernandes, A.R.; Sá-correia, I. *Saccharomyces cerevisiae* multidrug transporter Qdr2p (Yil121wp): Localization and function as a quinidine resistance determinant. *Antimicrob. Agents Chemother.* **2004**, *48*, 2531–2537. [CrossRef] [PubMed]

122. Tenreiro, S.; Vargas, R.C.; Teixeira, M.C.; Magnani, C.; Sá-Correia, I. The yeast multidrug transporter Qdr3 (Ybr043c): Localization and role as a determinant of resistance to quinidine, barban, cisplatin, and bleomycin. *Biochem. Biophys. Res. Commun.* **2005**, *327*, 952–959. [CrossRef] [PubMed]

123. Santos, R.; Costa, C.; Mil-Homens, D.; Romão, D.; de Carvalho, C.C.C.R.; Pais, P.; Mira, N.P.; Fialho, A.M.; Teixeira, M.C. The multidrug resistance transporters CgTpo1_1 and CgTpo1_2 play a role in virulence and biofilm formation in the human pathogen *Candida glabrata*. *Cell. Microbiol.* **2017**, *19*, 1–13. [CrossRef] [PubMed]

124. Lin, X.; Qi, Y.; Yan, D.; Liu, H.; Chen, X.; Liu, L. *CgMED3* changes membrane sterol composition to help *Candida glabrata* tolerate. *Appl. Environ. Microbiol.* **2017**, *83*, 1–15. [CrossRef] [PubMed]

125. Li, R.; Kumar, R.; Tati, S.; Puri, S.; Edgerton, M. *Candida albicans* Flu1-Mediated efflux of salivary histatin 5 reduces its cytosolic concentration and fungicidal activity. *Antimicrob. Agents Chemother.* **2013**, *57*, 1832–1839. [CrossRef] [PubMed]

126. Kuchler, K.; Sterne, R.E.; Thorner, J. *Saccharomyces cerevisiae STE6* gene product: A novel pathway for protein export in eukaryotic cells. *EMBO J.* **1989**, *8*, 3973–3984. [CrossRef] [PubMed]

127. Raymond, M.; Dignard, D.; Alarco, A.M.; Mainville, N.; Magee, B.B.; Thomas, D.Y. A Ste6p/P-glycoprotein homologue from the asexual yeast *Candida albicans* transports the a-factor mating pheromone in *Saccharomyces cerevisiae*. *Mol. Microbiol.* **1998**, *27*, 587–598. [CrossRef] [PubMed]

128. Hsueh, Y.P.; Shen, W.C. A homolog of Ste6, the a-factor transporter in *Saccharomyces cerevisiae*, is required for mating but not for monokaryotic fruiting in *Cryptococcus neoformans*. *Eukaryot. Cell* **2005**, *4*, 147–155. [CrossRef] [PubMed]

129. Gibbons, J.G.; Beauvais, A.; Beau, R.; McGary, K.L.; Latgé, J.P.; Rokas, A. Global transcriptome changes underlying colony growth in the opportunistic human pathogen *Aspergillus fumigatus*. *Eukaryot. Cell* **2012**, *11*, 68–78. [CrossRef] [PubMed]

130. Kragl, C.; Schrettl, M.; Abt, B.; Sarg, B.; Lindner, H.H.; Haas, H. EstB-mediated hydrolysis of the siderophore triacetylfusarinine C optimizes iron uptake of *Aspergillus fumigatus*. *Eukaryot. Cell* **2007**, *6*, 1278–1285. [CrossRef] [PubMed]

131. Schrettl, M.; Kim, H.S.; Eisendle, M.; Kragl, C.; Nierman, W.C.; Heinekamp, T.; Werner, E.R.; Jacobsen, I.; Illmer, P.; Yi, H.; et al. SreA-mediated iron regulation in *Aspergillus fumigatus*. *Mol. Microbiol.* **2008**, *70*, 27–43. [CrossRef] [PubMed]

132. Weiss, G. Iron and immunity: A double-edged sword. *Eur. J. Clin. Investig.* **2002**, *32*, 70–78. [CrossRef]

133. Albertsen, M.; Bellahn, I.; Krämer, R.; Waffenschmidt, S. Localization and function of the yeast multidrug transporter Tpo1p. *J. Biol. Chem.* **2003**, *278*, 12820–12825. [CrossRef] [PubMed]

134. Tomitori, H.; Kashiwagi, K.; Asakawa, T.; Kakinuma, Y.; Michael, A.J.; Igarashi, K. Multiple polyamine transport systems on the vacuolar membrane in yeast. *Biochem. J.* **2001**, *353*, 681–688. [CrossRef] [PubMed]

135. Teixeira, M.C.; Cabrito, R.; Hanif, Z.M.; Vargas, R.C.; Tenreiro, S.; Sá-Correia, I. Yeast response and tolerance to polyamine toxicity involving the drug: H^+ antiporter Qdr3 and the transcription factors Yap1 and Gcn4. *Microbiology* **2011**, *157*, 945–956. [CrossRef] [PubMed]

136. Tyms, A. Polyamines and the growth of bacteria and viruses. In *The Physiology of Polyamines*; Bachrach, U., Heime, Y., Eds.; CRC Press: Boca Raton, FL, USA, 1989.

137. Tenreiro, S.; Nunes, P.A.; Viegas, C.A.; Neves, M.S.; Teixeira, M.C.; Cabral, G.; Sá-Correia, I. *AQR1* gene (ORF YNL065w) encodes a plasma membrane transporter of the major facilitator superfamily that confers resistance to short-chain monocarboxylic acids and quinidine in *Saccharomyces cerevisiae*. *Biochem. Biophys. Res. Commun.* **2002**, *292*, 741–748. [CrossRef] [PubMed]

138. Velasco, I.; Tenreiro, S.; Calderon, I.L.; André, B. *Saccharomyces cerevisiae* Aqr1 is an internal-membrane transporter involved in excretion of amino acids. *Eukaryot. Cell* **2004**, *3*, 1492–1503. [CrossRef] [PubMed]

139. Romão, D.; Cavalheiro, M.; Mil-Homens, D.; Santos, R.; Pais, P.; Costa, C.; Takahashi-Nakaguchi, A.; Fialho, A.M.; Chibana, H.; Teixeira, M.C. A New determinant of *Candida glabrata* virulence: The acetate exporter CgDtr1. *Front. Cell. Infect. Microbiol.* **2017**, *7*, 473. [CrossRef] [PubMed]

140. Bernardo, R.T.; Cunha, D.V.; Wang, C.; Pereira, L.; Silva, S.; Salazar, S.B.; Schröder, M.S.; Okamoto, M.; Takahashi-Nakaguchi, A.; Chibana, H.; et al. The CgHaa1-Regulon Mediates Response and Tolerance to Acetic Acid Stress in the Human Pathogen *Candida glabrata*. *G3: Genes Genomes Genet.* **2017**, *7*, 1–18. [CrossRef] [PubMed]

141. Smriti; Krishnamurthy, S.; Dixit, B.L.; Gupta, C.M.; Milewski, S.; Prasad, R. ABC transporters Cdr1p, Cdr2p and Cdr3p of a human pathogen *Candida albicans* are general phospholipid translocators. *Yeast* **2002**, *19*, 303–318. [CrossRef]

142. Fox, L.M.; Cox, D.G.; Lockridge, J.L.; Wang, X.; Chen, X.; Scharf, L.; Trott, D.L.; Ndonye, R.M.; Veerapen, N.; Besra, G.S.; et al. Recognition of lyso-phospholipids by human natural killer T lymphocytes. *PLoS Biol.* **2009**, *7*, e1000228. [CrossRef] [PubMed]

143. Cox, D.; Fox, L.; Tian, R.; Bardet, W.; Skaley, M.; Mojsilovic, D.; Gumperz, J.; Hildebrand, W. Determination of cellular lipids bound to human CD1d molecules. *PLoS ONE* **2009**, *4*, e5325. [CrossRef] [PubMed]

144. De Libero, G.; Mori, L. Recognition of lipid antigens by T cells. *Nat. Rev. Immunol.* **2005**, *5*, 485–496. [CrossRef] [PubMed]

145. Netea, M.G.; Joosten, L.A.B.; van der Meer, J.W.M.; Kullberg, B.-J.; van de Veerdonk, F.L. Immune defence against *Candida* fungal infections. *Nat. Rev. Immunol.* **2015**, *15*, 630–642. [CrossRef] [PubMed]

146. Brown, G.D. Innate antifungal immunity: The key role of phagocytes. *Annu. Rev. Immunol.* **2011**, *29*, 1–21. [CrossRef] [PubMed]

147. Miramón, P.; Kasper, L.; Hube, B. Thriving within the host: *Candida* spp. interactions with phagocytic cells. *Med. Microbiol. Immunol.* **2013**, *202*, 183–195. [CrossRef] [PubMed]

148. Seider, K.; Brunke, S.; Schild, L.; Jablonowski, N.; Wilson, D.; Majer, O.; Barz, D.; Haas, A.; Kuchler, K.; Schaller, M.; et al. the facultative intracellular pathogen *Candida glabrata* subverts macrophage cytokine production and phagolysosome maturation. *J. Immunol.* **2011**, *187*, 3072–3086. [CrossRef] [PubMed]

149. Orsi, C.F.; Colombari, B.; Ardizzoni, A.; Peppoloni, S.; Neglia, R.; Posteraro, B.; Morace, G.; Fadda, G.; Blasi, E. The ABC transporter-encoding gene *AFR1* affects the resistance of *Cryptococcus neoformans* to microglia-mediated antifungal activity by delaying phagosomal maturation. *FEMS Yeast Res.* **2009**, *9*, 301–310. [CrossRef] [PubMed]

150. Goulart, L.; Silva, L.K.R.E.; Chiapello, L.; Silveira, C.; Crestani, J.; Masih, D.; Vainstein, M.H. *Cryptococcus neoformans* and *Cryptococcus gattii* genes preferentially expressed during rat macrophage infection. *Med. Mycol.* **2010**, *48*, 932–941. [CrossRef] [PubMed]

151. Fontes, A.C.L.; Oliveira, D.B.; Santos, J.R.A.; Carneiro, H.C.S.; De Queiroz Ribeiro, N.; De Oliveira, L.V.N.; Barcellos, V.A.; Paixão, T.A.; Abrahão, J.S.; Resende-Stoianoff, M.A.; et al. A subdose of fluconazole alters the virulence of *Cryptococcus gattii* during murine cryptococcosis and modulates type I interferon expression. *Med. Mycol.* **2017**, *55*, 203–212. [CrossRef] [PubMed]

152. Paul, S.; Diekema, D.; Moye-Rowley, W.S. Contributions of *Aspergillus fumigatus* ATP-binding cassette transporter proteins to drug resistance and virulence. *Eukaryot. Cell* **2013**, *12*, 1619–1628. [CrossRef] [PubMed]

153. White, T.C. Increased mRNA Levels of *ERG16*, *CDR*, and *MDR1* correlate with increases in azole resistance in *Candida albicans* isolates from a patient infected with human immunodeficiency virus. *Antimicrob. Agents Chemother.* **1997**, *41*, 1482–1487. [PubMed]

154. Hiller, D.; Sanglard, D.; Morschhauser, J. Overexpression of the *MDR1* gene is sufficient to confer increased resistance to toxic compounds in *Candida albicans*. *Antimicrob. Agents Chemother.* **2006**, *50*, 1365–1371. [CrossRef] [PubMed]

155. Becker, J.M.; Henry, L.K.; Jiang, W.; Koltin, Y. Reduced virulence of *Candida albicans* mutants affected in multidrug resistance. *Infect. Immun.* **1995**, *63*, 4515–4518. [PubMed]

156. Yamada-Okabe, T.; Yamada-Okabe, H. Characterization of the *CaNAG3*, *CaNAG4*, and *CaNAG6* genes of the pathogenic fungus *Candida albicans*: Possible involvement of these genes in the susceptibilities of cytotoxic agents. *FEMS Microbiol. Lett.* **2002**, *212*, 15–21. [CrossRef] [PubMed]

157. Cavalheiro, M.; Teixeira, M.C. *Candida* biofilms: Threats, challenges, and promising strategies. *Front. Med.* **2018**, *5*, 28. [CrossRef] [PubMed]

158. Rajendran, R.; Mowat, E.; McCulloch, E.; Lappin, D.F.; Jones, B.; Lang, S.; Majithiya, J.B.; Warn, P.; Williams, C.; Ramage, G. Azole resistance of *Aspergillus fumigatus* biofilms is partly associated with efflux pump activity. *Antimicrob. Agents Chemother.* **2011**, *55*, 2092–2097. [CrossRef] [PubMed]

159. Ferrari, S.; Sanguinetti, M.; Torelli, R.; Posteraro, B.; Sanglard, D. Contribution of *CgPDR1*-regulated genes in enhanced virulence of azole-resistant *Candida glabrata*. *PLoS ONE* **2011**, *6*, e17589. [CrossRef] [PubMed]

160. Vale-Silva, L.A.; Moeckli, B.; Torelli, R.; Posteraro, B.; Sanglard, D. Upregulation of the adhesin gene *EPA1* mediated by *PDR1* in *Candida glabrata* leads to enhanced host colonization. *mSphere* **2016**, *1*, 1–16. [CrossRef] [PubMed]

161. Monteiro, P.T.; Pais, P.; Costa, C.; Manna, S.; Sa-Correia, I.; Teixeira, M.C. The PathoYeastract database: An information system for the analysis of gene and genomic transcription regulation in pathogenic yeasts. *Nucleic Acids Res.* **2017**, *45*, D597–D603. [CrossRef] [PubMed]

162. Alonso-Monge, R.; Román, E.; Arana, D.M.; Prieto, D.; Urrialde, V.; Nombela, C.; Pla, J. The Sko1 protein represses the yeast-to-hypha transition and regulates the oxidative stress response in *Candida albicans*. *Fungal Genet. Biol.* **2010**, *47*, 587–601. [CrossRef] [PubMed]

163. Zhang, X.; De Micheli, M.; Coleman, S.T.; Sanglard, D.; Moye-Rowley, W.S. Analysis of the oxidative stress regulation of the *Candida albicans* transcription factor, Cap1p. *Mol. Microbiol.* **2000**, *36*, 618–629. [CrossRef] [PubMed]

164. Davis, D.; Wilson, R.B.; Mitchell, A.P. *RIM101*-dependent and-independent pathways govern pH responses in *Candida albicans*. *Mol. Cell. Biol.* **2000**, *20*, 971–978. [CrossRef] [PubMed]

165. Davis, D.; Edwars John, J.; Mitchell, A.P.; Ibrahim, A.S. *Candida albicans RIM101* pH response pathway is required for host-pathogen interactions. *Infect. Immun.* **2000**, *68*, 5953–5959. [CrossRef] [PubMed]

166. Nicholls, S.; Leach, M.D.; Priest, C.L.; Brown, A.J.P. Role of the heat shock transcription factor, Hsf1, in a major fungal pathogen that is obligately associated with warm-blooded animals. *Mol. Microbiol.* **2009**, *74*, 844–861. [CrossRef] [PubMed]

167. Leach, M.D.; Cowen, L.E. to sense or die: mechanisms of temperature sensing in fungal pathogens. *Curr. Fungal Infect. Rep.* **2014**, *8*, 185–191. [CrossRef]

168. Bruno, V.M.; Kalachikov, S.; Subaran, R.; Nobile, C.J.; Kyratsous, C.; Mitchell, A.P. Control of the *C. albicans* cell wall damage response by transcriptional regulator Cas5. *PLoS Pathog.* **2006**, *2*, e21. [CrossRef] [PubMed]

169. Vasicek, E.M.; Berkow, E.L.; Bruno, V.M.; Mitchell, A.P.; Wiederhold, N.P.; Barker, K.S.; Rogers, P.D. Disruption of the transcriptional regulator Cas5 results in enhanced killing of *Candida albicans* by Fluconazole. *Antimicrob. Agents Chemother.* **2014**, *58*, 6807–6818. [CrossRef] [PubMed]

170. Nobile, C.J.; Fox, E.P.; Nett, J.E.; Sorrells, T.R.; Mitrovich, Q.M.; Hernday, A.D.; Tuch, B.B.; Andes, D.R.; Johnson, A.D. A Recently evolved transcriptional network controls biofilm development in *Candida albicans*. *Cell* **2013**, *148*, 126–138. [CrossRef] [PubMed]

171. Mukherjee, P.K.; Chandra, J.; Kuhn, D.M.; Ghannoum, M.A. Mechanism of fluconazole resistance in *Candida albicans* biofilms: Phase-specific role of efflux pumps and membrane sterols. *Infect. Immun.* **2003**, *71*, 4333–4340. [CrossRef] [PubMed]

172. Lepak, A.; Nett, J.; Lincoln, L.; Marchillo, K.; Andes, D. Time course of microbiologic outcome and gene expression in *Candida albicans* during and following in vitro and in vivo exposure to fluconazole. *Antimicrob. Agents Chemother.* **2006**, *50*, 1311–1319. [CrossRef] [PubMed]

173. Fox, E.P.; Bui, C.K.; Nett, J.E.; Hartooni, N.; Mui, M.C.; Andes, D.R.; Nobile, C.J.; Johnson, A.D. An expanded regulatory network temporally controls *Candida albicans* biofilm formation. *Mol. Microbiol.* **2015**, *96*, 1226–1239. [CrossRef] [PubMed]

174. Nobile, C.J.; Mitchell, A.P. Genetics and genomics of *Candida albicans* biofilm formation. *Cell. Microbiol.* **2006**, *8*, 1382–1391. [CrossRef] [PubMed]

175. Lane, S.; Birse, C.; Zhou, S.; Matson, R.; Liu, H. DNA array studies demonstrate convergent regulation of virulence factors by Cph1, Cph2, and Efg1 in *Candida albicans*. *J. Biol. Chem.* **2001**, *276*, 48988–48996. [CrossRef] [PubMed]

176. Ha, K.C.; White, T.C. Effects of azole antifungal drugs on the transition from yeast cells to hyphae in susceptible and resistant isolates of the pathogenic yeast *Candida albicans*. *Antimicrob. Agents Chemother.* **1999**, *43*, 763–768. [PubMed]

177. Mogavero, S.; Tavanti, A.; Senesi, S.; Rogers, P.D.; Morschhäuser, J. Differential requirement of the transcription factor Mcm1 for activation of the *Candida albicans* multidrug efflux pump *MDR1* by its regulators Mrr1 and Cap1. *Antimicrob. Agents Chemother.* **2011**, *55*, 2061–2066. [CrossRef] [PubMed]

178. Znaidi, S.; Nesseir, A.; Chauvel, M.; Rossignol, T.; D'Enfert, C. A comprehensive functional portrait of two heat shock factor-type transcriptional regulators involved in *Candida albicans* morphogenesis and virulence. *PLoS Pathog.* **2013**, *9*, e1003519. [CrossRef] [PubMed]

179. Silver, P.M.; Oliver, B.G.; White, T.C. Role of *Candida albicans* Transcription factor Upc2p in Drug resistance and sterol metabolism. *Eukaryot. Cell* **2004**, *3*, 1391–1397. [CrossRef] [PubMed]

180. Lavoie, H.; Hogues, H.; Mallick, J.; Sellam, A.; Nantel, A.; Whiteway, M. Evolutionary tinkering with conserved components of a transcriptional regulatory network. *PLoS Biol.* **2010**, *8*, e1000329. [CrossRef] [PubMed]

181. Dogra, S.; Krishnamurthy, S.; Gupta, V.; Dixit, B.L.; Gupta, C.M.; Sanglard, D.; Prasad, R. Asymmetric distribution of phosphatidylethanolamine in *C. albicans*: Possible mediation by *CDR1*, a multidrug transporter belonging to ATP binding cassette (ABC) superfamily. *Yeast* **1999**, *15*, 111–121. [CrossRef]

182. Noble, J.A.; Tsai, H.-F.; Suffis, S.D.; Su, Q.; Myers, T.G.; Bennett, J.E. *STB5* is a negative regulator of azole resistance in *Candida glabrata*. *Antimicrob. Agents Chemother.* **2013**, *57*, 959–967. [CrossRef] [PubMed]

183. Roetzer, A.; Klopf, E.; Gratz, N.; Marcet-Houben, M.; Hiller, E.; Rupp, S.; Gabaldón, T.; Kovarik, P.; Schüller, C. Regulation of *Candida glabrata* oxidative stress resistance is adapted to host environment. *FEBS Lett.* **2011**, *585*, 319–327. [CrossRef] [PubMed]

184. Chen, K.H.; Miyazaki, T.; Tsai, H.F.; Bennett, J.E. The bZip transcription factor Cgap1p is involved in multidrug resistance and required for activation of multidrug transporter gene *CgFLR1* in *Candida glabrata*. *Gene* **2007**, *386*, 63–72. [CrossRef] [PubMed]

185. Merhej, J.; Thiebaut, A.; Blugeon, C.; Pouch, J.; Ali Chaouche, M.E.A.; Camadro, J.-M.; Le Crom, S.; Lelandais, G.; Devaux, F. A network of paralogous stress response transcription factors in the human pathogen *Candida glabrata*. *Front. Microbiol.* **2016**, *7*, 645. [CrossRef] [PubMed]

Concept Paper

Strengthening the One Health Agenda: The Role of Molecular Epidemiology in *Aspergillus* Threat Management

Eta E. Ashu [1] and Jianping Xu [1,2,*]

[1]   Department of Biology, McMaster University, 1280 Main St. W, Hamilton, Ontario, ON L8S 4K1, Canada;
      ashue@mcmaster.ca
[2]   Public Research Laboratory, Hainan Medical University, Haikou, Hainan 571199, China
*   Correspondence: jpxu@mcmaster.ca; Tel.: +1-905-525-9140 (ext. 27934); Fax: +1-905-522-6066

Received: 28 June 2018; Accepted: 16 July 2018; Published: 19 July 2018

Abstract: The United Nations' One Health initiative advocates the collaboration of multiple sectors within the global and local health authorities toward the goal of better public health management outcomes. The emerging global health threat posed by *Aspergillus* species is an example of a management challenge that would benefit from the One Health approach. In this paper, we explore the potential role of molecular epidemiology in *Aspergillus* threat management and strengthening of the One Health initiative. Effective management of *Aspergillus* at a public health level requires the development of rapid and accurate diagnostic tools to not only identify the infecting pathogen to species level, but also to the level of individual genotype, including drug susceptibility patterns. While a variety of molecular methods have been developed for *Aspergillus* diagnosis, their use at below-species level in clinical settings has been very limited, especially in resource-poor countries and regions. Here we provide a framework for *Aspergillus* threat management and describe how molecular epidemiology and experimental evolution methods could be used for predicting resistance through drug exposure. Our analyses highlight the need for standardization of loci and methods used for molecular diagnostics, and surveillance across *Aspergillus* species and geographic regions. Such standardization will enable comparisons at national and global levels and through the One Health approach, strengthen *Aspergillus* threat management efforts.

Keywords: molecular epidemiology; One Health; *Aspergillus fumigatus*; invasive fungal diseases; threat management

1. The Genus *Aspergillus*

Fungal infections affect over a billion people and cause approximately 1.5 million deaths each year worldwide [1]. Regrettably, due to increases in the number of at-risk populations, fungal infections are projected to rise [1,2]. It is estimated that death can be averted in over 80% of fungal disease patients through improved diagnostics, treatment surveillance, and effective antifungal therapies [1]. However, to achieve such success, an inter-disciplinary approach is needed. An emerging example of the inter-disciplinary approach is the One Health initiative. The World Health Organization defines One Health as 'an approach to designing and implementing programs, polices, legislation, and research in which multiple sectors communicate and work together to achieve better public health outcomes'.

Species in the ascomycete genus *Aspergillus* have emerged as key agents of the fungal infections around the world [3–19]. For example, *Aspergilli* are the leading cause of chronic severe and allergic fungal infections, and the second leading cause of acute invasive fungal infections [1]. The genus *Aspergillus* was first described at the end of the 18th century by a Catholic priest and botanist named Pier Antonio Micheli. Viewing the microscopic spore-bearing structure of *Aspergillus*, Micheli was reminded of a holy water sprinkler—an aspergillum [20–22]. Since then, the number of species in genus *Aspergillus* has grown to encompass eight subgenera and over 250 species [23,24]. Of these species, approximately 15% are of known clinical importance [25,26]. DNA sequence-based methods are revealing an increasing number of cryptic species of *Aspergillus* associated with human diseases [27–30]. For example, surveys carried out in the United States, Brazil, and Spain revealed the percentage of phylogenetically divergent lineages representing cryptic *Aspergilli* species among clinical samples to be between 11–19%, a percentage which is notably higher than those seen in other clinically important filamentous fungi, including those belonging to the orders Mucorales, Microascales, and Hypocreales [31–33]. This is particularly important given that, in addition to being pathogenic, up to 40% of these cryptic *Aspergilli* can be resistant to antifungal drugs [30,32]. Of greater importance is the fact that some of these cryptic *Aspergilli* are resistant to multiple antifungals which can exacerbate infections caused by these species. Indeed, fungal infections, including those caused by *Aspergilli*, have become a menace to global public health.

Aspergilli cause a wide range of infections, commonly referred to as aspergillosis. Allergic bronchopulmonary, chronic pulmonary, and invasive aspergillosis (IA) are the three most common types of *Aspergillus* infections. Allergic bronchopulmonary and chronic pulmonary aspergillosis results from immune hypersensitivity and scarring due to an *Aspergillus* respiratory tract infection. On the other hand, IA can affect a wider range of body organs belonging to the urinary, digestive, and nervous systems. A significant proportion of *Aspergillus* infections are asymptomatic. However, in patients with symptomatic infections, most symptoms are non-specific, and include low-grade fevers, generalized malaise, wheezing, headaches, and haemoptysis [34,35]. Approximately eight million people world-wide are estimated to have aspergillosis [1]. Invasive aspergillosis is the most lethal type of aspergillosis and is estimated to affect >300,000 people globally every year, with a mortality rate as high as 90% in at-risk populations [1,36]. Allergic bronchopulmonary and chronic pulmonary aspergillosis affect approximately 4.8 and 3 million people annually, respectively [1].

Although aspergillosis cases are predominantly sporadic, outbreaks are not uncommon. Specifically, there have been at least 75 documented aspergillosis outbreaks between January 1966 and December 2015 [37–45]. Interestingly, a recent study showed a non-construction-related outbreak that was associated with high airborne spore concentrations in hospital areas with low efficiency air filters [44]. These results highlight the threat posed if environmental spore concentrations reach critical levels within community or home settings. Multiple *Aspergillus* species including *A. fumigatus*, *A. flavus*, *A. terreus*, *A. niger*, *A. glaucus*, *A. oryzae*, and *A. ustus* are known to have caused outbreaks. Among these, *A. fumigatus*, and *A. flavus* are the most frequently identified species [38], and are responsible for approximately 87% of all aspergillosis case reports [26].

Changes in the antifungal susceptibility patterns of these two species have further increased the threat posed by *Aspergilli*. Since its emergence in 1997, resistance to triazole in *A. fumigatus* has steadily increased and is a current global health menace [46,47]. Furthermore, in *A. fumigatus*, there are emerging reports of increased resistance to polyenes such as amphotericin B (AMB), an antifungal to which very little resistance has been reported thus far [48,49]. For example, a recent study carried out in Brazil showed that 27% of a clinical sample of *A. fumigatus* isolates was resistant to AMB (minimum inhibitory concentration (MIC) $\geq$ 2 mg/L) [48]. Similarly, our group also very recently found that 96% of a combined environmental and clinical sample from Hamilton, Canada was resistant to amphotericin. The recent emergence of voriconazole (VRC) resistance in *A. flavus* will likely cause significant problems in the management of aspergillosis caused by *A. flavus*. Triazoles, especially VRC, are first-line drugs used in the treatment of aspergillosis [50–53]. Although not yet reported, multi-drug resistant aspergillosis outbreaks similar to those caused by *Candida auris* and *Acinetobacter baumannii* will likely emerge in the near future [54–58].

With the increasing number of clinically important *Aspergillus* species and the changing antifungal susceptibility patterns of key *Aspergilli* such *A. fumigatus*, and *A. flavus,* there is a pressing need to develop novel and effective *Aspergillus* threat management strategies. Below, we propose a framework that can be used in *Aspergillus* threat management. When put in context, this framework can also be used in the management of *Candida* and other clinically relevant fungi. We encourage essential stakeholders to engage in discussions aimed at *Aspergillus* threat management.

2. Molecular Epidemiology in *Aspergillus* Threat Management

A variety of host, pathogen, host-pathogen interaction, and environmental factors have been identified as contributors to the increased threat caused by *Aspergilli*. Of interest among pathogen-related factors is the recent global rise in resistance to antifungal drugs. Triazole resistance in *A. fumigatus* has now been reported in every continent but Antarctica [47]. Generally speaking, pathogen threat management has three interdependent components: preparedness, response, and prevention. In Figure 1, we suggest a non-exclusive framework that could be used in the management of *Aspergillus* threats, including those caused by resistant strains. This review however only focuses on the molecular epidemiology components of preparedness and prevention; specifically, on molecular diagnostics and surveillance (Figure 1).

2.1. The Usefulness of Molecular Epidemiology in Aspergillus Surveillance

In epidemiology, surveillance is defined as the collection and analysis of data necessary to develop, implement, and evaluate preventative health measures. Over the last two decades, molecular epidemiology has emerged as a very important tool in the surveillance of diverse human pathogens [59,60]. This burgeoning branch of epidemiology merges traditional epidemiology and molecular biology in order to better characterize virulence, pathogen transmission patterns, and outbreak incidence. In molecular epidemiology, marker genes are used to elucidate the genotypes and the relationship between strains and populations. In addition, some of these marker genes are becoming indispensable for understanding virulence determinants and the distribution of aspergillosis.

Over the years, a wide range of molecular markers has been used to study the molecular epidemiology of *Aspergilli*. These markers include multilocus sequences, microsatellites, PCR-restriction fragment length polymorphisms, Southern hybridization of restriction enzyme-digested DNA, randomly amplified polymorphic DNA, and mating type genes [61,62]. For instance, using microsatellite markers, Guinea and colleagues investigated an aspergillosis outbreak in a major heart surgery unit of a hospital in Spain and showed that such markers were a valuable tool in IA outbreak source investigation [41]. It is however, important to note that aspergillosis outbreaks most often do not have a single source and can consist of a series of unrelated events. As such, pinpointing the source of aspergillosis outbreaks can be difficult [63]. In contrast, molecular markers have been used with more success in determining the sources of infections in non-outbreak cases. For example, a recent study highlighted that the home environment can be an important source of infection for isolated cases of triazole-resistant *A. fumigatus* [64].

In addition to its value in infection source investigation, molecular epidemiology can be used to track *Aspergilli* transmission patterns. For example, using microsatellite markers, a recent global study showed that resistant populations of *A. fumigatus* are significantly differentiated geographically [65]. This result suggests that it may be possible to track triazole resistant *A. fumigatus* strains across national and regional borders. However, significant caution should be applied here. Compared to the large global population of *Aspergilli*, relatively few isolates and genotypes have been analyzed to date, and our current understanding of the molecular variation between and within these populations may not be representative of the true global diversity. In regard to tracking *Aspergilli*, geographic sub-structuring can vary by country and region, hence tracking transmission patterns of clinically relevant *Aspergilli* within or between certain countries might prove to be easier than in others [39,66]. For example, little to no geographic population structuring has been reported in *A. fumigatus* samples from India and Netherlands, whereas Cameroonian *A. fumigatus* samples show significant evidence for geographic sub-structuring [66]. As a result, tracking clinically relevant *A. fumigatus* strains would be a more feasible task in Cameroon than it would be in India or Netherlands.

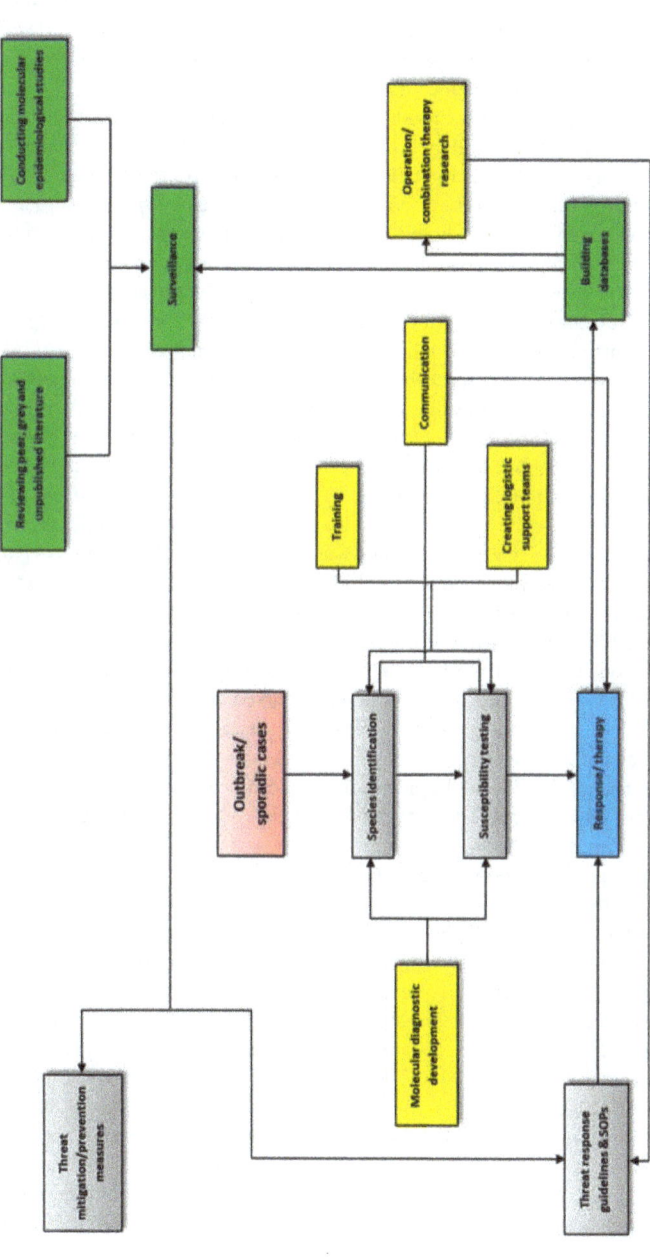

Figure 1. Framework for the management of *Aspergillus* outbreaks and critically important sporadic cases. Activities pertaining to threat preparedness are highlighted in yellow, while those pertaining to prevention are in green. Training within the context of this review refers to training laboratory technicians to perform rapid and accurate diagnosis of *Aspergilli*, including resistant strains and cryptic species. It also entails cross-training other laboratory technicians to carry out the designated emergency technician's routine duties. Communication here covers having a standardized plan to act as quickly as possible in relaying information on species and susceptibility diagnosis to the appropriate people, in order to ensure an adequate response and the safety of groups or persons at risk. Logistic support teams will be responsible for ensuring diagnostic supplies, admission documentation, travel arrangements, quality control, and all components essential for rapid and accurate diagnosis are in place in the event of an *Aspergillus* threat. SOP: standard operating procedure.

Aside from transmission pattern tracking and outbreak incidence investigation, molecular markers can also be used in virulence and antifungal drug-resistance characterization. Mating type loci genotyping has been successfully used to characterize the degree of disease severity caused by *A. fumigatus* strains [67–69], making it possible for these markers to be used in mapping the distribution of hyper-virulent strains. Similarly, certain mutations in the *cyp51A* gene are known to cause triazole drug resistance. Being able to map the distribution of hyper-virulent and drug resistant *Aspergilli* is essential for emergency preparedness and infection control efforts.

With reference to the wide range of markers used in genotyping *Aspergilli*, whilst molecular epidemiology holds promise for *Aspergilli* surveillance, there are still two pertinent issues that need to be addressed in order for the surveillance data to be fully implemented at the global scale. Firstly, we need a consensus on which markers should be used for the surveillance of *Aspergilli* of public health importance. Secondly, we need to establish and curate a reliable genotype database on the consensus gene markers to which similar information on new isolates can be added and compared. At present, there is a multilocus sequence type database for *A. fumigatus* based on seven loci (https: //pubmlst.org/afumigatus/) [70]. However, such a database is not available for other *Aspergillus* species. In addition, while the current multilocus sequence typing (MLST) method for *A. fumigatus* is highly reproducible and can distinguish strains at the genus and species levels, it has a relatively low discriminatory power among strains within the same species [70]. Instead, due to their high polymorphism and greater discrimination power, a group of nine microsatellite markers is more commonly used for genotyping *A. fumigatus*, making the MLST database of *A. fumigatus* of limited use in epidemiological and population genetic investigations [71]. However, a shared database for strain genotypes based on the microsatellite markers is not available. In the short/medium term, we recommend that future work should use the seven consensus MLST loci for genotyping all *Aspergilli* strains, and for *A. fumigatus*, the additional nine microsatellite loci should be used for genotyping. Furthermore, a publicly available database should be set up to include both types of genetic data. Ultimately, with increasing availability and affordability of next-generation sequencing, whole-genome sequencing (WGS) should be considered for epidemiological monitoring for the long term. Data from WGS will not only be highly discriminatory and commonly archived, but also can help identify genetic polymorphisms associated with virulence and antifungal drug susceptibility. Such data will significantly strengthen national and global *Aspergillus* threat management efforts.

Despite the rather successful use of molecular epidemiology in characterizing hyper-virulence, pathogen transmission patterns, and outbreak incidences of *Aspergilli*, very little has been done in leveraging molecular epidemiological methods in predicting drug resistance in endemic wild-type genotypes. For example, wild-type genotypes that are strongly associated with the emergence of resistance have been shown in the human immunodeficiency virus type 1 (HIV-1) [72]. This is particularly important, as infections with wild-type genotypes with higher propensities to become drug resistant could result in inappropriate antifungal therapy, and ultimately lead to treatment failure. In Figure 2, we suggest an in-vitro experimental evolutionary model that can be used in predicting drug resistance likelihood through drug exposure. The above experiment is aimed at measuring two primary outcomes: the likelihood of becoming resistant and the time it takes to attain resistance breakpoint or epidemiological cutoff (resistance time). Data from such experiments can be used to develop statistical learning models that can predict the likelihood of resistance and resistance time in a matter of minutes on a laptop. For example, supposing that acquiring resistance is through a stochastic process related to spontaneous mutation(s), then the likelihood of a genotype becoming resistant can be obtained using the following function $L(G/O) = P(O/G)$, where O is the observed outcome, G is the parameter set (genotype) that define the stochastic process, L is likelihood, and P is probability. Information on a wild-type genotype's propensity to become resistant and resistance time can be very useful in determining a treatment course and formulating disease control strategies. Similarly, given the role of sexual reproduction in the emergence of genotypes of clinical importance [63], developing tools capable of predicting highly fit and super mating genotypes is critical, as such genotypes can easily spread resistance genes through sexual reproduction and rapidly expanding their distribution across geographic areas. Indeed, a highly fit multi-triazole resistant *A. fumigatus* genotype which very likely acquired resistance through sexual reproduction has expanded across thousands of kilometers in India [73,74].

In conclusion, from a practical perspective, researchers involved in *Aspergillus* surveillance should monitor the following: (i) genetic structure of the local, regional, national, and global populations of human pathogenic *Aspergillus*; (ii) the distribution and overall prevalence of hyper-virulent strains; (iii) the distribution and overall prevalence of antifungal drug resistant strains including those with a higher propensity to become resistant; and (iv) the transmission patterns of the latter two categories of genotypes. Furthermore, given current increases in non-*fumigatus* aspergillosis, significant attention should be paid to monitoring non-*fumigatus Aspergilli*, especially cryptic species.

2.2. The Usefulness of Molecular Epidemiology in Threat Preparedness

Recent increases in globalization have led to more frequent movement of goods and people than ever before in human history. The constant movement of people and goods across geographic scales has significant implications for the spread of clinically important pathogens like *Aspergilli*. Indeed, both anthropogenic and non-anthropogenic activities have impacted the spread of clinically important *Aspergilli* across geographic boarders [65,76]. Such impacts highlight the need for preparedness even in countries with low aspergillosis incidences. There is an age-old adage that says "Luck favors the prepared mind". Threat preparedness in the context of this review implies being able to effectively anticipate and take the right steps to manage threats caused by *Aspergilli*. Among other things, this entails being able to accurately and rapidly diagnose *Aspergilli*, especially drug-resistant *Aspergilli*. Thus far, multiple diagnostic methods—including electrospray ionization and matrix-assisted laser desorption ionization mass spectrometry, nucleic acid sequence-based amplification, polymerase chain reactions (real-time, multiplex and nested), microsphere-based Luminex, loop-mediated isothermal amplification, and enzyme linked immunosorbent assays—have been used to identify *Aspergilli* in a wide range of specimens including whole blood, serum, plasma, bronchoalveolar lavages, and exhale breath condensate [77–81]. These methods, including their respective advantages and disadvantages, have been extensively reviewed and are not discussed in detail here. Instead, our focus is on providing suggestions necessary for leveraging molecular diagnostics in preparedness against the threat caused by *Aspergilli*.

Specifically, we would like to highlight two critical issues that need to be addressed by many countries in their efforts to manage the threat caused by *Aspergilli*. Firstly, we note that more research needs to be done in developing rapid diagnostic molecular markers for clinically important non-*fumigatus Aspergilli*, including the divergent lineages/cryptic species. This is particularly important because although some of these species represent only a small proportion of clinically diagnosed *Aspergilli*, they form significant proportions of drug-resistant *Aspergilli*. For example, although *A. lentulus* accounted for only ~3% (3/86) of a set of 86 *Aspergillus* isolates obtained from Italy and Netherlands, they represented approximately 21% (3/14) of all ($\geq$ 2 mg/L) voriconazole-resistant strains [82]. Another important example to note is *A. calidoustus*, an emerging pathogen in lung transplants which can also colonize water distribution systems, including those in health care settings [83,84]. Thus, the need for accurate and rapid diagnosis of non-*fumigatus Aspergilli* and cryptic species is becoming increasingly important.

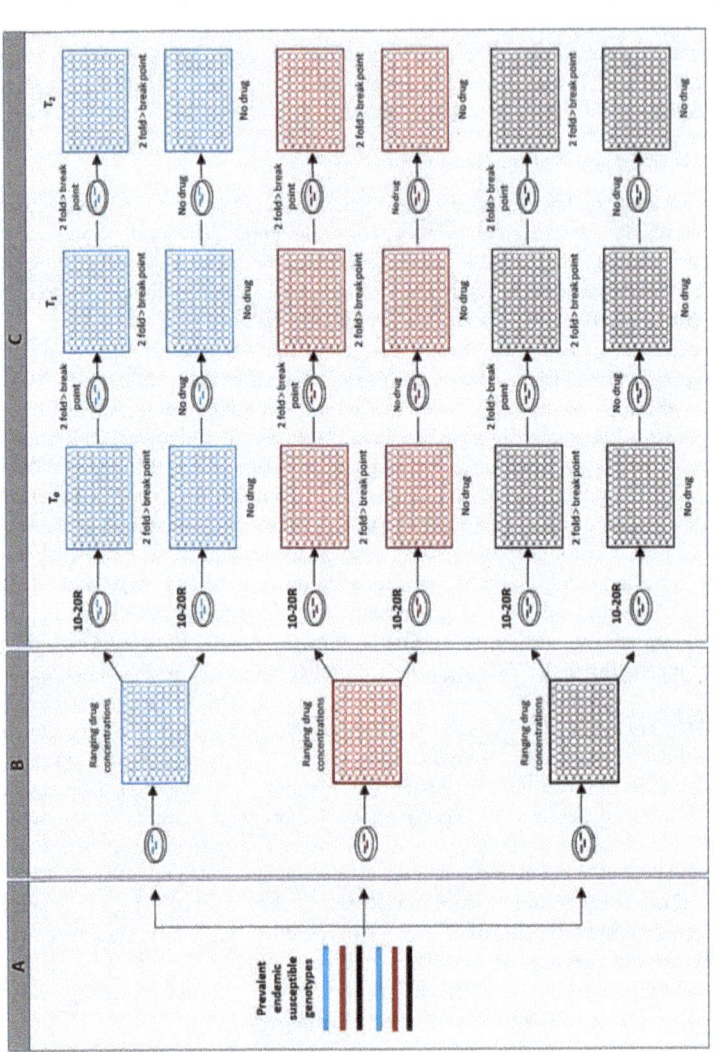

Figure 2. Experimental design to predict resistance through drug exposure in susceptible genotypes. The procedures depicted in panels A and B entail identifying prevalent endemic susceptible genotypes and determining their minimum inhibitory concentrations (MICs) to specific antifungal drugs. All genotypes used for the evolutionary experiment in panel C should have similar starting MICs, preferably very low. The evolutionary experiment consists of continuously exposing 10–20 replicates (R) of the selected genotypes to an antifungal drug as described in the Clinical & Laboratory Standards Institute (CLSI) protocol while including a no-drug control group [75]. At the end of 48 h, MICs are recorded, and new petri dishes are inoculated with microtiter plate content from the previous round of selection. The steps are repeated until genotypes reach resistance break point or epidemiological cutoff values.

Secondly, we highlight the need for standardization of molecular diagnostic procedures and markers at country and regional levels. Generally speaking, of all molecular diagnostic platforms, PCR platforms seem to show the best potential as they are able to identify *Aspergilli* to the species level while also being able to help identify antifungal susceptibility patterns. Thus far, plasma samples have resulted in the highest sensitivity (91%), while whole blood produced the highest specificity (96%) [85]. The highest diagnostic sensitivity and specificity in bronchoalveolar lavages samples thus far were 91% and 92%, respectively [86]. In cerebrospinal fluid, an *Aspergillus*-specific nested PCR assay showed sensitivity and specificity values of 100% and 93%, respectively [87]. Despite the great potential of PCR platforms, the specificity and sensitivity of PCR can be notably affected by two key factors: (i) method of DNA extraction; and (ii) specimen type. It was recently shown that factors including bead beating, white cell lysis, elution, and specimen volume can notably affect the quality and quantity of extracted DNA, and consequently PCR sensitivity [88]. In spite of the significant progress made by the European *Aspergillus* PCR initiative in quelling variation associated with DNA extraction from whole blood [89], the blood fraction best suited for PCR assays is still disputed. Furthermore, PCR performance is still to be evaluated in a wide range of patients, with other types of immunosuppressive conditions as most specimens used for PCR assays this far were performed on samples from hematology-oncology patients. Similarly, the differences in bronchoalveolar lavage procedures between patients and medical centers can affect PCR interpretation [90]. Indeed, there is room for improvements in all molecular diagnostic platforms.

Regardless of these shortfalls, molecular diagnostic platforms such as PCR can still be leveraged in the management of *Aspergilli* threats in two key ways. Firstly, molecular diagnostic platforms can be used in combinations to obtain optimal results. For example, a recent study showed that combining a lateral flow device with PCR yielded 100% diagnostic sensitivity and specificity [91]. Similarly, an earlier diagnosis and a lower incidence of IA were associated with a combination of galactomannan (GM) and PCR-based *Aspergillus* detection [92]. Secondly, PCR and similar molecular diagnostic platforms can be preemptively used in the surveillance of high risk patients in order to provide rapid usable results if needed. For instance, an improved 30-day survival rate was observed in a group that received PCR-guided prophylactic treatment compared to that for those treated on the basis of symptoms alone [93]. Similarly, combined surveillance of serum GM and PCR was shown to help decrease the incidence of IA in at-risk populations [92]. Surveillance of at-risk populations, health care facilities and adjacent environmental areas using molecular diagnostic platforms depicts how interdependent components of pathogen threat management can be. However, combining different platforms for diagnosis, prevention, and treatment requires standardization of procedures across different immunosuppressed populations and the mastery of diverse mycological procedures. Similarly, routine molecular diagnostic surveillance in high risk populations will require a significant amount of human resources. In the context of *Aspergillus* threat management, more specifically preparedness, standard operating procedures for the rapid and accurate diagnosis of both common and uncommon clinically important *Aspergilli* still need to be developed and adopted in many countries. Furthermore, the necessary human resources, cost effectiveness, and research capacity required to achieve the latter should be considered in most clinical microbiology laboratories.

Within individual *Aspergillus* species, having a good understanding of the molecular epidemiology is vital in formulating integrative molecular diagnostic plans. For instance, given that almost all multi-triazole resistant *A. fumigatus* strains in India belonged to a single microsatellite genotype (14/20/9/31/9/10/8/10/28) [74], screening for this genotype in addition to the common *A. fumigatus* species-level gene markers discussed here could diagnose an *A. fumigatus* isolate to a below-species level, including whether it is multi-triazole resistant. Such a plan could also be used in other geographic areas if one or a few genotypes dominate the drug-resistant population.

3. Conclusions

In this paper, we highlight the importance of surveillance and molecular diagnostics in *Aspergillus* threat preparedness and prevention. Given how interconnected that the world has become, standardization of loci and methods used for molecular diagnostics and surveillance is critically important for managing the current global *Aspergillus* threat. Although not discussed in detail here, combination therapy recommendations and vaccine research are two other key *Aspergillus* threat preparedness and prevention components worth mentioning.

Effectively managing the current threat caused by *Aspergilli* will require the significant incorporation of the molecular epidemiology component into *Aspergillus* threat preparedness and prevention. From a public health policy perspective, incorporation of genetic information into *Aspergillus* threat management is still a relatively new concept. For example, the European Organization for Research and Treatment of Cancer/Invasive Fungal Infections Cooperative Group and the National Institute of Allergy and Infectious Diseases Mycoses Study Group (EORTC/MSG) only recently suggested the inclusion of molecular diagnostics for the case definition of IA [94]. However, the recommended diagnosis can only identify the infecting pathogen to species level. A recent study identified that different genetic populations within *A. fumigatus* have different rates of triazole resistance [65]. Thus, identifying infecting pathogens to the genotype level will have significant treatment value. Furthermore, molecular markers targeting drug resistance mutations are being developed. The adoption by clinical microbiology labs around the globe of these fine-scale molecular methods will lead to better-targeted treatments and improved patient outcome at the population level.

Author Contributions: J.X. conceived the paper. E.E.A. produced the initial draft. Both E.E.A. and J.X. contributed to finalizing the text.

Funding: This research was funded by the Natural Science and Engineering Research Council (NSERC) of Canada (grant number CRDPJ474638-14) and by Hainan Medical University's Visiting Professorship Program.

Acknowledgments: We thank Heather Yoell for proofreading the manuscript.

Conflicts of Interest: The authors declare that they have no significant competing financial, professional or personal interests.

References

1. Bongomin, F.; Gago, S.; Oladele, R.O.; Denning, D.W. Global and multi-national prevalence of fungal diseases—Estimate precision. *J. Fungi* **2017**, *3*, 57. [CrossRef] [PubMed]
2. Vallabhaneni, S.; Mody, R.K.; Walker, T.; Chiller, T. The global burden of fungal diseases. *Infect. Dis. Clin.* **2016**, *30*, 1–11. [CrossRef] [PubMed]
3. Pegorie, M.; Denning, D.W.; Welfare, W. Estimating the burden of invasive and serious fungal disease in the United Kingdom. *J. Infect.* **2017**, *74*, 60–71. [CrossRef] [PubMed]
4. Gamaletsou, M.N.; Drogari-Apiranthitou, M.; Denning, D.W.; Sipsas, N.V. An estimate of the burden of serious fungal diseases in Greece. *Eur. J. Clin. Microbiol. Infect. Dis.* **2016**, *35*, 1115–1120. [CrossRef] [PubMed]
5. Gangneux, J.-P.; Bougnoux, M.-E.; Hennequin, C.; Godet, C.; Chandenier, J.; Denning, D.W.; Dupont, B. LIFE program, the Société française de mycologie médicale SFMM-study group. An estimation of burden of serious fungal infections in France. *J. Mycol. Med.* **2016**, *26*, 385–390. [CrossRef] [PubMed]
6. Oladele, R.O.; Denning, D.W. Burden of serious fungal infection in Nigeria. *West Afr. J. Med.* **2014**, *33*, 107–114. [PubMed]
7. Ben, R.; Denning, D.W. Estimating the burden of fungal diseases in Israel. *Isr. Med. Assoc. J. IMAJ* **2015**, *17*, 374–379. [PubMed]
8. Taj-Aldeen, S.J.; Chandra, P.; Denning, D.W. Burden of fungal infections in Qatar. *Mycoses* **2015**, *58* (Suppl. 5), 51–57. [CrossRef] [PubMed]
9. Lagrou, K.; Maertens, J.; Van Even, E.; Denning, D.W. Burden of serious fungal infections in Belgium. *Mycoses* **2015**, *58* (Suppl. 5), 1–5. [CrossRef] [PubMed]
10. Osmanov, A.; Denning, D.W. Burden of serious fungal infections in Ukraine. *Mycoses* **2015**, *58* (Suppl. 5), 94–100. [CrossRef] [PubMed]

11. Mortensen, K.L.; Denning, D.W.; Arendrup, M.C. The burden of fungal disease in Denmark. *Mycoses* **2015**, *58* (Suppl. 5), 15–21. [CrossRef] [PubMed]

12. Khwakhali, U.S.; Denning, D.W. Burden of serious fungal infections in Nepal. *Mycoses* **2015**, *58* (Suppl. 5), 45–50. [CrossRef] [PubMed]

13. Badiane, A.S.; Ndiaye, D.; Denning, D.W. Burden of fungal infections in Senegal. *Mycoses* **2015**, *58* (Suppl. 5), 63–69. [CrossRef] [PubMed]

14. Klimko, N.; Kozlova, Y.; Khostelidi, S.; Shadrivova, O.; Borzova, Y.; Burygina, E.; Vasilieva, N.; Denning, D.W. The burden of serious fungal diseases in Russia. *Mycoses* **2015**, *58* (Suppl. 5), 58–62. [CrossRef] [PubMed]

15. Faini, D.; Maokola, W.; Furrer, H.; Hatz, C.; Battegay, M.; Tanner, M.; Denning, D.W.; Letang, E. Burden of serious fungal infections in Tanzania. *Mycoses* **2015**, *58* (Suppl. 5), 70–79. [CrossRef] [PubMed]

16. Sinkó, J.; Sulyok, M.; Denning, D.W. Burden of serious fungal diseases in Hungary. *Mycoses* **2015**, *58* (Suppl. 5), 29–33. [CrossRef] [PubMed]

17. Mandengue, C.E.; Denning, D.W. The burden of serious fungal infections in Cameroon. *J. Fungi* **2018**, *4*, 44. [CrossRef] [PubMed]

18. Guto, J.A.; Bii, C.C.; Denning, D.W. Estimated burden of fungal infections in Kenya. *J. Infect. Dev. Ctries.* **2016**, *10*, 777–784. [CrossRef] [PubMed]

19. Corzo-León, D.E.; Armstrong-James, D.; Denning, D.W. Burden of serious fungal infections in Mexico. *Mycoses* **2015**, *58* (Suppl. 5), 34–44. [CrossRef] [PubMed]

20. George Agrios. *Plant Pathology*, 5th ed.; Academic Press: New York, NY, USA, 2005; ISBN 978-0-12-044565-3.

21. Bennett, J.W. An overview of the genus *Aspergillus*. In *Aspergillus: Molecular Biology and Genomics*; Katsuya, G., Masayuki, M., Eds.; Horizon Scientific Press: Norfolk, UK, 2010; pp. 1–17.

22. Ashu, E.; Forsythe, A.; Vogan, A.; Xu, J. Filamentous Fungi in Fermented Foods. In *Fermented Foods, Part I: Biochemistry and Biotechnology*; Didier, M., Ramesh, R., Eds.; CRC Press: Boca Raton, FL, USA, 2016; pp. 60–90, ISBN 978-1-4987-4081-4.

23. Dyer, P.S.; O'Gorman, C.M. A fungal sexual revolution: *Aspergillus* and *Penicillium* show the way. *Curr. Opin. Microbiol.* **2011**, *14*, 649–654. [CrossRef] [PubMed]

24. Geiser, D.M.; Klich, M.A.; Frisvad, J.C.; Peterson, S.W.; Varga, J.; Samson, R.A. The current status of species recognition and identification in *Aspergillus*. *Stud. Mycol.* **2007**, *59*, 1–10. [CrossRef] [PubMed]

25. Sugui, J.A.; Kwon-Chung, K.J.; Juvvadi, P.R.; Latgé, J.-P.; Steinbach, W.J. *Aspergillus fumigatus* and related species. *Cold Spring Harb. Perspect. Med.* **2015**, *5*, a019786. [CrossRef] [PubMed]

26. Guarro, J.; Xavier, M.O.; Severo, L.C. Differences and similarities amongst pathogenic *Aspergillus* species. In *Aspergillosis: From Diagnosis to Prevention*; Alessandro, C.P., Ed.; Springer: Dordrecht, The Netherlands, 2009; pp. 7–32. ISBN 978-90-481-2407-7.

27. Varga, J.; Houbraken, J.; Van Der Lee, H.A.L.; Verweij, P.E.; Samson, R.A. *Aspergillus calidoustus* sp. nov., causative agent of human infections previously assigned to *Aspergillus ustus*. *Eukaryot. Cell* **2008**, *7*, 630–638. [CrossRef] [PubMed]

28. Balajee, S.A.; Gribskov, J.L.; Hanley, E.; Nickle, D.; Marr, K.A. *Aspergillus lentulus* sp. nov., a new sibling species of *A. fumigatus*. *Eukaryot. Cell* **2005**, *4*, 625–632. [CrossRef] [PubMed]

29. Gautier, M.; Normand, A.-C.; Ranque, S. Previously unknown species of *Aspergillus*. *Clin. Microbiol. Infect.* **2016**, *22*, 662–669. [CrossRef] [PubMed]

30. Alastruey-Izquierdo, A.; Alcazar-Fuoli, L.; Cuenca-Estrella, M. Antifungal susceptibility profile of cryptic species of *Aspergillus*. *Mycopathologia* **2014**, *178*, 427–433. [CrossRef] [PubMed]

31. Balajee, S.A.; Kano, R.; Baddley, J.W.; Moser, S.A.; Marr, K.A.; Alexander, B.D.; Andes, D.; Kontoyiannis, D.P.; Perrone, G.; Peterson, S.; et al. Molecular identification of *Aspergillus* species collected for the transplant-associated infection surveillance network. *J. Clin. Microbiol.* **2009**, *47*, 3138–3141. [CrossRef] [PubMed]

32. Alastruey-Izquierdo, A.; Mellado, E.; Peláez, T.; Pemán, J.; Zapico, S.; Alvarez, M.; Rodríguez-Tudela, J.L.; Cuenca-Estrella, M.; Group, F.S. Population-based survey of filamentous fungi and antifungal resistance in Spain (FILPOP Study). *Antimicrob. Agents Chemother.* **2013**, *57*, 3380–3387. [CrossRef] [PubMed]

33. Negri, C.E.; Gonçalves, S.S.; Xafranski, H.; Bergamasco, M.D.; Aquino, V.R.; Castro, P.T.O.; Colombo, A.L. Cryptic and rare *Aspergillus* species in Brazil: Prevalence in clinical samples and in vitro susceptibility to triazoles. *J. Clin. Microbiol.* **2014**, *52*, 3633–3640. [CrossRef] [PubMed]

34. Maturu, V.N.; Agarwal, R. Itraconazole in chronic pulmonary aspergillosis: In whom, for how long, and at what dose? *Lung India Off. Organ Indian Chest Soc.* **2015**, *32*, 309–312. [CrossRef]

35. Agarwal, R. Allergic bronchopulmonary aspergillosis. *Chest* **2009**, *135*, 805–826. [CrossRef] [PubMed]

36. Dagenais, T.R.T.; Keller, N.P. Pathogenesis of *Aspergillus fumigatus* in invasive aspergillosis. *Clin. Microbiol. Rev.* **2009**, *22*, 447–465. [CrossRef] [PubMed]

37. Vonberg, R.-P.; Gastmeier, P. Nosocomial aspergillosis in outbreak settings. *J. Hosp. Infect.* **2006**, *63*, 246–254. [CrossRef] [PubMed]

38. Weber, D.J.; Peppercorn, A.; Miller, M.B.; Sickbert-Benett, E.; Rutala, W.A. Preventing healthcare-associated *Aspergillus* infections: Review of recent CDC/HICPAC recommendations. *Med. Mycol.* **2009**, *47*, S199–S209. [CrossRef] [PubMed]

39. Balajee, S.A.; Tay, S.T.; Lasker, B.A.; Hurst, S.F.; Rooney, A.P. Characterization of a novel gene for strain typing reveals substructuring of *Aspergillus fumigatus* across North America. *Eukaryot. Cell* **2007**, *6*, 1392–1399. [CrossRef] [PubMed]

40. Chang, C.C.; Cheng, A.C.; Devitt, B.; Hughes, A.J.; Campbell, P.; Styles, K.; Low, J.; Athan, E. Successful control of an outbreak of invasive aspergillosis in a regional haematology unit during hospital construction works. *J. Hosp. Infect.* **2008**, *69*, 33–38. [CrossRef] [PubMed]

41. Guinea, J.; García de Viedma, D.; Peláez, T.; Escribano, P.; Muñoz, P.; Meis, J.F.; Klaassen, C.H.W.; Bouza, E. Molecular epidemiology of *Aspergillus fumigatus*: An in-depth genotypic analysis of isolates involved in an outbreak of invasive aspergillosis. *J. Clin. Microbiol.* **2011**, *49*, 3498–3503. [CrossRef] [PubMed]

42. Peláez, T.; Muñoz, P.; Guinea, J.; Valerio, M.; Giannella, M.; Klaassen, C.H.W.; Bouza, E. Outbreak of invasive aspergillosis after major heart surgery caused by spores in the air of the intensive care unit. *Clin. Infect. Dis.* **2012**, *54*, e24–e31. [CrossRef] [PubMed]

43. Pettit, A.C.; Kropski, J.A.; Castilho, J.L.; Schmitz, J.E.; Rauch, C.A.; Mobley, B.C.; Wang, X.J.; Spires, S.S.; Pugh, M.E. The index case for the fungal meningitis outbreak in the United States. *N. Engl. J. Med.* **2012**, *367*, 2119–2125. [CrossRef] [PubMed]

44. Vena, A.; Muñoz, P.; Pelaez, T.; Guinea, J.; Valerio, M.; Bouza, E. Non-construction related *Aspergillus* outbreak in non-hematological patients related to high concentrations of airborne spores in non-HEPA filtered areas. *Open Forum Infect. Dis.* **2015**, *2*, 352. [CrossRef]

45. Kabbani, D.; Goldraich, L.; Ross, H.; Rotstein, C.; Husain, S. Outbreak of invasive aspergillosis in heart transplant recipients: The role of screening computed tomography scans in asymptomatic patients and universal antifungal prophylaxis. *Transpl. Infect. Dis.* **2017**, *20*, e12808. [CrossRef] [PubMed]

46. Rivero-Menendez, O.; Alastruey-Izquierdo, A.; Mellado, E.; Cuenca-Estrella, M. Triazole resistance in *Aspergillus* spp.: A worldwide problem? *J. Fungi* **2016**, *2*, 21. [CrossRef] [PubMed]

47. Garcia-Rubio, R.; Cuenca-Estrella, M.; Mellado, E. Triazole resistance in *Aspergillus* species: An emerging problem. *Drugs* **2017**, *77*, 599–613. [CrossRef] [PubMed]

48. Reichert-Lima, F.; Lyra, L.; Pontes, L.; Moretti, M.L.; Pham, C.D.; Lockhart, S.R.; Schreiber, A.Z. Surveillance for azoles resistance in *Aspergillus* spp. highlights a high number of amphotericin B-resistant isolates. *Mycoses* **2018**, *61*. [CrossRef] [PubMed]

49. Ashu, E.; Korfanty, G.; Samarasinghe, H.; Pum, N.; Man, Y.; Yamamura, D.; Xu, J. Widespread presence of amphotericin B resistant *Aspergillus fumigatus* in Hamilton, Canada. *Infect. Drug Resist.* **2018**, under review.

50. Paul, R.A.; Rudramurthy, S.M.; Meis, J.F.; Mouton, J.W.; Chakrabarti, A. A novel Y319H substitution in CYP51C associated with azole resistance in *Aspergillus flavus*. *Antimicrob. Agents Chemother.* **2015**, *59*, 6615–6619. [CrossRef] [PubMed]

51. Liu, W.; Sun, Y.; Chen, W.; Liu, W.; Wan, Z.; Bu, D.; Li, R. The T788G mutation in the *CYP51C* gene confers voriconazole resistance in *Aspergillus flavus* causing aspergillosis. *Antimicrob. Agents Chemother.* **2012**, *56*, 2598–2603. [CrossRef] [PubMed]

52. Sharma, C.; Kumar, R.; Kumar, N.; Masih, A.; Gupta, D.; Chowdhary, A. Investigation of multiple resistance mechanisms in voriconazole-resistant *Aspergillus flavus* clinical isolates from a chest hospital surveillance in Delhi, India. *Antimicrob. Agents Chemother.* **2018**, *62*. [CrossRef] [PubMed]

53. Nami, S.; Baradaran, B.; Mansoori, B.; Kordbacheh, P.; Rezaie, S.; Falahati, M.; Mohamed Khosroshahi, L.; Safara, M.; Zaini, F. The utilization of RNA silencing technology to mitigate the voriconazole resistance of *Aspergillus Flavus*; lipofectamine-based delivery. *Adv. Pharm. Bull.* **2017**, *7*, 53–59. [CrossRef] [PubMed]

54. Chowdhary, A.; Sharma, C.; Meis, J.F. *Candida auris*: A rapidly emerging cause of hospital-acquired multidrug-resistant fungal infections globally. *PLoS Pathog.* **2017**, *13*, e1006290. [CrossRef] [PubMed]

55. Sarma, S.; Upadhyay, S. Current perspective on emergence, diagnosis and drug resistance in *Candida auris*. *Infect. Drug Resist.* **2017**, *10*, 155–165. [CrossRef] [PubMed]

56. Dettori, M.; Piana, A.; Deriu, M.G.; Lo Curto, P.; Cossu, A.; Musumeci, R.; Cocuzza, C.; Astone, V.; Contu, M.A.; Sotgiu, G. Outbreak of multidrug-resistant *Acinetobacter baumannii* in an intensive care unit. *New Microbiol.* **2014**, *37*, 185–191. [PubMed]

57. Ghaith, D.M.; Zafer, M.M.; Al-Agamy, M.H.; Alyamani, E.J.; Booq, R.Y.; Almoazzamy, O. The emergence of a novel sequence type of MDR *Acinetobacter baumannii* from the intensive care unit of an Egyptian tertiary care hospital. *Ann. Clin. Microbiol. Antimicrob.* **2017**, *16*, 34. [CrossRef] [PubMed]

58. Zarrilli, R.; Casillo, R.; Di Popolo, A.; Tripodi, M.-F.; Bagattini, M.; Cuccurullo, S.; Crivaro, V.; Ragone, E.; Mattei, A.; Galdieri, N.; et al. Molecular epidemiology of a clonal outbreak of multidrug-resistant *Acinetobacter baumannii* in a university hospital in Italy. *Clin. Microbiol. Infect.* **2007**, *13*, 481–489. [CrossRef] [PubMed]

59. Eybpoosh, S.; Haghdoost, A.A.; Mostafavi, E.; Bahrampour, A.; Azadmanesh, K.; Zolala, F. Molecular epidemiology of infectious diseases. *Electron. Phys.* **2017**, *9*, 5149–5158. [CrossRef] [PubMed]

60. Villari, P.; Iacuzio, L.; Torre, I.; Scarcella, A. Molecular epidemiology as an effective tool in the surveillance of infections in the neonatal intensive care unit. *J. Infect.* **1998**, *37*, 274–281. [CrossRef]

61. De Valk, H.A.; Klaassen, C.H.W.; Meis, J.F.G.M. Molecular typing of *Aspergillus* species. *Mycoses* **2008**, *51*, 463–476. [CrossRef] [PubMed]

62. Varga, J. Molecular typing of *Aspergilli*: Recent developments and outcomes. *Med. Mycol.* **2006**, *44*, 149–161. [CrossRef]

63. Ashu, E.E.; Xu, J. The roles of sexual and asexual reproduction in the origin and dissemination of strains causing fungal infectious disease outbreaks. *Infect. Genet. Evol.* **2015**, *36*, 199–209. [CrossRef] [PubMed]

64. Lavergne, R.-A.; Chouaki, T.; Hagen, F.; Toublanc, B.; Dupont, H.; Jounieaux, V.; Meis, J.F.; Morio, F.; Le Pape, P. Home environment as a source of life-threatening azole-resistant *Aspergillus fumigatus* in immunocompromised patients. *Clin. Infect. Dis.* **2017**, *64*, 76–78. [CrossRef] [PubMed]

65. Ashu, E.E.; Hagen, F.; Chowdhary, A.; Meis, J.F.; Xu, J. Global population genetic analysis of *Aspergillus fumigatus*. *MSphere* **2017**, *2*, e00019-17. [CrossRef] [PubMed]

66. Ashu, E.E.; Korfanty, G.A.; Xu, J. Evidence of unique genetic diversity in *Aspergillus fumigatus* isolates from Cameroon. *Mycoses* **2017**, *60*, 739–748. [CrossRef] [PubMed]

67. Alvarez-Perez, S.; Blanco, J.L.; Alba, P.; Garcia, M.E. Mating type and invasiveness are significantly associated in *Aspergillus fumigatus*. *Med. Mycol.* **2010**, *48*, 273–277. [CrossRef] [PubMed]

68. Cheema, M.S.; Christians, J.K. Virulence in an insect model differs between mating types in *Aspergillus fumigatus*. *Med. Mycol.* **2011**, *49*, 202–207. [CrossRef] [PubMed]

69. Monteiro, M.C.; Garcia-Rubio, R.; Alcazar-Fuoli, L.; Peláez, T.; Mellado, E. Could the determination of *Aspergillus fumigatus* mating type have prognostic value in invasive aspergillosis? *Mycoses* **2018**, *61*, 172–178. [CrossRef] [PubMed]

70. Bain, J.M.; Tavanti, A.; Davidson, A.D.; Jacobsen, M.D.; Shaw, D.; Gow, N.A.; Odds, F.C. Multilocus sequence typing of the pathogenic fungus *Aspergillus fumigatus*. *J. Clin. Microbiol.* **2007**, *45*, 1469–1477. [CrossRef] [PubMed]

71. De Valk, H.A.; Meis, J.F.G.M.; Curfs, I.M.; Muehlethaler, K.; Mouton, J.W.; Klaassen, C.H.W. Use of a novel panel of nine short tandem repeats for exact and high-resolution fingerprinting of *Aspergillus fumigatus* isolates. *J. Clin. Microbiol.* **2005**, *43*, 4112–4120. [CrossRef] [PubMed]

72. García-Lerma, J.G.; Nidtha, S.; Blumoff, K.; Weinstock, H.; Heneine, W. Increased ability for selection of zidovudine resistance in a distinct class of wild-type HIV-1 from drug-naive persons. *Proc. Natl. Acad. Sci. USA* **2001**, *98*, 13907–13912. [CrossRef] [PubMed]

73. Abdolrasouli, A.; Rhodes, J.; Beale, M.A.; Hagen, F.; Rogers, T.R.; Chowdhary, A.; Meis, J.F.; Armstrong-James, D.; Fisher, M.C. Genomic context of azole resistance mutations in *Aspergillus fumigatus* determined using whole-genome sequencing. *MBio* **2015**, *6*, e00536-15. [CrossRef] [PubMed]

74. Chang, H.; Ashu, E.; Sharma, C.; Kathuria, S.; Chowdhary, A.; Xu, J. Diversity and origins of Indian multi-triazole resistant strains of *Aspergillus fumigatus*. *Mycoses* **2016**, *59*, 450–466. [CrossRef] [PubMed]

75. Clinical and Laboratory Standards Institute. *Reference Method for Broth Dilution Antifungal Susceptibility Testing of Filamentous Fungi*, 2nd ed.; Approved Standard, CLSI M38-A2; Clinical and Laboratory Standards Institute: Wayne, PA, USA, 2008.

76. Dunne, K.; Hagen, F.; Pomeroy, N.; Meis, J.F.; Rogers, T.R. Intercountry transfer of triazole-resistant *Aspergillus fumigatus* on Plant Bulbs. *Clin. Infect. Dis.* **2017**, *65*, 147–149. [CrossRef] [PubMed]

77. Lamoth, F. *Aspergillus fumigatus*-related species in clinical practice. *Front. Microbiol.* **2016**, *7*, 683. [CrossRef] [PubMed]

78. Tang, Q.; Tian, S.; Yu, N.; Zhang, X.; Jia, X.; Zhai, H.; Sun, Q.; Han, L. Development and evaluation of a loop-mediated isothermal amplification method for rapid detection of *Aspergillus fumigatus*. *J. Clin. Microbiol.* **2016**, *54*, 950–955. [CrossRef] [PubMed]

79. Powers-Fletcher, M.V.; Hanson, K.E. Molecular diagnostic testing for *Aspergillus*. *J. Clin. Microbiol.* **2016**, *54*, 2655–2660. [CrossRef] [PubMed]

80. Etienne, K.A.; Kano, R.; Balajee, S.A. Development and validation of a microsphere-based luminex assay for rapid identification of clinically relevant *Aspergilli*. *J. Clin. Microbiol.* **2009**, *47*, 1096–1100. [CrossRef] [PubMed]

81. Bhimji, A.; Bhaskaran, A.; Singer, L.G.; Kumar, D.; Humar, A.; Pavan, R.; Lipton, J.; Kuruvilla, J.; Schuh, A.; Yee, K.; et al. *Aspergillus* galactomannan detection in exhaled breath condensate compared to bronchoalveolar lavage fluid for the diagnosis of invasive aspergillosis in immunocompromised patients. *Clin. Microbiol. Infect.* **2018**, *24*, 640–645. [CrossRef] [PubMed]

82. Mello, E.; Posteraro, B.; Vella, A.; De Carolis, E.; Torelli, R.; D'Inzeo, T.; Verweij, P.E.; Sanguinetti, M. Susceptibility testing of common and uncommon *Aspergillus* species against posaconazole and other mold-active antifungal azoles using the Sensititre method. *Antimicrob. Agents Chemother.* **2017**, *61*, e00168-17. [CrossRef] [PubMed]

83. Hageskal, G.; Kristensen, R.; Fristad, R.F.; Skaar, I. Emerging pathogen *Aspergillus calidoustus* colonizes water distribution systems. *Med. Mycol.* **2011**, *49*, 588–593. [CrossRef] [PubMed]

84. Egli, A.; Fuller, J.; Humar, A.; Lien, D.; Weinkauf, J.; Nador, R.; Kapasi, A.; Kumar, D. Emergence of *Aspergillus calidoustus* infection in the era of post-transplantation azole prophylaxis. *Transplantation* **2012**, *94*, 403–410. [CrossRef] [PubMed]

85. Springer, J.; White, P.L.; Hamilton, S.; Michel, D.; Barnes, R.A.; Einsele, H.; Löffler, J. Comparison of performance characteristics of *Aspergillus* PCR in testing a range of blood-based samples in accordance with international methodological recommendations. *J. Clin. Microbiol.* **2016**, *54*, 705–711. [CrossRef] [PubMed]

86. Sun, W.; Wang, K.; Gao, W.; Su, X.; Qian, Q.; Lu, X.; Song, Y.; Guo, Y.; Shi, Y. Evaluation of PCR on bronchoalveolar lavage fluid for diagnosis of invasive aspergillosis: A bivariate meta-analysis and systematic review. *PLoS ONE* **2011**, *6*, e28467. [CrossRef] [PubMed]

87. Reinwald, M.; Buchheidt, D.; Hummel, M.; Duerken, M.; Bertz, H.; Schwerdtfeger, R.; Reuter, S.; Kiehl, M.G.; Barreto-Miranda, M.; Hofmann, W.-K.; et al. Diagnostic performance of an *Aspergillus*-specific nested PCR assay in cerebrospinal fluid samples of immunocompromised patients for detection of central nervous system aspergillosis. *PLoS ONE* **2013**, *8*, e56706. [CrossRef] [PubMed]

88. White, P.L.; Bretagne, S.; Klingspor, L.; Melchers, W.J.G.; McCulloch, E.; Schulz, B.; Finnstrom, N.; Mengoli, C.; Barnes, R.A.; Donnelly, J.P.; et al. European *Aspergillus* PCR initiative *Aspergillus* PCR: One step closer to standardization. *J. Clin. Microbiol.* **2010**, *48*, 1231–1240. [CrossRef] [PubMed]

89. White, P.L.; Perry, M.D.; Loeffler, J.; Melchers, W.; Klingspor, L.; Bretagne, S.; McCulloch, E.; Cuenca-Estrella, M.; Finnstrom, N.; Donnelly, J.P.; et al. European *Aspergillus* PCR initiative critical stages of extracting DNA from *Aspergillus fumigatus* in whole-blood specimens. *J. Clin. Microbiol.* **2010**, *48*, 3753–3755. [CrossRef] [PubMed]

90. Alanio, A.; Bretagne, S. Challenges in microbiological diagnosis of invasive *Aspergillus* infections. *F1000Research* **2017**, *6*. [CrossRef] [PubMed]

91. White, P.L.; Parr, C.; Thornton, C.; Barnes, R.A. Evaluation of real-time PCR, galactomannan enzyme-linked immunosorbent assay (ELISA), and a novel lateral-flow device for diagnosis of invasive aspergillosis. *J. Clin. Microbiol.* **2013**, *51*, 1510–1516. [CrossRef] [PubMed]

92. Aguado, J.M.; Vázquez, L.; Fernández-Ruiz, M.; Villaescusa, T.; Ruiz-Camps, I.; Barba, P.; Silva, J.T.; Batlle, M.; Solano, C.; Gallardo, D.; et al. PCRAGA Study Group; Spanish Stem Cell Transplantation Group; Study Group of Medical Mycology of the Spanish Society of Clinical Microbiology and Infectious Diseases; Spanish Network for Research in Infectious Diseases. Serum galactomannan versus a combination of galactomannan and polymerase chain reaction-based *Aspergillus* DNA detection for early therapy of invasive aspergillosis in high-risk hematological patients: A randomized controlled trial. *Clin. Infect. Dis.* **2015**, *60*, 405–414. [CrossRef] [PubMed]

93. Hebart, H.; Klingspor, L.; Klingebiel, T.; Loeffler, J.; Tollemar, J.; Ljungman, P.; Wandt, H.; Schaefer-Eckart, K.; Dornbusch, H.J.; Meisner, C.; et al. A prospective randomized controlled trial comparing PCR-based and empirical treatment with liposomal amphotericin B in patients after allo-SCT. *Bone Marrow Transplant.* **2009**, *43*, 553–561. [CrossRef] [PubMed]
94. De Pauw, B.; Walsh, T.J.; Donnelly, J.P.; Stevens, D.A.; Edwards, J.E.; Calandra, T.; Pappas, P.G.; Maertens, J.; Lortholary, O.; Kauffman, C.A.; et al. Revised definitions of invasive fungal disease from the European Organization for Research and Treatment of Cancer/Invasive Fungal Infections Cooperative Group and the National Institute of Allergy and Infectious Diseases Mycoses Study Group (EORTC/MSG) consensus group. *Clin. Infect. Dis.* **2008**, *46*, 1813–1821. [CrossRef] [PubMed]

Article

Dual RNA-Seq Analysis of *Trichophyton rubrum* and HaCat Keratinocyte Co-Culture Highlights Important Genes for Fungal-Host Interaction

Monise Fazolin Petrucelli [1], Kamila Peronni [2], Pablo Rodrigo Sanches [3], Tatiana Takahasi Komoto [1], Josie Budag Matsuda [1], Wilson Araújo da Silva Jr. [2,4], Rene Oliveira Beleboni [1], Nilce Maria Martinez-Rossi [3], Mozart Marins [1] and Ana Lúcia Fachin [1,*]

1. Biotechnology Unit, University of Ribeirão Preto-UNAERP, São Paulo 2201, Brazil; mofazolin@gmail.com (M.F.P.); tattytk@hotmail.com (T.T.K.); josie@unidavi.edu.br (J.B.M.); rbeleboni@unaerp.br (R.O.B.); mmarins@gmb.bio.br (M.M.)
2. Laboratory of Molecular Genetics and Bioinformatics, Regional Hemotherapy Center of Ribeirão Preto, Ribeirão Preto 2501, Brazil; kcperoni@gmail.com (K.P.); wilsonjr@usp.br (W.A.d.S.J.)
3. Department of Genetics, Ribeirão Preto Medical School, University of São Paulo, Ribeirão Preto 14049-900, Brazil; psanches@gmail.com (P.R.S.); nmmrossi@usp.br (N.M.M.-R.)
4. Center for Medical Genomics at the Clinics Hospital of Ribeirão Preto Medical School, University of São Paulo, Ribeirão Preto 14049-900, Brazil
* Correspondence: afachin@unaerp.br; Fax: +55-16-3603-7030

Received: 30 May 2018; Accepted: 16 July 2018; Published: 19 July 2018

Abstract: The dermatophyte *Trichophyton rubrum* is the major fungal pathogen of skin, hair, and nails that uses keratinized substrates as the primary nutrients during infection. Few strategies are available that permit a better understanding of the molecular mechanisms involved in the interaction of *T. rubrum* with the host because of the limitations of models mimicking this interaction. Dual RNA-seq is a powerful tool to unravel this complex interaction since it enables simultaneous evaluation of the transcriptome of two organisms. Using this technology in an in vitro model of co-culture, this study evaluated the transcriptional profile of genes involved in fungus-host interactions in 24 h. Our data demonstrated the induction of glyoxylate cycle genes, *ERG6* and *TERG_00916*, which encodes a carboxylic acid transporter that may improve the assimilation of nutrients and fungal survival in the host. Furthermore, genes encoding keratinolytic proteases were also induced. In human keratinocytes (HaCat) cells, the *SLC11A1*, *RNASE7*, and *CSF2* genes were induced and the products of these genes are known to have antimicrobial activity. In addition, the *FLG* and *KRT1* genes involved in the epithelial barrier integrity were inhibited. This analysis showed the modulation of important genes involved in *T. rubrum*–host interaction, which could represent potential antifungal targets for the treatment of dermatophytoses.

Keywords: dermatophytes; *ERG6*; epithelial barrier; glyoxylate cycle; fungal-host interaction

1. Introduction

Dermatophytoses are superficial infections of keratinized tissues caused by a group of filamentous fungi called dermatophytes [1]. Although these infections are restricted to the superficial layers of the epidermis, they can become invasive and can lead to severe diseases in immunocompromised [2] and diabetic patients [3]. Data from the World Health Organization estimate that approximately 25% of the world's population have skin infections caused by fungi.

Most human dermatophytoses are caused by anthropophilic dermatophytes. Among these species, *Trichophyton rubrum* is the main cause of dermatophytoses in the world [4,5]. It is estimated that *T. rubrum* is the etiological agent of 69.5% of all cases of dermatophytosis caused by species of the genus *Trichophyton*, followed by *Trichophyton interdigitale*, *Trichophyton verrucosum* and *Trichophyton tonsurans* [6].

Despite the importance of these infections in clinical practice, knowledge of the molecular mechanisms involved in the dermatophyte-host interaction is limited, possibly because of the technical difficulties of the models mimicking this interaction, as well as the lack of genetic tools that allow for a more in-depth study of these organisms [7]. However, this scenario has been changing with the sequencing of mixed transcriptomes, also called dual RNA-seq, an approach widely used for the study of the complex interaction that exists between the host and pathogen [8] including bacteria [9], viruses [10], and fungi [11,12].

With the advent of this technology and the published sequence of the *T. rubrum* genome, the present study evaluated the transcriptional profile of *T. rubrum* co-cultured with human keratinocytes (HaCat) for 24 h by dual RNA-seq to identify important genes involved in the host defense and fungal pathogenicity in order to increase our understanding of the molecular aspects of this interaction. After 24 h of co-culture, we observed the induction of specific genes of the glyoxylate cycle and of a carboxylic acid transporter in *T. rubrum*, which may contribute to metabolic flexibility in nutrient-limited host niches, as well as of the *ERG6* gene involved in plasma membrane permeability, which may favor the assimilation of nutrients and fungal survival in the host. In addition, we found that the modulation of the *LAP2* and *DPPV* genes involved in the production of keratinolytic proteases that are important for the virulence of this dermatophyte. In contrast, in keratinocytes, genes involved in the repair of the epithelial barrier, in the increase of cell migration and the *RNASE7*, *SLC11A1* and *CSF2* genes (whose gene products have potential antimicrobial activity) were induced. Furthermore, the inhibition of *FLG* and *KRT1* genes whose products are directly involved in the maintenance of skin barrier integrity was observed.

2. Materials and Methods

2.1. Strains, Media and Growth Conditions

The *T. rubrum* strain CBS 118892 (CBS-KNAW Fungal Biodiversity Center, Utrech, The Netherlands) sequenced by the Broad Institute (Cambridge, MA, USA) was cultured on Sabouraud dextrose agar (Oxoid, Hampshire, UK) for 15 days at 28 °C.

2.2. Keratinocytes, Media and Growth Conditions

The immortalized human keratinocytes cell line HaCat was purchased from Cell Lines Service GmbH (Eppelheim, Germany). The cells were cultured in an RPMI medium (Sigma Aldrich, St. Louis, MO, USA) supplemented with 10% fetal bovine serum at 37 °C in a humidified atmosphere containing 5% CO_2. Antibiotics (100 U/mL penicillin and 100 µg/mL streptomycin) were added to the medium to prevent bacterial contamination.

2.3. Co-Culture Assay and Conditions

For co-culture assay, a ratio of 2.5×10^5 cells/mL of keratinocytes to 1×10^7 conidia/mL of *T. rubrum* solution was used, and the co-culture was performed as described in [13]. The assays were carried out in three independent experiments performed in triplicate. Cultured keratinocytes and *T. rubrum* conidia were used as controls and were cultured similarly to the co-infection in RPMI Medium (Sigma Aldrich). Scanning electron microscopy was performed with a JEOL JEM 100CXII electron microscope at the Multiuser Electron Microscopy Laboratory of the Department of Cell and Molecular Biology (Ribeirão Preto Medical School, São Paulo, Brazil) to determine whether the penetration of fungal hyphae into keratinocytes occurred within 24 h of co-culture. The cell viability of HaCat keratinocytes prior to *T. rubrum* inoculation and after 24 h of co-culture was determined by measuring the release of the enzyme lactate dehydrogenase (LDH) (TOX7 kit from Sigma-Aldrich) in the RPMI Medium (Sigma Aldrich) according to the manufacturer's instructions and described in [14].

The absorbance was read in a microplate reader (Elx 800 UV Bio-Tek Instruments, Inc., Winooski, VT, USA) at 490 nm.

2.4. RNA Isolation and Integrity Analysis

After 24 h of incubation, fungi and human cells were recovered by scraping and centrifuging at $1730 \times g$ for 10 min. For the disruption of the fungal cell wall, the samples (co-culture and controls) were treated with lysis solution (20 mg/mL of lysing enzymes from *Trichoderma harzianum* purchased from Sigma-Aldrich; 0.7 M KCl and 1 M $MgSO_4$, pH 6.8) for 1 h at 28 °C under gentle shaking, followed by centrifugation at $1000 \times g$ for 10 min, as described in [13]. Total RNA was extracted using the Illustra RNAspin Mini RNA Isolation Kit (GE Healthcare, Chicago, IL, USA) according to the manufacturer's instructions. After extraction, the absence of proteins and phenol in the RNA was analyzed in a MidSci Nanophotometer (Midwest Scientific, St. Louis, MO, USA) and the RNA integrity was assessed by microfluidic electrophoresis in an Agilent 2100 Bioanalyzer (Agilent Technologies, Santa Clara, CA, USA). Only RNA with an RNA integrity number (RIN) >7.0 was used. These RNAs were quantified in a Quantus™ Fluorometer (Promega Corporation, Madison, WI, USA) to verify if they had the adequate concentration for library construction.

2.5. Library Construction and Sequencing

The cDNA libraries for RNA sequencing were constructed in triplicate for each condition (cultured keratinocytes and *T. rubrum* conidia as control and co-culture). The libraries were constructed using the TrueSeq® RNA Sample Preparation Kit v2 (Illumina, San Diego, CA, USA) according to manufacturer's instructions and the libraries were validated according to the Library quantitative PCR (qPCR) Quantification Guide (Illumina). A pool of 11 pM of each library was distributed on the flowcell lanes and cluster amplification was performed in a cBot (Illumina) according to the manufacturer's instructions.

Single read and paired-end sequencing were performed in a Genome Analyzer IIx and Hiseq 2000 (Illumina), respectively, according to the manufacturer's instructions. The RNA-seq data are deposited in the GEO (Gene Expression Omnibus) database [15] under the accession number GSE110073

2.6. Sequence Data Analysis

The reads generated for each library were filtered using the FastQC software (https://www. bioinformatics.babraham.ac.uk/index.html) for removal of Illumina adapters and poor-quality reads. Only those with a Phred score > 20 were considered high-quality reads.

The high-quality reads were aligned to the *T. rubrum* reference genome of the Broad Institute's Dermatophyte Comparative Database and to the *Homo sapiens* reference genome HG19 [16].

After alignment, the triplicate of each library was normalized according to each library size and the number of reads was calculated using the summarize Overlaps function in the Genomic Ranges Bioconductor package, obtaining the expression levels of the transcripts in the samples. For statistical evaluation of the gene expression data between the samples, the false discovery rate (FDR) procedure was applied using the DEseq package [17] implemented in the R/Bioconductor software. Genes exhibiting statistical significance <0.05 and a $\log_2$ fold change ratio ≥ 1 or ≤ -1 were defined as differentially expressed genes (DEGs). The functional categorization of *T. rubrum* and keratinocyte DEGs in co-culture was performed according to Gene Ontology [18] using the Blast2GO algorithm [19] for *T. rubrum* and the website http://www.geneontology.org/ for human keratinocyte DEGs. For functional enrichment, the BayGO algorithm [20] and Enrichr enrichment tool [21,22], were used for the *T. rubrum* and keratinocyte DEGs, respectively. A *p*-value < 0.05 indicated the over-represented categories.

2.7. qPCR Validation

A set of 14 genes, including the *T. rubrum* and keratinocyte genes, were selected for validation by qPCR. For the reaction, 1 µg of the total RNA used for sequencing was treated with DNAse 1 Amplification Grade® (Sigma Aldrich) to remove any genomic DNA contamination. The High-Capacity cDNA Reverse Transcription® Kit (Applied Biosystems, Foster city, CA, USA) was used for cDNA conversion according to the manufacturer's instructions. Quantitative Real Time (RT)-PCR experiments were performed in triplicate using the SYBR Taq Ready Mix Kit (Sigma Aldrich) in a Mx3300 qPCR System (Stratagene, San Diego, CA, USA). The cycling conditions were initial denaturation at 94 °C for 10 min, followed by 40 cycles at 94 °C for 2 min, at 60 °C for 60 s and at 72 °C for 1 min. A dissociation curve was constructed at the end of each PCR cycle to verify single product amplification. Gene expression levels were calculated using the $2^{-\Delta\Delta C_T}$ comparative method. GAPDH [23] and β-actin [24] were used as normalizer genes for keratinocytes and 18S [25] and β-tubulin [26] as normalizer genes for *T. rubrum*. The results are reported as the mean ± standard deviation of three experiments. Pearson's correlation test was used to evaluate the correlation between the qPCR and RNA-seq techniques. The primers used for qPCR validation are available in Table S4.

3. Results

3.1. Electron Microscopy of T. rubrum and HaCat Co-Culture

Figure 1B shows the penetration of a *T. rubrum* hypha into a HaCat cell after 24 h of co-culture. Thus, the period of co-culture was considered appropriate for the evaluation of the fungal-host interaction.

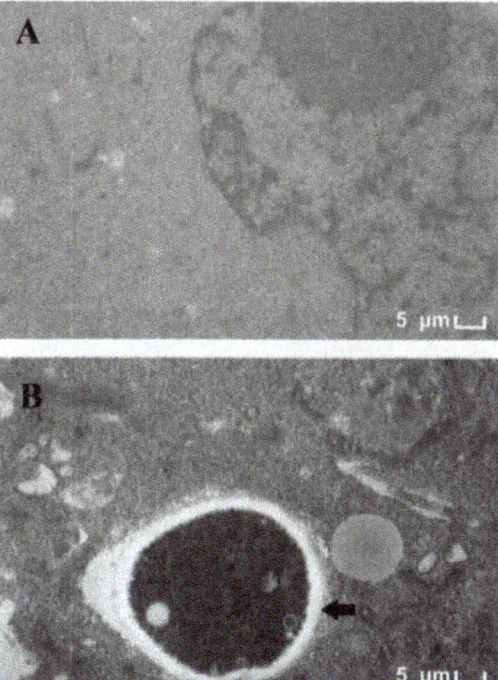

Figure 1. The transmission electron microscopy of the *Trichophyton rubrum*-HaCat co-culture after 24 h. (**A**) Human keratinocytes (HaCat) keratinocyte as the control (14kx); (**B**) Co-culture (14kx). The arrow indicates a fragment of *T. rubrum* hyphae inside the HaCat cells.

We performed the LDH assay with 24 h of co-culture to evaluate the keratinocyte cell viability. The percentage of LDH release was 18%. This LDH release may be due to the penetration of some fungal hyphae into keratinocyte cells (as observed in Figure 1B). LDH release was also evaluated at 0 h to assess cell viability prior to the addition of the fungus. The LDH release rate at 0 h was 1%. As a positive control, Triton X-100 (1%) was used in which 100% of the LDH release was obtained. Considering that we used 2.5×10^5 cells/mL prior to inoculation of the fungus and that the percentage of LDH was 18%, we can estimate that approximately 2×10^5 cells/mL are still viable in 24 h of co-cultivation.

3.2. Dual RNA-Seq Analysis of the Fungal-Host Interaction

Sequencing resulted in an average of 40, 34 and 47 million raw reads corresponding to the libraries of *T. rubrum* conidia, co-culture, and keratinocytes, respectively. Low-quality reads were then removed, and the resulting reads were aligned to the references genomes of *T. rubrum* and *Homo sapiens* HG19 (UCSC Genome Bioinformatics site, Santa Cruz, CA, USA). On average, 85% and 5% of the quality reads of the *T. rubrum* conidia and co-culture libraries, respectively, aligned to the *T. rubrum* reference genome (CBS 118892). These percentages were 84% and 85%, respectively, when the quality reads of the co-culture and keratinocyte cell line were aligned to the HG19 reference genome. The total number of filtered and aligned reads of each library is shown in Table S1.

3.3. Transcriptional Profile Analysis of Differentially Expressed Genes in the T. rubrum-Keratinocyte Co-Culture System

Tables 1 and 2 show the genes that are up-regulated and down-regulated in keratinocytes and *T. rubrum*, respectively. According to the distribution of the genes, those showing a *p*-value < 0.05 and $\log_2$ fold change ≥ 1 or ≥ -1 in each condition were considered differentially expressed (Figure S1). A total of 353 HaCat genes and 70 *T. rubrum* genes were differentially expressed during 24 h of co-culture (Tables S2 and S3).

Table 1. The major up- and down-regulated genes in HaCat cells after 24 h of co-culture.

ID	Gene Product Name	$\log_2$ Fold Change
SLC9A2	Sodium/hydrogen exchanger 2	5.01
ANGPTL4	Angiopoietin-related protein 4	4.71
DES	Desmin	4.53
C4orf47	UPF0602 protein C4orf47	4.51
KISS1R	KiSS-1 receptor	4.49
NSA2	Ribosome biogenesis protein NSA2 homolog	4.35
HIST1H3C	Histone cluster 1 H3 family member c	4.04
SEC11C	Signal peptidase complex catalytic subunit	3.87
KPNA7	Importin subunit alpha-8	3.83
CASP14	Caspase 14	3.74
SLC2A3	Facilitated glucose transporter member 3	3.73
ALDOC	Fructose-bisphosphate aldolase C	3.70
MT1B	Metallothionein-1B	3.62
SERPINE1	Plasminogen activator inhibitor 1	3.55
MAF	Transcription factor Maf	3.54
CA9	Carbonic anhydrase 9	3.36
TGM2	Transglutaminase 2	3.35
PADI1	Protein-arginine deiminase type-1	3.29
STC1	Stanniocalcin 1	3.14
BNIP3	BCL2 interacting protein 3	3.08
LSS	Lanosterol synthase	3.06
MT1H	Metallothionein 1H	3.05
MT1X	Metallothionein 1X	2.97
PLA2G2F	Group IIF secretory phospholipase A2	2.96
CALB1	Calbindin 1	2.93
POTEM	Putative POTE ankyrin domain family member M	−5.31

Table 1. *Cont.*

ID	Gene Product Name	Log$_2$ Fold Change
SNORA51	Small nucleolar RNA. H/ACA box	−4.90
ANP32A-IT1	ANP32A intronic transcript 1	−4.64
UCKL1	Uridine-cytidine kinase 1 like 1	−4.50
FNDC3B	Fibronectin type III domain containing	−4.37
KRT1	Keratin 1	−4.02
MMP12	Matrix metallopeptidase 12	−3.22
NSD1	Nuclear receptor binding SET domain	−3.06
CYCSP52	Cytochrome c. somatic pseudogene	−3.02
EME2	Essential meiotic structure-specific endonuclease subunit 2	−3.00
COL12A1	Collagen type XII alpha 1 chain	−2.88
SNORD45A	Small nucleolar RNA. C/D box	−2.82
FBXL19-AS1	FBXL19 antisense RNA 1 (head to head)	−2.80
TRIM26	Tripartite motif containing 26	−2.76
IARS	Isoleucyl-tRNA synthetase	−2.76
KIF14	Kinesin family member 14	−2.74
MEGF8	Multiple EGF like domains 8	−2.67
HNRNPL	Heterogeneous nuclear ribonucleoprotein	−2.66

Table 2. The major up- and down-regulated genes in *T. rubrum* after 24 h of co-culture.

ID	Gene Product Name	Log$_2$ Fold Change
TERG_12606	Dipeptidyl peptidase V (DPPV)	2.16
TERG_01280	Hypothetical protein	2.06
TERG_03102	Sterol 24-C-methyltransferase- ERG6	2.05
TERG_08104	Potassium/sodium efflux P-type ATPase	1.98
TERG_01281	Malate synthase	1.72
TERG_04399	Phthalate transporter	1.62
TERG_00215	MFS peptide transporter	1.47
TERG_00348	Galactose-proton symporter	1.47
TERG_02811	Hypothetical protein	1.42
TERG_12645	Hypothetical protein	1.40
TERG_07017	Oxidoreductase	1.35
TERG_08333	1-pyrroline-5-carboxylate dehydrogenase	1.34
TERG_02671	Hypothetical protein	1.34
TERG_02023	Extracellular matrix protein	1.32
TERG_08405	Leucine aminopeptidase 2	1.30
TERG_00916	Carboxylic acid transporter	1.29
TERG_11638	Isocitrate lyase	1.28
TERG_04952	ABC transporter	1.26
TERG_01406	Hypothetical protein	−2.91
TERG_07726	Hypothetical protein	−2.25
TERG_03174	MFS siderochrome iron transporter	−1.99
TERG_06355	Hypothetical protein	−1.90
TERG_07035	Hypothetical protein	−1.85
TERG_04156	Hypothetical protein	−1.77
TERG_05655	AN1 zinc finger protein	−1.73
TERG_01622	Hypothetical protein	−1.63
TERG_07477	Hypothetical protein	−1.57
TERG_06186	Protein disulfide-isomerase domain-containing protein	−1.57
TERG_03708	Hypothetical protein	−1.53
TERG_03855	Hypothetical protein	−1.50
TERG_00499	Hypothetical protein	−1.45
TERG_03175	Hypothetical protein	−1.45
TERG_04073	Glutathione synthetase	−1.41
TERG_12563	Hypothetical protein	−1.37
TERG_08139	NAD dependent epimerase/dehydratase	−1.34
TERG_06963	Hsp90-like protein	−1.33
TERG_01731	Hypothetical protein	−1.32
TERG_04006	Rho guanyl nucleotide exchange factor	−1.32

3.4. Functional Categorization of Differentially Expressed Genes

To evaluate the molecular and biological mechanisms involved in the fungal-host interaction, the DEGs were categorized according to biological processes and molecular functions. The most enriched categories considering a *p* < 0.05 are shown in Figure 2.

Most of the up-regulated *T. rubrum* genes (Figure 2A) belong to categories related to metabolic processes, membrane proteins, and substance transport, while the down-regulated genes are mainly involved in ATP binding. However, categories important to the fungus-host relationship, such as those including genes involved in the glyoxylate cycle and pathogenicity, should also be highlighted. Table 3 shows some functional categories that are important for the interaction of *T. rubrum* with HaCat keratinocytes. Within these categories, we selected some genes considered to play a fundamental role in the attack mechanisms and survival of the fungus when in contact with the host for validation and discussion: genes involved in protease secretion (*TERG_12606*; *TERG_08405*), metabolic flexibility for nutrient assimilation (*TERG_01281*; *TERG_11638*; *TERG_11639*; *TERG_00916*), and plasma membrane permeability (*TERG_03102*). On the other hand, up-regulated genes in keratinocytes (Figure 2B) are mainly found in the categories related to RNA binding, translation, and rRNA processing, while most of the down-regulated genes belong to the RNA binding category. Furthermore, Table 4 shows some functional categories that are important for the cell defense mechanisms of human keratinocytes during co-culture with *T. rubrum*, such as the genes involved in the innate immune response, epidermal cell differentiation, regulation of cell migration, and establishment of the skin barrier.

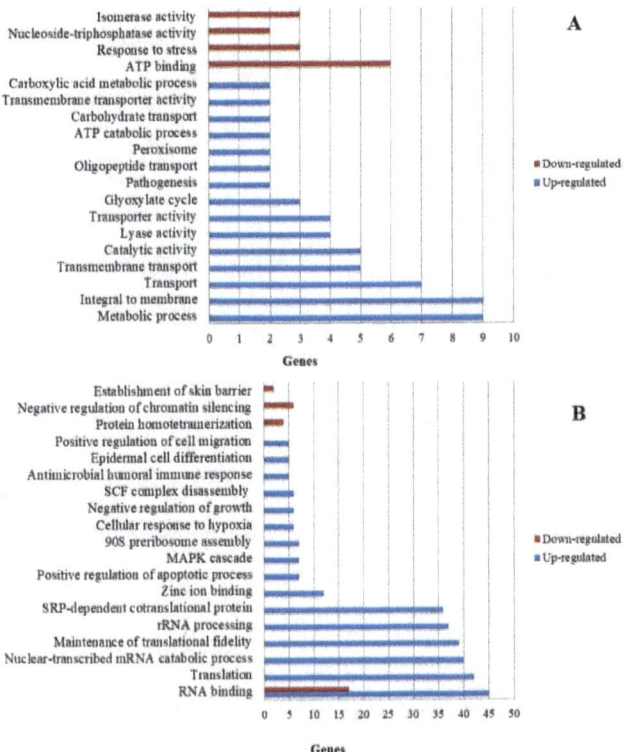

Figure 2. The gene Ontology-based functional categorization of differentially expressed genes. The main representative functional categories (*p* < 0.05) of genes differentially expressed in *T. rubrum* (**A**) and HaCat (**B**).

Table 3. Some functional categories and related genes important for the pathogenesis of *T. rubrum*.

ID	Gene Product Name	Log₂ Fold Change
Metabolic process		
TERG_03102	Sterol 24-C-methyltransferase	2.05
TERG_08104	Sodium transport ATPase	1.98
TERG_02811	Hypothetical protein	1.40
TERG_08333	Delta 1-pyrroline-5-carboxylate dehydrogenase	1.34
TERG_11638	Isocitrate lyase	1.26
TERG_01270	AMP-dependent ligase	1.13
TERG_07691	Nonspecific lipid-transfer protein	1.13
TERG_07222	Carbonic anhydrase	1.05
Transmembrane transport		
TERG_04399	Phthalate transporter (MFS transporter)	1.62
TERG_00348	Galactose-proton symporter (MFS transporter)	1.42
TERG_00916	Carboxylic acid transporter (MFS transporter)	1.28
TERG_04952	ABC transporter	1.25
TERG_04356	Amino acid permease	1.06
Pathogenesis		
TERG_12606	Dipeptidyl peptidase V	2.16
TERG_08405	Leucine Aminopeptidase 2	1.29
Glyoxylate cycle		
TERG_01281	Malate synthase	1.72
TERG_11638	Isocitrate lyase	1.26
TERG_11639	Isocitrate lyase	1.13

Table 4. Some functional categories and related genes important for human host defense.

ID	Gene Product Name	Log₂ Fold Change
Positive regulation of cell migration		
TCAF2	TRPM8 channel-associated factor 2	2.11
MMP9	Matrix metalloproteinase-9	2.06
LAMC2	Laminin subunit gamma-2	1.97
HBEGF	Proheparin-binding EGF-like growth factor	1.84
HAS2	Hyaluronan synthase 2	1.46
MAPK cascade involved in the innate immune response		
CSF2	Granulocyte-macrophage colony-stimulating factor	2.86
HBEGF	Proheparin-binding EGF-like growth factor	1.84
DUSP5	Dual specificity protein phosphatase 5	1.57
PSMB3	Proteasome subunit beta type-3	1.46
PPP5C	Serine/threonine-protein phosphatase 5	1.37
PSMB2	Proteasome subunit beta type-2	1.28
UBB	Polyubiquitin-B	1.24
Antimicrobial humoral immune response		
SERPINE1	Plasminogen activator inhibitor 1	3.55
SLC11A1	Natural resistance-associated macrophage protein 1	2.28
RNASE7	Ribonuclease 7	2.27
RPS19	40S ribosomal protein S19	1.66
RPL30	60S ribosomal protein L30	1.41
Epidermal cell differentiation		
CASP14	Caspase-14	3.74
ALDOC	Fructose-bisphosphate aldolase C	3.70
AKR1C1	Aldo-keto reductase family	2.80
LAMC2	Laminin subunit gamma-2	1.97
PGK1	Phosphoglycerate kinase 1	1.62
Establishment of the skin barrier		
KRT1	Keratin type II cytoskeletal 1	−4.02
FLG	Filaggrin	−1.86

3.5. Validation by qPCR

Pearson's correlation test was used to evaluate the correlation between dual RNA-seq and qPCR. For this purpose, 14 genes were chosen for validation, including 6 *T. rubrum* genes (*TERG_11638*; *TERG_01281*; *TERG_08405*; *TERG_12606*; *TERG_00916*; *TERG_03102*) and 8 HaCat genes (*HAS2*; *CSF2*; *SLC11A1*; *RNASE7*; *CASP14*; *MMP9*; *KRT1*; *FLG*). Figure 3 shows the comparison of the log₂ fold change values obtained with the two techniques. The gene expression results obtained by RNA-seq showed a strong correlation ($r = 0.80$, $p < 0.001$) with the gene modulation values obtained by qPCR. This finding suggests that sequencing provided reliable results, demonstrating the reproducibility and accuracy of the technique.

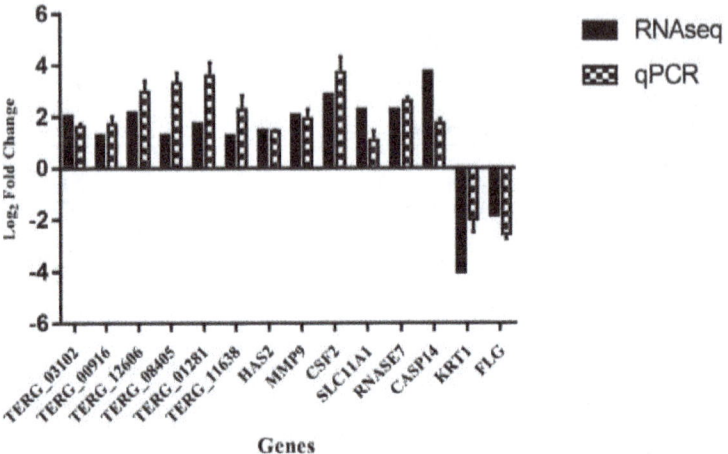

Figure 3. The comparison of gene modulation obtained by RNA-seq and quantitative PCR (qPCR). The error bars represent the standard error of three independent replicates. Pearson's test indicated a strong correlation between the two techniques ($r = 0.80$; $p < 0.01$).

4. Discussion

Through the analysis of mixed transcriptomes, this is the first study to sequence by dual RNA-seq the dermatophyte *T. rubrum* with a HaCat cells in an in vitro model of co-culture for 24 h.

Based on the sequencing data generated, only about 5% of the quality reads of the co-culture could be aligned to the *T. rubrum* reference genome (CBS 118892), indicating a predominance of human reads in this library. Indeed, a major challenge encountered in the sequencing of mixed transcriptomes is the difference in the amount of RNA between different cell types. Whereas a human cell contains about 20–25 pg RNA, a fungal cell contains 0.5–1 pg [8,27], a fact that may explain the smaller number of reads generated for *T. rubrum* compared to human keratinocytes. This obstacle was also observed in dual RNA-seq analysis of a *Magnaporthe oryzae* and *Oryza sativa* co-culture, in which the percentage of alignment of fungal reads to the *M. oryzae* reference genome ranged from 0.1–0.2% [28]. As the latest example, in [12] also obtained a low percentage (~1%) of reads corresponding to the pathogen *Phytophthora cinnamomi* in dual RNAseq with *Eucalyptus nitens*. In that study, the authors obtained 283 genes of *Phytophthora cinnamomi* in a genome comprising approximately 58.38 Mb (National Center for Biotechnology). Comparing these data with our study, we obtained about 5% of read alignment and 70 modulated *T. rubrum* genes considering a fold change ≥ 1 or ≤ -1 within a genome of 22.5 Mb.

However, we reached coverage of 90.7% of the 8.616 annotated genes in the *T. rubrum* genome considering the genes with at least one count read.

Seventy DEGs of *T. rubrum* were identified after 24 h of co-culture, which could be allocated to different categories according to biological function. Categories that were relevant for the understanding of the attack mechanisms of *T. rubrum* against keratinocytes included those containing *TERG_12606* and *TERG_08405* which encode important proteases for tissue invasion by the fungus, *TERG_03102* or the *ERG6* gene which is considered a promising target for the development of new antifungal agents [29], and *TERG_01281*, *TERG_11638* and *TERG_ 00916* which may be involved in the metabolic flexibility of *T. rubrum*, improving the adaptation and development in the host.

Regarding the functional categories containing the 353 HaCat DEGs, we highlight the following genes as important for the host defense mechanisms: *SLC11A1*, *RNASE7* and *CSF2* involved in innate immune response signaling; *MMP9* and *HAS2* involved in the regulation of epithelial cell migration; *KRT1* and *FLG* involved in maintaining skin barrier integrity, and *CASP14* involved in epidermal cells differentiation.

4.1. Genes Involved in Protease Secretion Are Important for the Pathogenicity of T. rubrum

During the course of infection, dermatophytes such as *T. rubrum* secrete endo- and exoproteases that degrade the keratin of the host tissue into oligopeptides and amino acids [30]. These compounds are used as a source of carbon, nitrogen, phosphorus, and sulfur for nutrition of the fungus [31].

The results of dual RNA-seq showed the induction of *TERG_12606* ($\log_2$ fold change: 2.16) and *TERG_08405* ($\log_2$ fold change: 1.29) (functional category: pathogenicity), which encode exoproteases (dipeptidyl peptidase V and leucine aminopeptidase 2, respectively). These findings corroborate the results of Reference [32] which evaluated gene expression by microarray in *T. rubrum* grown in a keratin-containing medium, and in [33] which evaluated the secretion of exoproteases, including dipeptidyl peptidase V, by *T. rubrum* in a keratin-containing medium. The secretion of endo- and exo-proteases by dermatophytes is one of the best-characterized virulence factors [32,33] and is of fundamental importance for invasion and dissemination of the fungus through the stratum corneum of the host [34].

4.2. The ERG6 Gene Is a Promising Target for Developing a New Antifungal Agent Against T. rubrum

In addition to the need of effective degradation of skin protein components for penetration of the fungus into tissue, the maintenance of fungal plasma membrane permeability and fluidity is essential for the correct assimilation of nutrients and the consequent growth and survival of *T. rubrum* in the host. In the present study, we observed the induction of *TERG_03102* ($\log_2$ fold change: 2.05) (functional category: metabolic process), which corresponds to the *ERG6* gene. This gene encodes the enzyme 24-C-methyltransferase, which participates in ergosterol biosynthesis [35]. Ergosterol is known to be responsible for fungal plasma membrane fluidity and permeability and it is important for the adequate function of membrane-anchored proteins [36].

The ergosterol biosynthesis pathway, which is absent in mammals, is the target of antifungal agents such as terbinafine. However, new genes of this pathway should be explored as potential targets because of reports of resistance of *T. rubrum* to this commercial antifungal drug [37]. One example of a promising potential target of new antifungals is the *ERG6* gene whose expression was found to be modulated in this study. Altered expression of this gene results in plasma membrane changes, impairing the transport of nutrients into the fungal cell [38]. The importance of this gene as a new therapeutic strategy has also been reported in [29]. In a comparative genomics study, these authors identified this gene in important human fungal pathogens such as *Candida albicans* and *Aspergillus fumigatus*.

4.3. Glyoxylate Cycle Genes and a Carboxylic Acid Transporter May Be Associated with Mechanisms of Metabolic Flexibility in the T. rubrum-Host Relationship

Additionally, regarding the importance of nutrient assimilation by the fungus for its development during infection, the metabolic flexibility of some pathogenic fungi is worth noting. This flexibility enables the fungus to obtain nutrients through the assimilation of alternative carbon sources in nutrient-limited host niches [39,40]. Knowledge of the genes that are induced to favor this metabolic flexibility is still limited. Thus, these genes are interesting targets for the development of more selective antifungals since the induction of alternative metabolic pathways is an exclusive property of pathogenic fungi [41].

In the present study, genes involved in metabolic flexibility were modulated: *TERG_01281* ($\log_2$ fold change: 1.72), *TERG_11638* ($\log_2$ fold change: 1.26) and *TERG_11639* ($\log_2$ fold change: 1.13) (functional category: glyoxylate cycle), which encode malate synthase and isocitrate lyases, respectively, are enzymes that participate in the glyoxylate cycle. In other clinical fungi, the activation of this cycle permits cell survival in low-glucose environments through the synthesis of glucose from lipids and other carbon sources [41]. We suggest this strategy could favor the growth and persistence of *T. rubrum* in the host since the fungus infects tissues rich in keratin and lipids. Furthermore, this cycle provides pathogenicity and virulence to other pathogens such as *C. albicans*, since the alternative assimilation of nutrients in nutrient-limited host niches favors pathogen survival and adaptation to the host [39].

The role of the glyoxylate cycle in the pathogenicity of *T. rubrum* is still not well established considering that this fungus causes superficial infections. However, we also showed the induction of genes encoding isocitrate lyase and malate synthase during the co-culture of HaCat keratinocytes with *T. rubrum*. The same genes were repressed in the presence of antifungal compounds licochalcone and caffeic acid in the co-culture for 24 h [42]. We also highlight the induction of *TERG_00916* ($\log_2$ fold change: 1.28), which encodes a carboxylic acid transporter (functional category: transport), and suggest that the fungus can use this transporter to facilitate the assimilation of carboxylic acids as an alternative carbon source during infection. The expression of two short-chain carboxylic acid transporters has been demonstrated in *C. albicans* when glucose availability in the host is low. These findings indicate the importance of these transporters in the early stages of infection, contributing to the virulence of the pathogen [43].

4.4. The Modulation of Genes Involved in the Maintenance of the Skin Barrier, Cell Migration, and Differentiation May Be Associated with the Defense Strategies of Human Keratinocytes

The degradation of keratin present in the epidermis through the secretion of proteases such as those modulated in this study (*TERG_12606* and *TERG_08405*) causes marked changes in the function and structure of the epithelial barrier [44]. Repression of the *FLG* ($\log_2$ fold change: −1.86) and *KRT1* ($\log_2$ fold change: −4.02) genes that encode filaggrin and keratin 1, respectively, was observed during the 24 h of co-culture of *T. rubrum* with HaCat cells. We suggest the repression of the *FLG* and *KRT1* genes to be related to the loss of skin barrier integrity, favoring the installation and tissue invasion by the fungus since the proteins encoded by these genes act together during the transition of keratinocytes to corneocytes that will compose the epithelial barrier [45,46]. These results corroborate the findings reported in [47] which identified the reduced expression of filaggrin in cases of tinea corporis caused by *T. rubrum*, and in [48] which observed the loss of skin barrier integrity in *KRT1*-deficient mice.

In the case of damage to the skin barrier, creating a portal of entry for exogenous microorganisms, epithelial cells respond rapidly to close the wound by increasing cell proliferation. In addition, the remodeling of affected tissue occurs and the migration of epithelial and immunocompetent cells to the site of infection is facilitated [49,50].

Among the genes allocated to the functional category of epidermal cell differentiation, the most modulated gene was *CASP14* ($\log_2$ fold change: 3.74), which encodes caspase 14 (Table 4). This is the only caspase not involved in apoptotic pathways [51,52] and an increase in its expression is associated with the differentiation of keratinocytes into corneocytes [53,54], demonstrating a low accumulation of filaggrin fragments in the stratum corneum and increased epithelial water loss in caspase 14-deficient mice. Thus, the induction of *CASP14* expression might be related to the increased differentiation of keratinocytes into corneocytes in an attempt to strengthen the epithelial barrier. Another possibility is that the increased expression of the *CASP14* gene is involved in the repair of damage caused by the repression of the *FLG* and *KRT1* genes as a host defense response during infection with *T. rubrum*.

Regarding the functional category containing genes involved in the regulation of cell migration, the induction of the *MMP9* gene ($\log_2$ fold change: 1.46), which encodes matrix metalloproteinase 9, should be highlighted (Table 4). In addition to the role of matrix metalloproteinases in the remodeling of damaged tissues through the degradation of extracellular matrix, studies have shown that matrix metalloproteinase 9 is necessary for the migration of inflammatory cells to the epidermis [55]. Considering the data available so far, the induction of this gene may indicate an important role in the regulation of the flow of immunocompetent cells through the epidermal compartment in infections caused by *T. rubrum*. Since this protein is produced in its inactive form [56], the present results do not permit to establish whether the matrix metalloproteinase 9 becomes active in keratinocytes during dermatophyte infections. Furthermore, the increased expression of this enzyme in its active form may be associated with an increase in inflammation and the occurrence of ulcers in some diseases such as ocular herpes [57] and leishmaniasis [58], in addition to facilitating the dissemination of the pathogen through tissues by excessive cleavage of collagen IV present in the basement membrane [59].

With respect to other genes involved in the regulation of cell migration, the induction of the *HAS2* gene was observed ($\log_2$ fold change: 1.46), which encodes hyaluronan synthase 2, an enzyme that participates in the synthesis of hyaluronic acid. This acid is one of the main components of the extracellular matrix and plays an important role in the repair of damaged tissues, contributing to the activation of inflammatory cells and the stimulation of chemokines and cytokines through its interaction with Toll-like receptors [60]. Studies also indicate a potential antifungal effect of hyaluronic acid, which inhibits the growth of *C. albicans* in vitro [61].

4.5. The Induction of Genes Involved in the Immune Response of Human Keratinocytes that Encode Compounds with Antimicrobial Activity

Among the functional categories studied, the most important to be evaluated during the fungal-host interaction are those containing the set of genes involved in the human cellular defense. These genes participate not only in the signaling and recruitment of immune system cells, but also in the production of compounds by the host that have a potential antimicrobial effect. These include genes allocated to the MAPK cascade involved in the innate immune response and antimicrobial humoral immune response categories (Table 4).

As an innate cellular defense mechanism, keratinocytes produce peptides with antimicrobial activity, such as cathelicidins, defensins, and ribonucleases [62]. We observed the induction of the *RNASE7* gene ($\log_2$ fold change: 2.27) that encodes ribonuclease 7. This ribonuclease is known for its marked antimicrobial activity against Gram-positive and -negative bacteria, *C. albicans* [63] and dermatophytes [64], suggesting its use as a new antifungal agent.

Compounds that can be used as new approaches to the treatment of fungal diseases are increasingly being explored because of the growing resistance of pathogenic fungi to conventional antifungal agents [65]. In addition to the *RNASE7* gene, we highlight the induction of the *CSF2* gene ($\log_2$ fold change: 2.86), which encodes the cytokine granulocyte-macrophage colony-stimulating factor (GM-CSF). Studies indicate the clinical use of this cytokine as an immunological adjuvant for the treatment of fungal diseases. GM-CSF has already been used to treat neutropenic patients undergoing chemotherapy, HIV-infected patients, and bone marrow transplant recipients [66,67]. The effects of this cytokine have been evaluated in species of the genera *Candida* [68] and *Aspergillus* [69], administered alone or in combination with other commercial antifungals.

The induction of the *CSF2* and *RNASE7* genes during co-culture of human keratinocytes with *T. rubrum* may indicate an important cellular defense response of the host when in contact with this fungus since both genes encode compounds with antimicrobial activity. Furthermore, the production of these compounds favors the recruitment of immunocompetent cells to the affected sites that are important for the host's innate immune mechanisms [63,70,71]

Another gene that was found to be induced in this study and that is also known for its antimicrobial activity is *SLC11A1*. This gene encodes an integral membrane protein [72] that mediates the transport of divalent ions, activating macrophages and exerting other pleiotropic effects on the innate immune system [73]. The available data indicate that this protein protects the host against intracellular pathogens such as *Salmonella* by controlling iron homeostasis inside macrophages, limiting the access of the pathogen to this essential element inside the host, and by concomitantly promoting an increase in the production of antimicrobial effector molecules [72].

Although more elucidated in macrophages, the increased expression of this gene was also observed in keratinocytes of patients with severe burns, suggesting that this gene participates in the innate immune response in the presence of tissue injury [74]. Tissue damage also occurs in dermatophytoses as a result of the secretion of keratinolytic proteases by the fungus. We, therefore, suggest that the induction of this gene during co-culture of keratinocytes with *T. rubrum* may be associated with a defense mechanism of the host, since the *SLC11A1* gene can also exert some signaling effects on the immune system such as macrophage activation, the regulation of interleukin 1-β, and the induction of iNOS, major histocompatibility (MHC) class II molecules, and tumor necrosis factor α (TNFα), among others [75,76]. However, more in-depth studies are necessary to elucidate this possible mechanism of defense

In summary, within the complex interaction between the fungus and host, we highlight the importance of the modulation of expression of *T. rubrum* genes that contribute to the acquisition and assimilation of nutrients. In this respect, genes responsible for the secretion of keratinolytic proteases (*TERG_12606*; *TERG_08405*) and metabolic adaptation (*TERG_01281*; *TERG_11638*; *TERG_11639*; *TERG_00916*) were found to be induced, as well as the *ERG6* gene that is responsible for maintaining the integrity and permeability of the plasma membrane. In contrast, in the presence of keratinocytes, genes encoding proteins with antimicrobial activity (*RNASE7*; *SLC11A1*; *CSF2*) and genes involved in the maintenance of the skin barrier (*MMP9*; *HAS2*; *CASP14*) are induced, while two genes essential for the stability and integrity of the skin barrier (*FLG*; *KRT1*) are repressed (Figure 4). Considering the limited knowledge, the use of dual RNA-seq allowed for a better understanding of some of the molecular mechanisms involved in the *T. rubrum*-host relationship.

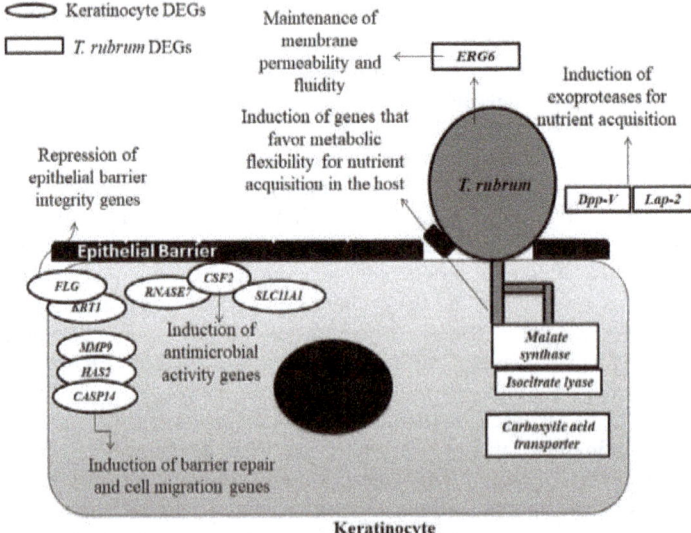

Figure 4. The schematic overview of the *T. rubrum*-keratinocyte interaction. The genes differentially expressed (DEGs) during host-pathogen interaction discussed in this paper are shown.

Supplementary Materials: The following are available online at http://www.mdpi.com/2073-4425/9/7/362/s1, Figure S1: The distribution of differentially expressed genes after 24 h of co-culture. The red points indicate differentially expressed genes. Table S1: The general features of dual RNA-seq sequences against reference genomes. CBS I, CBS II, CBS III: *T. rubrum* libraries; CO I, CO II, CO III: co-culture libraries; H I, H II, H III: human keratinocyte libraries. The libraries were constructed in triplicate, with I, II and III corresponding to the sample number of each condition. PE: paired-end sequence; SR: single read sequence. Table S2: The complete list of genes differentially expressed in keratinocytes after 24 h of co-culture. Table S3: The complete list of genes differentially expressed in *T. rubrum* after 24 h of co-culture. Table S4: The primers used for qPCR analysis.

Author Contributions: M.F.P. performed the laboratory experiments and bioinformatics analysis and wrote the manuscript. K.P. and W.A.d.S.J. constructed the libraries and performed the sequencing. P.R.S. performed the bioinformatics analysis. T.T.K. performed the co-culture experiments. J.B.M. helped with the laboratory experiments. R.O.B. discussed the manuscript. N.M.M.-R. supervised the bioinformatics analysis. M.M. supervised the research and contributed reagents and materials. A.L.F. designed the project, supervised the research, contributed reagents/materials/analysis tools, and wrote the manuscript. All authors have read and approved the final version of the manuscript.

Funding: This study was supported by grants from Fundação de Amparo à Pesquisa do Estado de São Paulo (FAPESP, 2016/22701-9) and CAPES (PhD fellowships granted to MFP and TTK).

Acknowledgments: We thank the Multiuser Electron Microscopy Laboratory of the Department of Cellular and Molecular Biology, Ribeirão Preto Medical School, for the electron microscopy experiment, and Nilce M. Martinez-Rossi for kindly providing the CBS *T. rubrum* strain.

Conflicts of Interest: The authors declare that they have no competing interests.

References

1. Bouchara, J.P.; Mignon, B.; Chaturvedi, V. Dermatophytes and dermatophytoses: a thematic overview of state of the art, and the directions for future research and developments. *Mycopathologia* **2017**, *182*, 1–4. [CrossRef] [PubMed]

2. Rodwell, G.E.; Bayles, C.L.; Towersey, L.; Aly, R. The prevalence of dermatophyte infection in patients infected with human immunodeficiency virus. *Int. J. Dermatol.* **2008**, *47*, 339–343. [CrossRef] [PubMed]

3. Romano, C.; Massai, L.; Asta, F.; Signorini, A.M. Prevalence of dermatophytic skin and nail infections in diabetic patients. *Mycoses* **2001**, *44*, 83–86. [CrossRef] [PubMed]

4. Aly, R. Ecology and epidemiology of dermatophyte infections. *J. Am. Acad. Dermatol.* **1994**, *31*, S21–S25. [CrossRef]

5. Havlickova, B.; Czaika, V.A.; Fredrich, M. Epidemiological trends in skin mycosis worldwide. *Mycosis* **2008**, *51*, 2–15. [CrossRef] [PubMed]

6. Hube, B.; Hay, R.; Brasch, J.; Veraldi, S.; Schaller, M. Dermatomycoses and inflammation: The adaptive balance between growth, damage, and survival. *J. Mycol. Med.* **2015**, *25*, e44–e58. [CrossRef] [PubMed]

7. Achterman, R.R.; White, T.C. A foot in the door for dermatophyte research. *PLoS Pathog.* **2012**, *8*, 6–9. [CrossRef] [PubMed]

8. Wolf, T.; Kämmer, P.; Brunke, S.; Linde, J. Two's company: Studying interspecies relationships with dual RNA-seq. *Curr. Opin. Microbiol.* **2018**, *42*, 7–12. [CrossRef] [PubMed]

9. Aprianto, R.; Slager, J.; Holsappel, S.; Veening, J.-W. Time-resolved dual RNA-seq reveals extensive rewiring of lung epithelial and pneumococcal transcriptomes during early infection. *Genome Biol.* **2016**, *17*, 198. [CrossRef] [PubMed]

10. Wesolowska-Andersen, A.; Everman, J.L.; Davidson, R.; Rios, C.; Herrin, R.; Eng, C.; Janssen, W.J.; Liu, A.H.; Oh, S.S.; Kumar, R.; et al. Dual RNA-seq reveals viral infections in asthmatic children without respiratory illness which are associated with changes in the airway transcriptome. *Genome Biol.* **2017**, *18*, 12. [CrossRef] [PubMed]

11. Tierney, L.; Linde, J.; Müller, S.; Brunke, S.; Molina, J.C.; Hube, B.; Schöck, U.; Guthke, R.; Kuchler, K. An interspecies regulatory network inferred from simultaneous RNA-seq of *Candida albicans* invading innate immune cells. *Front. Microbiol.* **2012**, *3*, 1–14. [CrossRef] [PubMed]

12. Meyer, F.E.; Shuey, L.S.; Naidoo, S.; Mamni, T.; Berger, D.K.; Myburg, A.A.; Van den Berg, N.; Naidoo, S. Dual RNA-sequencing of *Eucalyptus nitens* during *Phytophthora cinnamomi* challenge reveals pathogen and host factors influencing compatibility. *Front. Plant Sci.* **2016**, *7*, 1–15. [CrossRef] [PubMed]

13. Komoto, T.T.; Bitencourt, T.A.; Silva, G.; Beleboni, R.O.; Marins, M.; Fachin, A.L. Gene expression response of *Trichophyton rubrum* during coculture on keratinocytes exposed to antifungal agents. *Evid. Based Complement. Altern. Med.* **2015**, *2015*, 1–7. [CrossRef] [PubMed]

14. Santiago, K.; Bomfim, G.F.; Criado, P.R.; Almeida, S.R. Monocyte-derived dendritic cells from patients with dermatophytosis restrict the growth of *Trichophyton rubrum* and induce CD4-T cell activation. *PLoS ONE* **2014**, *9*, e110879. [CrossRef] [PubMed]

15. Edgar, R. Gene Expression Omnibus: NCBI gene expression and hybridization array data repository. *Nucleic Acids Res.* **2002**, *30*, 207–210. [CrossRef] [PubMed]

16. Langmead, B.; Salzberg, S.L. Fast gapped-read alignment with Bowtie 2. *Nat. Methods* **2012**, *9*, 357–359. [CrossRef] [PubMed]

17. Anders, S.; Huber, W. Differential expression analysis for sequence count data. *Genome Biol.* **2010**, *11*, 1–12. [CrossRef] [PubMed]

18. Blake, J.A.; Harris, M.A. The Gene Ontology (GO) Project: Structured vocabularies for molecular biology and their application to genome and expression analysis. In *Current Protocols in Bioinformatics*; John Wiley & Sons, Inc.: Hoboken, NJ, USA, 2008; pp. 1–9. ISBN 0471250953.

19. Gotz, S.; García-Gómez, J.M.; Terol, J.; Williams, T.D.; Nagaraj, S.H.; Nueda, M.J.; Robles, M.; Talón, M.; Dopazo, J.; Conesa, A. High-throughput functional annotation and data mining with the Blast2GO suite. *Nucleic Acids Res.* **2008**, *36*, 3420–3435. [CrossRef] [PubMed]

20. Vêncio, R.Z.N.; Koide, T.; Gomes, S.L.; de Pereira, C.A.B. BayGO: Bayesian analysis of ontology term enrichment in microarray data. *BMC Bioinf.* **2006**, *7*, 86. [CrossRef] [PubMed]

21. Chen, E.Y.; Tan, C.M.; Kou, Y.; Duan, Q.; Wang, Z.; Meirelles, G.; Clark, N.R.; Ma'ayan, A. Enrichr: Interactive and collaborative HTML5 gene list enrichment analysis tool. *BMC Bioinf.* **2013**, *14*, 128. [CrossRef] [PubMed]

22. Kuleshov, M.V.; Jones, M.R.; Rouillard, A.D.; Fernandez, N.F.; Duan, Q.; Wang, Z.; Koplev, S.; Jenkins, S.L.; Jagodnik, K.M.; Lachmann, A.; et al. Enrichr: A comprehensive gene set enrichment analysis web server 2016 update. *Nucleic Acids Res.* **2016**, *44*, W90–W97. [CrossRef] [PubMed]

23. Ma, R.; Zhang, D.; Hu, P.-C.; Li, Q.; Lin, C.-Y. HOXB7-S3 inhibits the proliferation and invasion of MCF-7 human breast cancer cells. *Mol. Med. Rep.* **2015**, 4901–4908. [CrossRef] [PubMed]

24. Dai, Z.; Ma, X.; Kang, H.; Gao, J.; Min, W.; Guan, H.; Diao, Y.; Lu, W.; Wang, X. Antitumor activity of the selective cyclooxygenase-2 inhibitor, celecoxib, on breast cancer in vitro and in vivo. *Cancer Cell Int.* **2012**, *19*, 1–8. [CrossRef] [PubMed]

25. Bitencourt, T.A.; Komoto, T.T.; Massaroto, B.G.; Miranda, C.E.S.; Beleboni, R.O.; Marins, M.; Fachin, A.L. Trans-chalcone and quercetin down-regulate fatty acid synthase gene expression and reduce ergosterol content in the human pathogenic dermatophyte *Trichophyton rubrum*. *BMC Complement. Altern. Med.* **2013**, *13*, 229. [CrossRef] [PubMed]

26. Jacob, T.R.; Peres, N.T.A.; Persinoti, G.F.; Silva, L.G.; Mazucato, M.; Rossi, A.; Martinez-Rossi, N.M. *Rpb2* is a reliable reference gene for quantitative gene expression analysis in the dermatophyte *Trichophyton rubrum*. *Med. Mycol.* **2012**, *50*, 368–377. [CrossRef] [PubMed]

27. Westermann, A.J.; Barquist, L.; Vogel, J. Resolving host–pathogen interactions by dual RNA-seq. *PLoS Pathog.* **2017**, *13*, 1–19. [CrossRef] [PubMed]

28. Kawahara, Y.; Oono, Y.; Kanamori, H.; Matsumoto, T.; Itoh, T.; Minami, E. Simultaneous RNA-seq analysis of a mixed transcriptome of rice and blast fungus interaction. *PLoS ONE* **2012**, *7*, e49423. [CrossRef] [PubMed]

29. Abadio, A.K.R.; Kioshima, E.S.; Teixeira, M.M.; Martins, N.F.; Maigret, B.; Felipe, M.S.S. Comparative genomics allowed the identification of drug targets against human fungal pathogens. *BMC Genom.* **2011**, *12*, 75. [CrossRef] [PubMed]

30. Baldo, A.; Monod, M.; Mathy, A.; Cambier, L.; Bagut, E.T.; Defaweux, V.; Symoens, F.; Antoine, N.; Mignon, B. Mechanisms of skin adherence and invasion by dermatophytes. *Mycoses* **2012**, *55*, 218–223. [CrossRef] [PubMed]

31. Peres, N.T.D.A.; Maranhão, F.C.A.; Rossi, A.; Martinez-Rossi, N.M. Dermatophytes: Host-pathogen interaction and antifungal resistance. *An. Bras. Dermatol.* **2010**, *85*, 657–667. [CrossRef] [PubMed]

32. Bitencourt, T.A.; Macedo, C.; Franco, M.E.; Assis, A.F.; Komoto, T.T.; Stehling, E.G.; Beleboni, R.O.; Malavazi, I.; Marins, M.; Fachin, A.L. Transcription profile of *Trichophyton rubrum* conidia grown on keratin reveals the induction of an adhesin-like protein gene with a tandem repeat pattern. *BMC Genom.* **2016**, *17*, 249. [CrossRef] [PubMed]

33. Monod, M.; Léchenne, B.; Jousson, O.; Grand, D.; Zaugg, C.; Stöcklin, R.; Grouzmann, E. Aminopeptidases and dipeptidyl-peptidases secreted by the dermatophyte *Trichophyton rubrum*. *Microbiology* **2005**, *151*, 145–155. [CrossRef] [PubMed]

34. Leng, W.; Liu, T.; Wang, J.; Li, R.; Jin, Q. Expression dynamics of secreted protease genes in *Trichophyton rubrum* induced by key host's proteinaceous components. *Med. Mycol.* **2009**, *47*, 759–765. [CrossRef] [PubMed]

35. Azam, S.S.; Abro, A.; Raza, S.; Saroosh, A. Structure and dynamics studies of sterol 24-C-methyltransferase with mechanism based inactivators for the disruption of ergosterol biosynthesis. *Mol. Biol. Rep.* **2014**, *41*, 4279–4293. [CrossRef] [PubMed]

36. Iwaki, T.; Iefuji, H.; Hiraga, Y.; Hosomi, A.; Morita, T.; Giga-Hama, Y.; Takegawa, K. Multiple functions of ergosterol in the fission yeast *Schizosaccharomyces pombe*. *Microbiology* **2008**, *154*, 830–841. [CrossRef] [PubMed]

37. Osborne, C.S.; Leitner, I.; Favre, B.; Neil, S.; Osborne, C.S.; Leitner, I.; Favre, B.; Ryder, N.S. Amino acid substitution in *Trichophyton rubrum* squalene epoxidase associated with resistance to terbinafine. **2005**, *49*, 2840–2844. [CrossRef] [PubMed]

38. Gaber, R.F.; Copple, D.M.; Kennedy, B.K.; Vidal, M.; Bard, M. The yeast gene *ERG6* is required for normal membrane function but is not essential for biosynthesis of the cell-cycle-sparking sterol. *Mol. Cell. Biol.* **1989**, *9*, 3447–3456. [CrossRef] [PubMed]

39. Mayer, F.L.; Wilson, D.; Hube, B. *Candida albicans* pathogenicity mechanisms. *Virulence* **2013**, *4*, 119–128. [CrossRef] [PubMed]

40. Cheah, H.L.; Lim, V.; Sandai, D. Inhibitors of the glyoxylate cycle enzyme ICL1 in *Candida albicans* for potential use as antifungal agents. *PLoS ONE* **2014**, *9*. [CrossRef] [PubMed]

41. Fleck, C.B.; Schöbel, F.; Brock, M. Nutrient acquisition by pathogenic fungi: Nutrient availability, pathway regulation, and differences in substrate utilization. *Int. J. Med. Microbiol.* **2011**, *301*, 400–407. [CrossRef] [PubMed]

42. Cantelli, B.A.M.; Bitencourt, T.A.; Komoto, T.T.; Beleboni, R.O.; Marins, M.; Fachin, A.L. Caffeic acid and licochalcone A interfere with the glyoxylate cycle of *Trichophyton rubrum*. *Biomed. Pharmacother.* **2017**. [CrossRef] [PubMed]

43. Vieira, N.; Casal, M.; Johansson, B.; MacCallum, D.M.; Brown, A.J.P.; Paiva, S. Functional specialization and differential regulation of short-chain carboxylic acid transporters in the pathogen *Candida albicans*. *Mol. Microbiol.* **2010**, *75*, 1337–1354. [CrossRef] [PubMed]

44. Lee, W.J.; Kim, J.Y.; Song, C.H.; Jung, H.D.; Lee, S.H.; Lee, S.J.; Kim, D.W. Disruption of barrier function in dermatophytosis and pityriasis versicolor. *J. Dermatol.* **2011**, *38*, 1049–1053. [CrossRef] [PubMed]

45. Brown, S.J.; Irvine, A.D. Atopic eczema and the filaggrin story. *Semin. Cutan. Med. Surg.* **2008**, *27*, 128–137. [CrossRef] [PubMed]

46. McGrath, J.A. Filaggrin and the great epidermal barrier grief. *Australas. J. Dermatol.* **2008**, *49*, 67–74. [CrossRef] [PubMed]

47. Jensen, J.-M.; Pfeiffer, S.; Akaki, T.; Schröder, J.-M.; Kleine, M.; Neumann, C.; Proksch, E.; Brasch, J. Barrier function, epidermal differentiation, and human β-defensin 2 expression in Tinea Corporis. *J. Investig. Dermatol.* **2007**, *127*, 1720–1727. [CrossRef] [PubMed]

48. Roth, W.; Kumar, V.; Beer, H.-D.; Richter, M.; Wohlenberg, C.; Reuter, U.; Thiering, S.; Staratschek-Jox, A.; Hofmann, A.; Kreusch, F.; et al. Keratin 1 maintains skin integrity and participates in an inflammatory network in skin through interleukin-18. *J. Cell Sci.* **2012**, *125*, 5269–5279. [CrossRef] [PubMed]

49. Parks, W.; Wilson, C.; López-Boado, Y. Matrix metalloproteinases as modulators of inflammation and innate immunity. *Nat. Rev. Immunol.* **2004**, *4*, 617–629. [CrossRef] [PubMed]

50. Purwar, R.; Kraus, M.; Werfel, T.; Wittmann, M. Modulation of keratinocyte-derived MMP-9 by IL-13: A possible role for the pathogenesis of epidermal inflammation. *J. Investig. Dermatol.* **2008**, *128*, 59–66. [CrossRef] [PubMed]

51. Hvid, M.; Johansen, C.; Deleuran, B.; Kemp, K.; Deleuran, M.; Vestergaard, C. Regulation of caspase 14 expression in keratinocytes by inflammatory cytokines—A possible link between reduced skin barrier function and inflammation? *Exp. Dermatol.* **2011**, *20*, 633–636. [CrossRef] [PubMed]

52. Gkegkes, I.D.; Aroni, K.; Agrogiannis, G.; Patsouris, E.S.; Konstantinidou, A.E. Expression of caspase-14 and keratin-19 in the human epidermis and appendages during fetal skin development. *Arch. Dermatol. Res.* **2013**, *305*, 379–387. [CrossRef] [PubMed]

53. Lippens, S.; Kockx, M.; Knaapen, M.; Mortier, L.; Polakowska, R.; Verheyen, A.; Garmyn, M.; Zwijsen, A.; Formstecher, P.; Huylebroeck, D.; et al. Epidermal differentiation does not involve the pro-apoptotic executioner caspases, but is associated with caspase-14 induction and processing. *Cell Death Differ.* **2000**, *7*, 1218–1224. [CrossRef] [PubMed]

54. Denecker, G.; Hoste, E.; Gilbert, B.; Hochepied, T.; Ovaere, P.; Lippens, S.; Van den Broecke, C.; Van Damme, P.; D'Herde, K.; Hachem, J.-P.; et al. Caspase-14 protects against epidermal UVB photodamage and water loss. *Nat. Cell Biol.* **2007**, *9*, 666–674. [CrossRef] [PubMed]

55. Ratzinger, G.; Stoitzner, P.; Ebner, S.; Lutz, M.B.; Layton, G.T.; Rainer, C.; Senior, R.M.; Shipley, J.M.; Fritsch, P.; Schuler, G.; et al. Matrix metalloproteinases 9 and 2 are necessary for the migration of langerhans cells and dermal dendritic cells from human and murine skin. *J. Immunol.* **2002**, *168*, 4361–4371. [CrossRef] [PubMed]

56. Visse, R.; Nagase, H. Matrix metalloproteinases and tissue inhibitors of metalloproteinases: Structure, function, and biochemistry. *Circ. Res.* **2003**, *92*, 827–839. [CrossRef] [PubMed]

57. Lee, S.; Zheng, M.; Kim, B.; Rouse, B.T. Role of matrix metalloproteinase-9 in angiogenesis caused by ocular infection with herpes simplex virus. *J. Clin. Investig.* **2002**, *110*, 1105–1111. [CrossRef] [PubMed]

58. Campos, T.M.; Passos, S.T.; Novais, F.O.; Beiting, D.P.; Costa, R.S.; Queiroz, A.; Mosser, D.; Scott, P.; Carvalho, E.M.; Carvalho, L.P. Matrix metalloproteinase 9 production by monocytes is enhanced by TNF and participates in the pathology of human Cutaneous Leishmaniasis. *PLoS Negl. Trop. Dis.* **2014**, *8*. [CrossRef] [PubMed]

59. Murphy, G.; Nagase, H. Progress in matrix metalloproteinase research. *Mol. Aspects Med.* **2009**, *29*, 290–308. [CrossRef] [PubMed]

60. Jiang, D.; Liang, J.; Noble, P.W. Hyaluronan in tissue injury and repair. *Annu. Rev. Cell Dev. Biol.* **2007**, *23*, 435–461. [CrossRef] [PubMed]

61. Sakai, A.; Akifusa, S.; Itano, N.; Kimata, K.; Kawamura, T.; Koseki, T.; Takehara, T.; Nishihara, T. Potential role of high molecular weight hyaluronan in the anti-*Candida* activity of human oral epithelial cells. *Med. Mycol.* **2007**, *45*, 73–79. [CrossRef] [PubMed]

62. Becknell, B.; Spencer, J.D. A Review of ribonuclease 7's structure, regulation, and contributions to host defense. *Int. J. Mol. Sci.* **2016**, *17*, 423. [CrossRef] [PubMed]

63. Harder, J.; Schröder, J.M. RNase 7, a novel innate immune defense antimicrobial protein of healthy human skin. *J. Biol. Chem.* **2002**, *277*, 46779–46784. [CrossRef] [PubMed]

64. Fritz, P.; Beck-Jendroschek, V.; Brasch, J. Inhibition of dermatophytes by the antimicrobial peptides human β-defensin-2, ribonuclease 7 and psoriasin. *Med. Mycol.* **2012**, *50*, 579–584. [CrossRef] [PubMed]

65. Mehra, T.; Köberle, M.; Braunsdorf, C.; Mailänder-Sanchez, D.; Borelli, C.; Schaller, M. Alternative approaches to antifungal therapies. *Exp. Dermatol.* **2012**, *21*, 778–782. [CrossRef] [PubMed]

66. Hubel, K.; Dale, D.C.; Liles, W.C. Therapeutic use of cytokines to modulate phagocyte function for the treatment of infectious diseases: Current status of granulocyte colony-stimulating factor, granulocyte-macrophage colony-stimulating factor, macrophage colony-stimulating factor, and interferon-gamma. *J. Infect. Dis.* **2002**, *185*, 1490–1501. [CrossRef] [PubMed]

67. Shi, Y.; Liu, C.H.; Roberts, A.I.; Das, J.; Xu, G.; Ren, G.; Zhang, Y.; Zhang, L.; Yuan, Z.R.; Tan, H.S.W.; et al. Granulocyte-macrophage colony-stimulating factor (GM-CSF) and T-cell responses: What we do and don't know. *Cell Res.* **2006**, *16*, 126–133. [CrossRef] [PubMed]

68. Liehl, E.; Hildebrandt, J.; Lam, C.; Mayer, P. Prediction of the role of granulocyte-macrophage colony-stimulating factor in animals and man from in vitro results. *Eur. J. Clin. Microbiol. Infect. Dis.* **1994**, *13*, 9–17. [CrossRef]

69. Bodey, G.P.; Anaissie, E.; Gutterman, J.; Vadhan-Raj, S. Role of granulocyte-macrophage colony-stimulating factor as adjuvant treatment in neutropenic patients with bacterial and fungal infection. *Eur. J. Clin. Microbiol. Infect. Dis.* **1994**, *13* (Suppl. 2), S18–S22. [CrossRef] [PubMed]

70. Hamilton, J.A. GM-CSF in inflammation and autoimmunity. *Trends Immunol.* **2002**, *23*, 403–408. [CrossRef]

71. Rademacher, F.; Simanski, M.; Harder, J. RNase 7 in cutaneous defense. *Int. J. Mol. Sci.* **2016**, *17*. [CrossRef] [PubMed]

72. Nairz, M.; Fritsche, G.; Crouch, M.L.V.; Barton, H.C.; Fang, F.C.; Weiss, G. Slc11a1 limits intracellular growth of *Salmonella enterica* sv. Typhimurium by promoting macrophage immune effector functions and impairing bacterial iron acquisition. *Cell. Microbiol.* **2009**, *11*, 1365–1381. [CrossRef] [PubMed]

73. Stober, C.B.; Brode, S.; White, J.K.; Popoff, J.-F.; Blackwell, J.M. Slc11a1, formerly Nramp1, is expressed in dendritic cells and influences major histocompatibility complex class II expression and antigen-presenting cell function. *Infect. Immun.* **2007**, *75*, 5059–5067. [CrossRef] [PubMed]

74. Noronha, S.A.; Noronha, S.M.; Lanziani, L.E.; Ferreira, L.M.; Gragnani, A. Innate and adaptive immunity gene expression of human keratinocytes cultured of severe burn injury. *Acta Cir. Bras.* **2014**, *29* (Suppl. 3), 60–67. [CrossRef] [PubMed]

75. Blackwell, J.M.; Searle, S.; Goswami, T.; Miller, E.N. Understanding the multiple functions of Nrampl. *Microbes Infect.* **2000**, *2*, 317–321. [CrossRef]

76. Blackwell, J.M.; Goswami, T.; Evans, C.A.W.; Sibthorpe, D.; Papo, N.; White, J.K.; Searle, S.; Miller, E.N.; Peacock, C.S.; Mohammed, H.; et al. SLC11A1 (formerly NRAMP1) and disease resistance. *Cell. Microbiol.* **2001**, *3*, 773–784. [CrossRef] [PubMed]

Article

Genome-Wide Comparative Analysis of *Aspergillus fumigatus* Strains: The Reference Genome as a Matter of Concern

Rocio Garcia-Rubio [1,†], Sara Monzon [2,†], Laura Alcazar-Fuoli [1], Isabel Cuesta [2] and Emilia Mellado [1,*]

1   Mycology Reference Laboratory, National Centre for Microbiology, Instituto de Salud Carlos III (ISCIII), Majadahonda, 28220 Madrid, Spain; rgarciar@isciii.es (R.G.-R.); lalcazar@isciii.es (L.A.-F.)

2   Bioinformatics Unit, Common Scientific Technical Units, ISCIII, Majadahonda, 28220 Madrid, Spain; smonzon@isciii.es (S.M.); isabel.cuesta@isciii.es (I.C.)

*   Correspondence: emellado@isciii.es; Tel.: +34-918-223-427

†   These authors contributed equally to this work.

Received: 28 June 2018; Accepted: 16 July 2018; Published: 19 July 2018

Abstract: *Aspergillus fumigatus* is a ubiquitous saprophytic mold and a major pathogen in immunocompromised patients. The effectiveness of triazole compounds, the *A. fumigatus* first line treatment, is being threatened by a rapid and global emergence of azole resistance. Whole genome sequencing (WGS) has emerged as an invaluable tool for the analysis of genetic differences between *A. fumigatus* strains, their genetic background, and antifungal resistance development. Although WGS analyses can provide a valuable amount of novel information, there are some limitations that should be considered. These analyses, based on genome-wide comparative data and single nucleotide variant (SNV) calling, are dependent on the quality of sequencing, assembling, the variant calling criteria, as well as on the suitable selection of the reference genome, which must be genetically close to the genomes included in the analysis. In this study, 28 *A. fumigatus* genomes sequenced in-house and 73 available in public data bases have been analyzed. All genomes were distributed in four clusters and showed a variable number of SNVs depending on the genome used as reference (Af293 or A1163). Each reference genome belonged to a different cluster. The results highlighted the importance of choosing the most suitable *A. fumigatus* reference genome to avoid misleading conclusions.

Keywords: *Aspergillus fumigatus*; whole genome sequencing; reference genomes; azole resistance; *cyp51A*

1. Introduction

Aspergillus fumigatus is a saprophytic filamentous fungus and the principal causative agent of human aspergillosis [1,2]. There are many types of diseases caused by *A. fumigatus* and their symptoms vary according to the site of infection and host health condition. Predominantly, this species causes invasive infections, such as invasive pulmonary aspergillosis (IPA), with high mortality rates in immunocompromised or immunosuppressed patients [3–5]. Currently, the treatment options are limited to three classes of antifungal drugs (azoles, echinocandins and polyenes) and azoles are drugs of first line for treating *Aspergillus*-caused diseases. However, azole drug efficacy is threatened by the worldwide spreading of resistance [6].

Despite all the technological improvements developed in recent years, many basic aspects about the biology of this opportunistic pathogen remained largely unknown. Genomics and whole genome sequencing (WGS) have emerged as useful tools to greatly enhance knowledge and understanding of infectious diseases and clinical microbiology [7–10]. In this context, genome-wide sequencing using high throughput sequencing (HTS) together with alignment comparison analysis has been described as a good approach to identify single nucleotide variants (SNVs) or polymorphisms (SNPs) for the analysis of genetic differences between *A. fumigatus* strains [11]. The genetic diversity between isolates can be used to explain many different aspects such as the strain background lineage, strain specific virulence phenotypes, the potential factors involved in antifungal resistant development, and more commonly, determining the genetic strain relatedness in epidemiological surveillance and outbreak studies [12–14].

Although WGS studies provide a valuable amount of novel data, there are some considerations that should be taken into account. To date, most WGS analyses are based on identifying SNVs in a set of strains in comparison to a reference genome, and consequently the results will be completely dependent on the chosen reference. Therefore, the identified SNVs will have a different significance depending on how genetically close the reference genome will be from the strains included in the analysis. Classically, most advances have been focused on unravelling the genomics of two different *A. fumigatus* reference strains [15], Af293 (AAHF00000000.1) and A1163 (ABDB00000000.1), or their derivatives. First, the Af293 strain was fully sequenced in 2005 [16], and soon after, the A1163 strain was the second *A. fumigatus* genome sequenced by the J. Craig Venter Institute [17]. This strain was a derivative of another clinical isolate, the CBS144-89/CEA10. The first genome comparison between both strains showed that despite high synteny and identity in most of the regions of their genomes, there were some hundreds of genes unique from each strain and not present in the other [18]. This heterogeneity could potentially lead to confusion, so care must be taken in drawing conclusions on different aspects from a single genetic background [15]. A recent study carried out in our laboratory [19] showed that the Af293 reference genome belonged to a specific lineage of *A. fumigatus* strains which harbors five polymorphisms (F46Y, M172V, N248T, D255E, E427K) in *cyp51A* gene which codes for the azole target. Based on these phylogenetic studies from whole genome sequencing, the other classic reference genome, the A1163 strain, belonged to a different and distant phylogenetic lineage, which implies that both genomes are significantly different. Also, and apart from the relevance of choosing the most suitable reference genome, the quality of sequencing and assembling will be of great importance for getting reliable results.

In this work, we describe the in-house genome sequencing and comparative analysis of 28 *A. fumigatus* genomes, including the *A. fumigatus* genomes of six strains routinely used in *Aspergillus* basic research laboratories. All these genomes were distributed in four separate clusters, corresponding to very different lineages. In addition, 73 *A. fumigatus* genomes, including azole susceptible and resistant strains from a worldwide distribution, were included in the study. The whole sequence analysis of these 101 *A. fumigatus* genomes provided information that led to an increase in the knowledge of *A. fumigatus* genetic background and the development of azole resistance.

2. Materials and Methods

2.1. Aspergillus fumigatus Isolates

A total of 169 genomes were evaluated including in-house sequenced *Aspergillus fumigatus* isolates and genomes downloaded from NCBI Sequence Read Archive (SRA) public database (www.ncbi. nlm.nih.gov/sra). The genomic samples were selected following this criteria: (i) all genomes had to be sequenced with Illumina platform; and (ii) had to be paired-end. After quality control filtering (>85% mapping rate against Af293 reference genome and >95% of genome coverage at more than 10× depth of coverage), a total of 101 *A. fumigatus* strain genome sequences representing 8 different geographical locations were selected for further analysis. Among them, 28 *A. fumigatus* strains were from the Mycology Reference Laboratory Collection (Table 1). Most of them were clinical azole susceptible strains. Some of them had specific *cyp51A* modifications: 12 strains harbored three changes (F46Y, M172V, and E427K) and 3 strains had five (F46Y, M172V, N248T, D255E and E427K) [19]. Af293 was included in the latter group of five *cyp51A* modifications and sequenced again in-house (named as AF293). Another group of *A. fumigatus* azole susceptible strains were isolates that have been classically used as reference or control strains in most *Aspergillus* laboratories. This set of strains included the A1163 (also known as CEA10, CBS144.89, FGSC A1163 or AF10), AF237 (called CM237), ATCC204305, *akuB*^{KU80} (FGSC A1160) [20], and ATCC46645. Two azole resistant strains without modifications in *cyp51A* were also included and sequenced in this work (CM7510 and CM7555).

Table 1. *Aspergillus fumigatus* strains whole genome sequenced in our laboratory. The assigned lineage was based on the phylogenetic study.

Samples	Origin	*cyp51A* Modifications	AZL SC	Mating Type	Source	Lineage
AF293	SP	5SNPs	S	M1.2	CL	3
akuB^{KU80}	SP	WT	S	M1.1	CL	1.1
ATCC204305	SP	I242V	S	M1.1	CL	1.2
ATCC46645	SP	WT	S	M1.1	CL	2.1
CEA10	SP	WT	S	M1.1	CL	1.1
CM2141	SP	WT	S	M1.1	CL	2.2
CM237	SP	WT	S	M1.1	CL	1.1
CM2495	SP	3SNPs	S	M1.1	CL	4
CM2730	SP	3SNPs	S	M1.2	CL	4
CM2733	SP	3SNPs	S	M1.1	CL	4
CM3248	SP	N248K	S	M1.1	CL	1.1
CM3249	SP	3SNPs	S	M1.1	CL	4
CM3249b	SP	3SNPs	S	M1.1	CL	4
CM3262	SP	3SNPs	S	M1.1	CL	4
CM3720	SP	3SNPs	S	M1.1	CL	4
CM4602	SP	3SNPs	S	M1.2	CL	4
CM4946	SP	3SNPs	S	M1.2	CL	4
CM5419	SP	WT	S	M1.2	CL	2.2
CM5757	SP	WT	S	M1.2	CL	1.1
CM6126	SP	WT	S	M1.2	CL	1.1
CM6458	SP	WT	S	M1.2	CL	1.3
CM7510	SP	WT	R	M1.1	CL	1.1
CM7555	SP	WT	R	M1.2	CL	1.3
CM7560	SP	3SNPs	S	M1.1	CL	4
CM7570	SP	3SNPs	S	M1.1	CL	4
CM7632	SP	5SNPs	S	M1.1	CL	3
TP12	SP	5SNPs	S	M1.2	CL	3
TP32	SP	3SNPs	S	M1.1	CL	4

SP (Spain), SNPS (single nucleotide polymorphisms), 3SNPs *cyp51A* modifications (F46Y, M172V, E427K) and 5SNPs *cyp51A* modifications (F46Y, M172V, N248T, D255E, E427K), AZL susceptibility (azole susceptibility), S (susceptible), R (resistant), CL (clinical).

The remaining 73 strain genomes were obtained from public data bases and belonged to strains from the United Kingdom, the Netherlands, Portugal, Denmark, India, Japan and Canada [13,21,22] (NCBI SRA public database) (Table 2). Some of the isolates from The Netherlands and India came from clinical samples (*n* = 8) and others from environmental soil sources (*n* = 8). Strains from Japan, UK and Canada were all from clinical origin while the Portuguese strains had an unknown origin (Table 2).

Table 2. *Aspergillus fumigatus* genomes from public data bases included in this study. The assigned lineage is based on the phylogenetic study.

Samples	Origin	*cyp51A* Modifications	AZL SC	Mating Type	Source	Lineage	Ref.
08-12-12-13	NT	TR$_{34}$/L98H, S297T, F495I	R	M1.1	CL	2.2	[13]
08-19-02-10	NT	TR$_{34}$/L98H	R	M1.2	ENV	2.2	[13]
08-19-02-30	NT	WT	S	M1.1	ENV	1.3	[13]
08-19-02-46	NT	TR$_{34}$/L98H	R	M1.1	ENV	2.2	[13]
08-19-02-61	NT	TR$_{34}$/L98H	R	M1.1	ENV	2.2	[13]
08-31-08-91	NT	TR$_{34}$/L98H	R	M1.1	CL	2.2	[13]
08-36-03-25	NT	TR$_{34}$/L98H, S297T, F495I	R	Unclear	CL	2.2	[13]
09-7500806	UK	WT	S	M1.2	CL	1.3	[13]
10-01-02-27	NT	TR$_{34}$/L98H	R	M1.2	CL	2.2	[13]
12-7504462	UK	WT	S	M1.2	CL	1.3	[13]
12-7504652	UK	WT	S	M1.1	CL	2.1	[13]
12-7505054	UK	WT	S	M1.1	CL	2.1	[13]

Table 2. *Cont.*

Samples	Origin	*cyp51A* Modifications	AZL SC	Mating Type	Source	Lineage	Ref.
12-7505220	UK	TR$_{34}$/L98H	R	M1.1	CL	2.2	[13]
12-7505446	UK	TR$_{34}$/L98H	R	M1.2	CL	2.2	[13]
Af293	UK	5SNPs	S	M1.2	CL	3	[13]
AF41	UK	WT	S	M1.2	CL	1.1	[13]
Af65	UK	WT	S	M1.2	CL	2.1	[13]
AF72	UK	G54E	R	M1.2	CL	1.1	[13]
AF90	UK	M220V	R	M1.1	CL	1.1	[13]
Afu1042-09	IN	TR$_{34}$/L98H	R	M1.1	CL	2.2	[13]
Afu124-E11	IN	TR$_{34}$/L98H	R	M1.1	ENV	2.2	[13]
Afu166-E11	IN	TR$_{34}$/L98H	R	M1.1	ENV	2.2	[13]
Afu218-E11	IN	TR$_{34}$/L98H	R	M1.1	ENV	2.2	[13]
Afu257-E11	IN	TR$_{34}$/L98H	R	M1.1	ENV	2.2	[13]
Afu343-P-11	IN	TR$_{34}$/L98H	R	Unclear	CL	2.2	[13]
Afu591-12	IN	TR$_{34}$/L98H	R	M1.1	CL	2.2	[13]
Afu942-09	IN	TR$_{34}$/L98H	R	M1.1	CL	2.2	[13]
F12041	UK	G138C	R	M1.2	CL	2.1	[21] *
F12219	UK	G54R	R	M1.2	CL	1.1	[21] *
F12636	UK	G54E	R	M1.1	CL	2.1	[21] *
F13535	UK	G138C	R	M1.2	CL	2.1	[21] *
F13619	UK	H147Y, G448S	R	M1.1	CL	2.1	[21] *
F13952	UK	G138C	R	M1.2	CL	2.1	[21] *
F14403	UK	G54R	R	M1.1	CL	1.1	[21] *
F14513G	UK	G138C	R	M1.2	CL	2.1	[21] *
F14532	UK	M220T	R	M1.1	CL	2.1	[21] *
F14946G	UK	WT	S	M1.2	CL	1.1	[21] *
F15390	UK	M220T	R	M1.1	CL	2.1	[21] *
F15927	CN	3SNPs	S	M1.1	CL	4	[21] *
F16134	DN	M220K	R	M1.2	CL	1.1	[21] *
F16216	UK	TR$_{34}$/L98H	R	M1.2	CL	2.1	[21] *
F16311	UK	WT	S	M1.1	CL	1.1	[21] *
F17582	UK	WT	S	M1.2	CL	2.1	[21] *
F17729	UK	WT	S	M1.2	CL	1.1	[21] *
F17729W	UK	WT	S	M1.2	CL	1.1	[21] *
F17764	CN	WT	S	M1.1	CL	1.1	[21] *
F18085	UK	WT	S	M1.1	CL	1.1	[21] *
F5211G	UK	WT	S	M1.1	CL	2.1	[21] *
F7763	UK	5SNPs	S	M1.1	CL	3	[21] *
IFM55369	JP	WT	S	M1.1	CL	1.2	[22]
IFM58026	JP	N248K	S	M1.2	CL	1.1	[22]
IFM58029	JP	WT	S	M1.2	CL	1.2	[22]
IFM58401	JP	WT	S	M1.2	CL	1.1	[22]
IFM59056	JP	WT	S	M1.1	CL	1.1	[22]
IFM59073	JP	WT	S	M1.2	CL	1.2	[22]
IFM59359	JP	WT	S	Unclear	CL	1.2	[22]
IFM59361	JP	WT	S	M1.1	CL	1.3	[22]
IFM59365	JP	WT	S	M1.2	CL	1.1	[22]
IFM59777	JP	WT	S	M1.1	CL	1.2	[22]
IFM60514	JP	WT	S	M1.2	CL	1.1	[22]
IFM61118	JP	N248K	S	M1.1	CL	1.1	[22]
IFM61407	JP	WT	S	Unclear	CL	1.1	[22]
IFM61578	JP	P216L	S	M1.2	CL	1.1	[22]
IFM61610	JP	WT	S	M1.2	CL	1.1	[22]
IFM62516	JP	P329P	S	M1.1	CL	1.2	[22]
SL143435	PT	N248K	S	M1.2	UNK	1.1	*
SL143436	PT	WT	S	M1.2	UNK	1.1	*
SL143437	PT	WT	S	M1.1	UNK	1.3	*
SL143438	PT	N248K	S	M1.2	UNK	1.1	*
SL143439	PT	WT	S	M1.1	UNK	1.3	*
SL143440	PT	3SNPs	S	M1.1	UNK	4	*
SL143441	PT	WT	S	M1.1	UNK	1.1	*
SL146112	PT	N248K	S	M1.2	UNK	1.1	*

NT (the Netherlands), UK (the United Kingdom), IN (India), CN (Canada), JP (Japan), PT (Portugal). SNPs (single nucleotide polymorphisms), 3SNPs *cyp51A* modifications (F46Y, M172V, E427K) and 5SNPs *cyp51A* modifications (F46Y, M172V, N248T, D255E, E427K), AZL SC (azole susceptibility), S (susceptible), R (resistant), CL (clinical), ENV (environmental), UNK (unknown), * NCBI SRA.

2.2. DNA Extraction and Aspergillus fumigatus Identification

First, conidia from each selected strain were cultured in 3 mL of glucose-yeast extract-peptone (GYEP) broth (2% glucose, 0.3% yeast extract, 1% peptone; OXOID LTD, Basingstoke, Hampshire, England) and grown overnight at 37 °C, after which mycelium mats were harvested and DNA was extracted [23]. The *A. fumigatus* Spanish isolates included in this work were identified to the species level on the basis of PCR amplification and sequencing of the internal transcribed spacer (ITS) region and the partial amplification of the *β-tubulin* gene [24].

2.3. DNA Quality and Quantity Assessment

The DNA quality and quantity was assessed using a spectrophotometer (Thermo Scientific NanoDrop One Spectrophotometer, Waltham, MA, USA). Since genomic DNA (gDNA) must be clean and pure for preparing WGS libraries, when the 260/280 or 260/230 rates were lower than 1.8 or higher than 2.2, the DNA was rejected. The strain was extracted again from a new mycelium mat and its quality was assessed. Genomic DNA (gDNAs) were stored at −20 °C until further use.

2.4. Screening of cyp51A Changes: PCR Amplification and Sequence Analysis

The full coding sequences of *cyp51A* including its promoter, were amplified and sequenced using the PCR conditions described before [25]. To exclude the possibility that any change identified in the sequences was due to PCR-induced errors, each isolate was independently analyzed twice. Isolates were screened for the presence of tandem repeat insertions in the *cyp51A* promoter region, as well as for the presence of other *cyp51A* modifications. All the nucleotide sequences were analyzed using the DNASTAR Lasergene package (DNASTAR Inc., Madison, WI, USA).

2.5. DNA Library Preparation and Illumina Whole-Genome Sequencing

Genomic DNA was extracted as previously described. Genomic DNA were quantified afterwards using the QuantiFluor® dsDNA System and the QuantiFluor® ST Fluorometer (Promega, Madison, WI, USA) and their quality was determined with the Agilent 2100 Bioanalyzer (Agilent Technologies, Inc., Santa Clara, CA, USA). The preparation of fragmented gDNA libraries was performed using Nextera® XT Library Prep Kit (Illumina Inc., San Diego, CA, USA), according to the manufacturer's protocols. The mean fragment length of the libraries ranged from 800 to 1800 bp. Sequencing was conducted in paired-end 2 × 150 bp on a NextSeq 500 system, according to the manufacturer's protocols (Illumina Inc., San Diego, CA, USA).

2.6. Whole-Genome Sequencing Alignment

The Illumina reads were trimmed using Trimmomatic (version 0.32) [26]. The sequencing adapters and sequences with low quality scores on $3'$ ends (Phred score [Q], <20) were trimmed. Raw Illumina WGS reads were quality checked performing a quality control with FastQC (version 0.11.3; Babraham Institute). Data sets were analyzed against two different *A. fumigatus* reference genomes, the Af293 (GenBank accession number AAHF00000000.1) and the A1163 (GenBank accession number ABDB00000000.1) using WGS-outbreaker v1.0 (Instituto de Salud Carlos III, Madrid, Spain) (https://github.com/BU-ISCIII/WGS-Outbreaker) with default parameters. The pipeline comprised all steps needed for SNV analysis using whole genome sequencing data. Mapping against genome reference was performed with bwa mem (version 0.7.12-r1039) [27], duplicated reads removed using Picard (version 1.140) (http://broadinstitute.github.io/picard), and the bedtools coverage v2.26 program [28] was used to perform further quality controls. Hereafter, in order to identify genetic variations among strains, single nucleotide variant (SNV) detection (variant calling) and SNV matrix generation were performed using GATK version 3.8.0 [29] with best practices parameters. ENSEMBL variant effect predictor script (version 88) was used for variant annotation. The whole genome sequencing project has been deposited in NCBI SRA (project accession number SRP151231).

2.7. Phylogenetic Analysis and Single Nucleotide Variant Comparison

Final step of WGS-Outbreaker pipeline comprised Maximum-likelihood trees construction using RaxML software (version 8.2.9) [30] with GTRCAT model and 100 bootstrap replicates. Phylogenetic trees were visualized and annotated using ggtree R package [31]. SNV comparisons were performed using a custom R script, mapping all genomes to Af293 and A1163 reference genomes.

2.8. Analysis of Genetic Diversity

Genetic diversity was calculated using the Molecular Evolutionary Genetics Analysis software (MEGA, version 7) [32]. The *A. fumigatus* population was analyzed according to the formed groups (clusters and subclusters) from the previous phylogenic studies. In MEGA, evolutionary distances between sequences can be estimated by computing the number or proportion of nucleotide differences between sequences using the FASTA alignment against each reference genome [33]. Two different models have been used in this work, the p-distance model and another based on the number of differences; (i) the p-distance model calculates the distance as the proportion of different nucleotide sites compared. It is obtained by dividing the number of nucleotide differences (transitions and transversions) by the total number of nucleotides compared. As recommended [34], when the genetic distance was estimated, the complete-deletion option was used, since it normalizes the number of differences based on the number of valid sites compared, not taking into account the alignment gaps and missing data that sequences could contain. However, this model does not make any correction for multiple substitutions at the same site, substitution rate biases (for example, differences in the transitional and transversional rates), or differences in evolutionary rates among sites; (ii) The second model, based on the number of differences, estimates the genetic distances according to the number of sites at which the compared sequences differ (transitions and transversions). In this model, the complete-deletion option was also used.

2.9. Determining Modifications in Genes of Interest

In order to see if the population structure could be based on particular genomic modifications, some genes that have already been described as important in *A. fumigatus* biology were analyzed in depth in each of the *A. fumigatus* population clusters formed from the SNV comparisons. One of the genes was the *cyp51A* gene including its promoter (AFUA_4G06890 in Af293, and AFUB_063960 in A1163). This region was used to determine the azole resistant phenotype based on the resistance mechanism. Since both reference genomes, Af293 and A1163, have opposite mating types, the mating locus genes were included in the analysis: Mat1.1.1 (AFUB_042900), Mat1.2.1 (AFUA_3G06170), and Mat1.2.4 (AFUA_3G06160). The presence or absence of these loci was also tested in all genomes using srst2 software (version 0.1.8) [35]. Every gene of interest was analyzed independently in all genomes.

2.10. Visualization of Depth of Coverage

An exploratory approach using Circos (v 0.69.3) [36], a tool to represent visual data, was used to plot the depth of coverage of 20 whole-sequenced genomes selected to represent each cluster and subcluster in comparison to the Af293 reference genome. Mean coverage data in 10,000 length bins was calculated from bam files with bedtools (version 2.2.17) (http://bedtools.readthedocs.io/en/latest/).

2.11. Mating Type Analysis

A mapping approach was used in order to identify genes associated with each mating type. Since Af293 was Mat1.2 and A1163 was Mat1.1, all samples were mapped against both references using the workflow displayed in Figure S1: (i) samples that had been determined with srst2 as Mat1.2 and mapped against A1163 were selected, and unmapped reads were retained in this step; (ii) the unmapped reads were then mapped against Af293. Statistics of coverage and depth of coverage were calculated for all genes annotated in A1163 reference genome using the bam generated in this step; (iii) genes with at least 70% of coverage were selected; (iv) to discard genes that may be found due to the lack of presence in the A1163 assembly, the same procedure was performed for Mat1.1 samples; (v) finally, genes in common among all Mat1.2 samples minus one but not present in any Mat1.1 sample were exported as a result (Figure S1A). The reverse procedure was performed starting with Mat1.1 samples determining the genes only present in Mat1.1 samples (Figure S1B).

3. Results

3.1. Species Identification

The isolates were identified at the species level and also the azole resistance mechanism was analyzed by PCR amplification and sequencing of the *cyp51A* gene and its promoter as explained in Materials and Methods section. All the strains were identified as *A. fumigatus* sensu stricto. The azole susceptibility profile of most of the strains sequenced in the laboratory is described elsewhere [19]. For the remaining strains, the azole susceptibility profile was determined based on the known *cyp51A* modifications detected by WGS analysis. According to the azole resistance mechanism, the *A. fumigatus* strains included in this study harbored the following *cyp51A* modifications: (i) strains with *cyp51A* single point mutations (G54E/R, M220V/T/K, and G138C) or double point mutations (G448S together with H147Y) in the minority of cases (only one strain); or (ii) strains with tandem integrations and *cyp51A* modifications (TR34/L98H, TR34/L98H/S297T/F495I). Two azole resistant *cyp51A* wild type (WT) strains were also included. All these strain features (*cyp51A* modifications and azole susceptibility profile, as well as the geographical origin and mating type), were described in Tables 1 and 2. Afterwards, the DNAs were further analyzed using WGS.

3.2. Whole-Genome Sequencing Analysis

All genome sequences mapped to a range of 85–99% genome coverage with at least 10x of depth of coverage and a mapping rate > 94% against Af293 reference genome. Similarly, sequences were mapped to a range of 84–99% genome coverage with at least $10\times$ of depth of coverage against A1163 reference with a mapping rate > 88% (Table S1). The Af293 and the A1163 strains were resequenced in our laboratory (and named hereafter AF293 and CEA10) and had a total of 313 and 1009 SNVs, respectively, compared to their respective reference genomes.

3.3. Phylogenetic Analysis

Phylogenetic analysis using SNV data of Spanish genomes (Table 1) clearly showed that our strains were divided into four clusters independently of the genome used as reference (Figure 1, with Af293 as reference genome and Figure 2, with A1163 as reference genome). In this previous analysis, the different wild type reference strains used, *akuB*^{KU80} and CEA10 as derivative strains of A1163, AF293, CM237, ATCC204305 and ATCC46645 were located in separated clusters (Figures 1 and 2). Those clusters were further confirmed when the remaining 73 *A. fumigatus* genomes available from other countries were included.

Therefore, based on phylogenetic analysis, the complete *A. fumigatus* population was divided into four well defined clusters (I, II, III, and IV). Two of them could be further subdivided into subclusters: cluster I split in 3 subclusters (I.1, I.2, I.3), and cluster II in 2 (II.1, II.2). Clusters III and IV remained undivided, because any subcluster was identified. Each subcluster was considered as an independent population group, and was used to compare itself against each reference genome (Table 3). The two reference genomes used were distributed in different clusters when compared against the other: A1163 in cluster I and Af293 in cluster III. The number of clusters and the tree topology was maintained regardless of the reference genome chosen.

Figure 1. Whole genome phylogenetic analysis of 101 *Aspergillus fumigatus* genomes included in the study. Dendrogram was performed using Af293 as reference genome.

Table 3. Mean of single nucleotide variants (SNVs). Variants within clusters and subclusters compared to both reference genomes.

Phylogenetic Groups	SNVs vs. A1163 Reference Genome	SNVs vs. Af293 Reference Genome
Subcluster I.1	34,544	75,653
Subcluster I.2	40,124	65,121
Subcluster I.3	46,811	79,899
Subcluster II.1	81,478	98,423
Subcluster II.2	80,994	97,749
Cluster III	82,235	35,268
Cluster IV	158,154	159,742

Figure 2. Whole genome phylogenetic analysis of 101 *Aspergillus fumigatus* genomes included in the study. Dendrogram was performed using A1163 as reference genome.

Based on *cyp51A* modifications, azole susceptible *cyp51A*-WT strains together with azole resistant *cyp51A* single point mutation strains were grouped together in cluster I. In cluster II, there were azole susceptible and resistant strains, with both *cyp51A* single point mutations and TR$_{34}$/L98H mechanisms. In cluster III, a set of strains with five *cyp51A* modifications (F46Y, M172V, N248T, D255E, E427K) grouped together. While in cluster IV strains with three *cyp51A* modifications (F46Y, M172V, E427K) also grouped together [19]. The Netherlands, Portugal and Canada genomes were distributed in two clusters. The United Kingdom genomes were distributed in three, while Japanese genomes (*cyp51A*-WT) were only in cluster I and Indian genomes (*cyp51A*-TR$_{34}$) in cluster II. Except for the tightly clustered isolates from India in cluster II.2, there was no relationship between the geographical origin of the isolates and the cluster where they grouped.

3.4. Single Nucleotide Variant Analysis against Both References

To explore the genomic differences between the clusters formed from phylogenetic studies, total SNVs between populations were determined. There were a total of 93,609 and 71,844 SNV positions identified against the reference genomes, Af293 and A1163, respectively. No noticeable differences were found in the SNV ranges when all strains were mapped against Af293 (313–165,788) or A1163 (1043–163,367). However, the number of specific SNVs identified in each strain was notably different depending on the genome that was used as reference (Table 4). In fact, some particular SNVs previously detected in our laboratory by PCR (L98H, F46Y, M172V, N248T, D255E, E427K) were specifically searched for in both reference whole genomes. The L98H variant was found in all genomes that harbored this mutation since none of the reference genomes had this change in its own genome. However, the remaining 5 SNVs (F46Y, M172V, N248T, D255E, E427K) were only detected when the reference genome chosen was the A1163 genome, as Af293 specifically harbored these 5 same polymorphisms.

Table 4. Single nucleotide variant differences compared to both reference genomes. Variants of each *A. fumigatus* genome included in this study.

Samples	A1163	Af293	Samples	A1163	Af293	Samples	A1163	Af293
08-12-12-13	80,791	91,956	CM2495	151,089	148,524	F17582	79,598	97,024
08-19-02-10	75,401	89,762	CM2730	157,505	159,177	F17729	44,054	79,106
08-19-02-30	42,487	87,831	CM2733	154,361	160,494	F17729W	43,663	78,416
08-19-02-46	78,449	96,243	CM3248	22,834	74,568	F17764	25,812	70,013
08-19-02-61	80,244	96,416	CM3249	157,242	159,900	F18085	51,241	78,553
08-31-08-91	88,972	98,911	CM3249b	154,510	157,780	F5211G	82,091	99,726
08-36-03-25	83,921	102,006	CM3262	161,695	160,926	F7763	84,764	55,332
09-7500806	45,286	74,134	CM3720	156,600	158,781	IFM55369	42,849	64,003
10-01-02-27	72,487	93,299	CM4602	155,381	161,767	IFM58026	22,186	68,777
12-7504462	48,927	75,301	CM4946	158,692	161,082	IFM58029	36,521	63,106
12-7504652	80,203	93,323	CM5419	89,354	99,771	IFM58401	26,669	63,489
12-7505054	86,952	101,471	CM5757	36,027	74,420	IFM59056	30,841	61,481
12-7505220	75,479	100,229	CM6126	55,871	86,902	IFM59073	43,230	68,775
12-7505446	86,194	100,736	CM6458	42,482	71,670	IFM59359	39,309	67,118
Af293	83,486	313	CM7510	33,559	79,283	IFM59361	42,795	73,228
AF293	78,045	659	CM7555	51,657	77,679	IFM59365	21,830	74,248
AF41	34,879	72,914	CM7560	158,657	153,769	IFM59777	40,117	65,300
Af65	90,143	103,614	CM7570	160,908	162,292	IFM60514	39,629	75,765
AF72	35,271	73,632	CM7632	81,519	25,490	IFM61118	21,905	70,230
AF90	40,262	73,395	F12041	78,812	93,712	IFM61407	32,025	76,034
Afu1042-09	80,346	96,348	F12219	35,440	73,872	IFM61578	26,201	67,946
Afu124-E11	81,115	97,509	F12636	80,359	98,501	IFM61610	41,690	74,102
Afu166-E11	80,948	97,350	F13535	78,741	93,777	IFM62516	41,229	67,364
Afu218-E11	80,571	96,955	F13619	85,412	107,985	SL143435	22,037	81,031
Afu257-E11	80,499	97,008	F13952	78,466	93,926	SL143436	52,604	84,188
Afu343-P11	73,797	101,305	F14403	27,476	74,454	SL143437	61,289	91,394
Afu591-12	80,781	97,039	F14513G	78,647	94,213	SL143438	22,676	81,649
Afu942-09	80,827	97,149	F14532	79,931	102,436	SL143439	39,562	87,955
akuB^{KU80}	1043	78,152	F14946G	44,442	79,719	SL143440	162,993	165,788
ATCC 204305	37,614	60,180	F15390	79,743	101,967	SL143441	38,498	72,994
ATCC 46645	84,994	103,941	F15927	161,157	164,076	SL146112	21,782	81,137
CEA10	1009	75,854	F16134	19,660	77,718	TP12	83,363	24,982
CM2141	88,704	107,238	F16216	78,073	90,736	TP32	163,367	162,032
CM237	48,402	83,874	F16311	51,387	78,630	F17582	79,598	97,024

Although the tree topology was maintained regardless of the reference genome chosen, the number of SNVs within clusters changed significantly when comparing each reference against each cluster, showing that both reference genomes were genetically very different (Table 3). The number of SNVs of clusters II and IV were similar against both references, while remarkable differences in SNVs were present in clusters I (closer to A1163) and III (closer to Af293) (Table 3). Considering subclusters, the number of SNVs in subclusters I.1, I.2 and I.3 were almost half when the reference genome was the

A1163, than the SNVs obtained using Af293 as reference genome. Another remarkable result was the small number of SNVs found in cluster III when compared to Af293 genome. The SNV differences found against both references were slighter but also considerable in cluster II, higher when compared to A1163. Cluster IV had the greatest number of SNVs, independent of the reference genome used.

Moreover, in order to explore any genomic differences between clusters and subclusters, some features of the SNVs found in each population were determined (Table 5). This table shows that all populations had a similar amount of SNVs, excluding cluster IV, which had much more variants, specifically missense and synonymous, compared to the other clusters.

Table 5. Types of SNVs and features. Variants found within each cluster compared to both genomes.

Variants	A1163 Reference Genome				Af293 Reference Genome			
	I	II	III	IV	I	II	III	IV
Frameshift variants	256	512	477	828	446	587	133	819
Inframe deletions	149	278	264	511	246	330	94	527
Inframe insertions	191	369	291	561	325	447	91	556
Intergenic variants	6405	11,107	12,819	6730	13,428	16,394	3,785	8792
Intron variants	1568	3668	3418	9066	3120	4322	960	8670
Missense variants	4342	10,251	10,008	23,107	8766	11,973	2562	22,746
Protein altering	8	19	12	18	16	22	4	21
Start lost	9	16	23	57	21	24	6	57
Stop gained	95	230	264	298	141	218	43	288
Stop lost	13	27	32	61	25	33	9	61
Stop retained	10	22	21	52	17	24	5	52
Synonymous variants	3675	9026	9005	26,080	8189	11,058	2365	25,418

3.5. Genetic Diversity

Evolutionary distances were estimated by computing the mean proportion or number of nucleotide differences between each subcluster and each reference genome using two different models provided by MEGA: the p-model and the number of differences model.

The pairwise distance matrices referred to the distance between each genome and the others in pairs and was calculated and expressed in proportion and number of differences (Tables S2–S5). The matrices obtained from the mean distance between groups in which each subcluster was compared to each other, were indicated also as the proportion and number of differences of all these comparisons (Tables S6 and S7). Cluster IV was identified as the most distant cluster, according to phylogenetic results, with 123,259, 118,271 and 114,815 differences in SNVs against cluster I, II and III, respectively. Moreover, while cluster III was more similar to cluster II with 36,031 differences, cluster I was closer to cluster II than to III with 38,197 and 50,040 differences, respectively. These data were calculated using Af293 genome as reference. Data using A1163 were not shown since no significant differences were found.

Homogeneity within a group or cluster could be inferred through the analysis of the mean distance within groups (Tables S8 and S9). Cluster IV was the most homogeneous since the number of differences was lower than in the remaining groups, followed by subcluster II.1, independently of the genome that was chosen as reference.

3.6. Genome-Wide Visualization of Depth of Coverage

The depth of coverage from a selected group of whole-sequenced genomes was plotted in a Circos image to make an exploratory analysis, and to visualize similarities and differences in genome structure (Figure 3). None of the 20 *A. fumigatus* genomes included in this study displayed chromosomal or copy number variations. However, small deletions were observed in multiple chromosomes: regions smaller than 100 kbp seemed to be deleted in chromosomes 3, 4, 5, 6, 7 and 8. Also, what seemed to be large-scale deletions (bigger than 300 kbp) were detected in the edges of chromosomes 1 and 7 in genomes of cluster I and II. It is noteworthy that there was a huge red region in chromosome 4 present in all genomes marked as an expected deletion, even in the two different Af293 genomes downloaded and sequenced in-house although the reference genome used was the same Af293 strain genome.

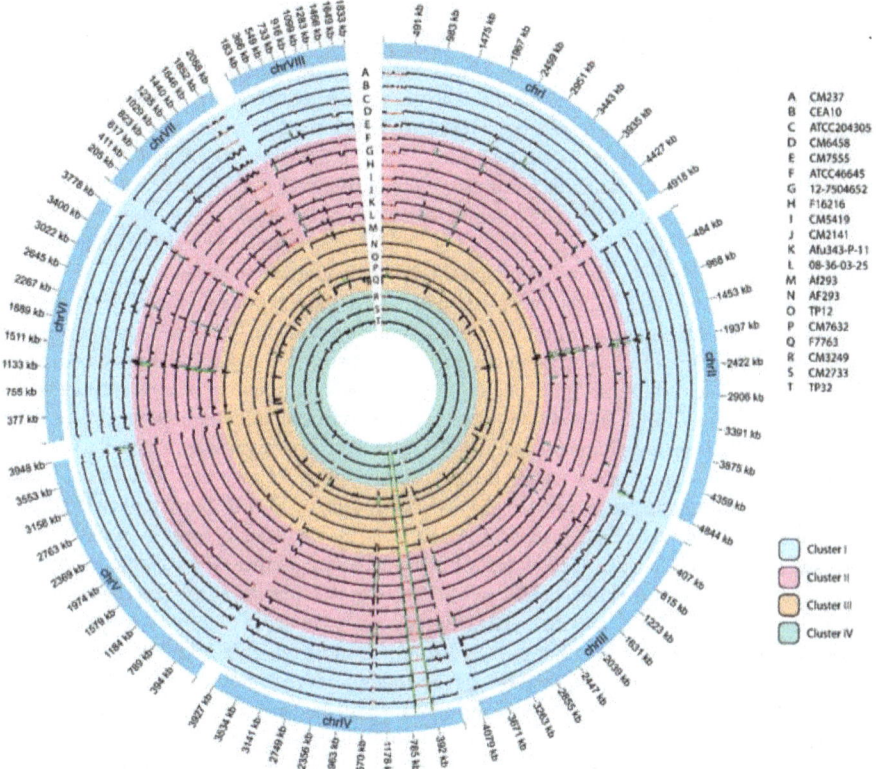

Figure 3. Circos representation of whole-genome depth of coverage of 20 *A. fumigatus* genomes, averaged over 10,000-bp bins. Red marks showed null coverage, black marks a range of coverage between 1–500, and green marks showed a coverage higher than 500. Color legend shows the selected genomes of each cluster: I (blue), II (pink), III (orange) and IV (green).

3.7. Genome Comparisons Based on Their Mating Type

There were no differences in the strain genetic background between genomes harboring the mating type 1 or 2. In fact, strain genomes of both mating types were present in all clusters. In order to detect if there was any differential presence or absence of certain genes based on their particular mating type, all genomes were divided according to their mating type and mapped to both reference genomes (Af293, Mat1.2 and A1163, Mat1.1). No differences in the presence of genes were found in Mat1.1 strain genomes other than those encoded by the Mat1.2 locus itself (Mat1.2.1 and Mat1.2.4). The same situation was found within strains with Mat1.2 genomes; there were no differences in genes other than the gene Mat1.1.1 itself encoded by the Mat1.1 locus.

4. Discussion

Different typing techniques have demonstrated remarkable genomic diversity among strains of *A. fumigatus* [13,37]. *Aspergillus fumigatus* is found in multiple environmental niches which contribute to its genotypic and phenotypic heterogeneity among isolates [38]. Therefore, understanding the association between genomic diversity and different biological and phenotypic features, such as the spectrum of pathogenicity and antifungal resistance, among others, is a challenge for multiple *Aspergillus* laboratories. In this context, WGS is emerging as a promising tool for strain characterization [18,39]. Since WGS considers the entire genetic material of every isolate, it provides the highest discriminatory power that any technique can reach, allowing for the differentiation of closely related strains [39]. In addition, WGS could emerge as a useful tool for gene detection and other sequence-based investigations [9]. This source of an unprecedented amount of novel and comparative data could aid in the development of a huge database which would contain information about the biological differences between *A. fumigatus* strains, such as their origin, diversity and population structure. Previous WGS studies with a suitable depth of coverage had shown variation in genome structure with small-scale insertions and deletions, and recombination events among *A. fumigatus* isolates [40]. However, due to the complexity of biological systems, some difficulties need to be overcome in order to correlate complex biological traits with genomic features.

To date, most of the WGS studies are based on genome-scale comparisons in which the number of SNVs of each strain are obtained comparing its genome to a previously selected reference genome. In this work, we have made all comparison independently using two *A. fumigatus* reference genomes, Af293 and A1163. The differential grouping of the *A. fumigatus* population in four clusters is driven by the different SNVs found. In fact, the number of SNVs in each cluster and subcluster differed depending on the reference genome used to determine the number of variants. Particularly, the small number of SNVs found in cluster III when it was compared to Af293 reference genome is due to the fact that the genomes included in this cluster were more similar to Af293 and had twice the SNVs obtained with A1163 as a reference. Because cluster III was the smallest, formed by only five genomes, the WGS analysis would be improved by the addition of more genomes when available. In addition, it is important to highlight that the specific five SNVs in *cyp51A* present in all genomes belonging to cluster III (F46Y, M172V, N248T, D255E, and E427K) would have not been detected when Af293 genome was used as reference since those changes are also found in its genome. Other important changes could have been missed during the analysis depending on which genome was used as a reference, highlighting the importance of choosing an appropriate reference genome for each case.

The highest number of SNVs was found in cluster IV, which means that was the furthest and most different population compared to both references (Table 3). In fact, this result was in agreement with those computed using MEGA. Cluster IV was the most distant (in number of differences as well as in proportion) from all the subclusters (Tables S6 and S7). When the differences within subclusters were analyzed (Tables S8 and S9), this cluster had the greatest homogeneity. In conclusion, cluster IV is the most different *A. fumigatus* population compared to the others but also the most homogeneously composed. The results obtained by MEGA to study the evolutionary distances among genomes were in line with the results obtained from WGS phylogenetic studies, since the evolutionary distances supported the formed clusters and subclusters (Figures 1 and 2, Tables S6 and S7).

Regarding the Circos visual representation of the depth of coverage (Figure 3), it is noteworthy that the described large-scale deletions were detected in the edges of chromosomes 1 and 7, so these possible deletion events could be confined to telomeric or subtelomeric regions. Notably, all cluster III genomes did not have these deletions, as well as the reference genome used, which belonged to this same cluster. Remarkably, a huge deletion in chromosome 4 was present in all genomes. This might be due to a misassembly or to sequencing errors since two different Af293 genomes (downloaded and sequenced in-house) were included in the analysis and also had this deletion in spite of being the same strain that the genome used as reference. Similar results have been found by other authors although no comments were made about it, probably because it was difficult to realize without sequencing in-house the Af293 [13]. Areas with higher depth of coverage might be a consequence of possible duplication events, pseudogene presence or artefacts. However, since these regions appeared specifically in chromosomes 2, 4 and 6 in most of the genomes of cluster II, it would be worthy to explore them further. A specific mapping approach should be developed to analyze all these regions and to determine if there are real duplications and also if they represent a hallmark for cluster characterization. Further research is in progress to clarify which genes are in these differential regions and which are their biological processes or functions in which they are involved.

Experimental investigations on *A. fumigatus* biology and virulence have historically used only a few particular strains, of which Af293 and A1163 are the most used. More than in any other comparisons, the genetic background must be highly considered in the genomes used as references, since it can be responsible for the variability in many biological aspects between isolates and thus responsible for different phenotypes. The first genome comparison study between Af293 and A1163 was accomplished in 2007, showing that both genomes were highly syntenic in most regions, having core genes highly conserved and with levels of identity of 99.8% at the genome level [18]. These results were confirmed later by Fedorova et al. [17]. Specifically, the Af293 genome contained 143 [17] to 208 unique genes [18], while the A1163, 218 [17] to 320 [18]. In both cases, more than 60% of the strain-specific genes were related to cellular metabolism, secondary metabolism, signaling and transcriptional regulation [17,18]. These sequence divergences could explain the phenotypic differences described by other authors who reported a high heterogeneity among these two isolates regarding abiotic stimuli physiological responses, levels of growth, and virulence [15,38,41], as well as in immunogenic responses [42]. The variability in the total amount of unique genes in each genome could be due to sequencing errors or, less probably, to new mutations generated during the repeated subculture [13]. These results are in agreement with our study, in which the Af293 and the A1163 reference genomes belonged to different phylogenetic clusters which means that there are considerable differences between both genomes. Therefore, the appropriate election of the most suitable reference genome in each context will ultimately result in a more appropriate comparison and lead to solid conclusions.

Apart from the differences already discussed between Af293 and A1163, comparative analyses of the mating loci showed that both strains belonged to opposite mating types [40,43]. Mating type genes are not true idiomorphs, as they occupy adjacent positions on the chromosome. Mat1.1 strains have only one gene in this locus that encodes a transcriptional factor with an alpha box domain (AFUB_042900, Mat1.1.), while Mat1.2 strains contain a truncated copy of the high mobility group (HMG) box (Afu3g06170, Mat1.2.1) and another gene (Afu3g06160, Mat1.2.4) needed for heterokaryon formation as part of the mating process [44–46]. In this study we also checked if there were gene differences (presence or absence) other than those responsible for the mating type (Mat1.1, Mat1.2.1 and Mat1.2.4) among both groups of *A. fumigatus* genomes using an innovative approach (Figure S1). No gene absences were found in Mat1.1 strain genomes other than those encoded by the Mat1.2 locus itself (Mat1.2.1 and Mat1.2.4). Some previous works have reported differential gene presence between strains that coincidentally had different mating types [17,18]. However, the differences found must be due to their own different genetic background and not because of their opposite mating type.

One of the main concerns of invasive aspergillosis treatment is the global emergence of azole resistance. Regarding azole resistance mechanisms, Cyp51A protein is the target of these drugs and the *cyp51A* gene is a hotspot for mutations that confer azole resistance. The most common *cyp51A* modifications could be grouped into two categories: *A. fumigatus* strains that harbor *cyp51A* single point mutations (G54, G138, M220, or G448) [47–50], and isolates with specific point mutations in *cyp51A* gene together with various size tandem repeat (TR) integrations in the promoter of the gene (TR34/L98H, TR34/L98H/S297T/F495I, TR46/Y121F/T289A, or TR53) [24,51–53]. Moreover, all these *cyp51A* modifications have been described to evolve from two different azole resistance acquisition routes: the first set of strains come from the clinical setting as a consequence of the in-host drug adaptation after azole exposure in the patient [54], while the second set of isolates with TR insertions are hypothesized to develop from the azole exposure in the environmental setting [13,37,55–58]. Thanks to the high amount of isolates harboring different azole resistance mechanisms included in the study, a clear tendency in terms of grouping can be observed. In agreement with previous studies [13,37,55,56,58–60], all genomes included in our work that harbor a TR integration with single point mutations in *cyp51A* gene were grouped in cluster II, while isolates with punctual modifications in this gene were spread across clusters I and II. The same happened with *cyp51A*-WT strains that were distributed across those clusters. Therefore, our study supported the idea that isolates with punctual modifications in *cyp51A* (G54, G138, M220, or G448) and also *cyp51A*-WT strains have a greater genetic diversity than TR azole resistant isolates (TR34/L98H, TR34/L98H/S297T/F495I), which reinforces the previously suggested idea that TR resistance mechanisms have developed from a reduced set of clonally related strains with shorter genetic distances among them [59,60]. In this study, only genomes from strains which harbor the TR34/L98H resistance mechanism (with or without the S297T and F495I modifications) have been included. It would be most interesting to include isolates harboring TR46/Y121F/T289A, and TR53 azole resistance mechanisms. Furthermore, it is noteworthy that cluster III and IV were formed specifically by particular isolates that harbor 3 or 5 *cyp51A* modifications (F46Y, M172V, E427K and F46Y, M172V, N248T, D255E, E427K, respectively). This result was in agreement with previous studies developed in our group [19]. Further characterization of clado-specific genes would enhance our understanding of antifungal resistance mechanisms and help to unravel the environmental route resistance development.

Despite all these known azole resistance mechanisms, there is an increasing number of *A. fumigatus* azole resistant isolates for which the underlying mechanism remains unknown [61,62] or which are independent of *cyp51A* modifications [63]. The availability of *A. fumigatus* genomes to compare closely related strains but with differences in their azole susceptibility phenotype will increase the chance of finding genetic differences that could explain the phenotypic variations.

In conclusion, a great amount of novel and useful information can be derived from WGS studies. Here we particularly demonstrated that the selection of strains and reference genomes is crucial for comparative analysis at genomic and phenotypic level. Further compilation of *A. fumigatus* data generated by WGS studies could greatly enhance the understanding of molecular mechanisms involved in antifungal drug resistance and its development, as well as in many other biological functions that remain unknown. As the number of *A. fumigatus* genomes have increasingly been publically available, researchers will be able to increase the possibilities for data analysis which will ultimately allow them to come up with possible conclusions about an infinite number of hypotheses.

Supplementary Materials: The following are available online at http://www.mdpi.com/2073-4425/9/7/363/s1, Figure S1: Workflow for obtaining genes present in all Mat1.1 samples but not present in Mat1.2 (A); and on the contrary (B). Table S1: Mapping rate and coverage at 10× of 101 *A. fumigatus* strains against both reference genomes. Table S2: MEGA genetic distance estimations: pairwise distance matrix calculating the proportion of differences (p-model) and using Af293 as reference genome. Table S3: MEGA genetic distance estimations: pairwise distance matrix calculating the number of differences and using Af293 as reference genome. Table S4: MEGA genetic distance estimations: pairwise distance matrix calculating the proportion of differences (p-model) and using A1163 as reference genome. Table S5: MEGA genetic distance estimations: pairwise distance matrix calculating the number of differences and using A1163 as reference genome. Table S6: MEGA genetic distance estimations: mean distances within subclusters calculating the proportion and number of differences and using Af293 as reference genome. Table S7: MEGA genetic distance estimations: mean distances within subclusters calculating the proportion and number of differences and using A1163 as reference genome. Table S8: MEGA genetic distance estimations: matrix of the mean distances between subclusters calculating the proportion and number of differences and using Af293 as reference genome. Table S9: MEGA genetic distance estimations: matrix of the mean distances between subclusters calculating the proportion and number of differences and using A1163 as reference genome.

Author Contributions: Conceptualization, E.M.; Data curation, S.M. and I.C.; Formal analysis, R.G.-R. and S.M.; Funding acquisition, E.M.; Investigation, R.G.-R and S.M.; Project administration, E.M.; Resources, E.M.; Software, S.M. and I.C.; Supervision, E.M.; Validation, I.C. and E.M.; Visualization, R.G.-R. and S.M.; Writing-original draft, R.G.-R. and E.M.; Writing-review & editing, R.G.-R., L.A.-F. and E.M.

Funding: This work has been supported by Fondo de Investigacion Sanitaria (FIS PI15/00019) and also by Plan Nacional de I+D+i 2013–2016 and Instituto de Salud Carlos III, Subdirección General de Redes y Centros de Investigación Cooperativa, Ministerio de Economía, Industria y Competitividad, Spanish Network for Research in Infectious Diseases (REIPI RD16/CIII/0004/0003), co-financed by European Development Regional Fund ERDF "A way to achieve Europe", Operative program Intelligent Growth 2014–2020.

Conflicts of Interest: The authors declare no conflicts of interest

References

1. Kontoyiannis, D.P.; Marr, K.A.; Park, B.J.; Alexander, B.D.; Anaissie, E.J.; Walsh, T.J.; Ito, J.; Andes, D.R.; Baddley, J.W.; Brown, J.M.; et al. Prospective surveillance for invasive fungal infections in hematopoietic stem cell transplant recipients, 2001–2006: Overview of the Transplant-Associated Infection Surveillance Network (TRANSNET) Database. *Clin. Infect. Dis.* **2010**, *50*, 1091–1100. [CrossRef] [PubMed]
2. Pappas, P.G.; Alexander, B.D.; Andes, D.R.; Hadley, S.; Kauffman, C.A.; Freifeld, A.; Anaissie, E.J.; Brumble, L.M.; Herwaldt, L.; Ito, J.; et al. Invasive fungal infections among organ transplant recipients: Results of the Transplant-Associated Infection Surveillance Network (TRANSNET). *Clin. Infect. Dis.* **2010**, *50*, 1101–1111. [CrossRef] [PubMed]
3. Denning, D.W. Invasive aspergillosis. *Clin. Infect. Dis.* **1998**, *26*, 781–805. [CrossRef] [PubMed]
4. Latgé, J.P. *Aspergillus fumigatus* and aspergillosis. *Clin. Microbiol. Rev.* **1999**, *12*, 310–350. [PubMed]
5. Kosmidis, C.; Denning, D.W. The clinical spectrum of pulmonary aspergillosis. *Thorax* **2015**, *70*, 270–277. [CrossRef] [PubMed]
6. Verweij, P.E.; Chowdhary, A.; Melchers, W.J.G.; Meis, J.F. Azole Resistance in *Aspergillus fumigatus*: Can We Retain the Clinical Use of Mold-Active Antifungal Azoles? *Clin. Infect. Dis.* **2016**, *62*, 362–368. [CrossRef] [PubMed]
7. Köser, C.U.; Ellington, M.J.; Peacock, S.J. Whole-genome sequencing to control antimicrobial resistance. *Trends Genet.* **2014**, *30*, 401–407. [CrossRef] [PubMed]
8. Peacock, S. Bring microbial sequencing to hospitals. *Nature* **2014**, *509*, 557–559. [CrossRef] [PubMed]
9. Kwong, J.C.; McCallum, N.; Sintchenko, V.; Howden, B.P. Whole genome sequencing in clinical and public health microbiology. *Pathology* **2015**, *47*, 199–210. [CrossRef] [PubMed]
10. Zoll, J.; Snelders, E.; Verweij, P.E.; Melchers, W.J.G. Next-Generation Sequencing in the Mycology Lab. *Curr. Fungal Infect. Rep.* **2016**, *10*, 37–42. [CrossRef] [PubMed]
11. Araujo, R. Towards the genotyping of fungi: Methods, benefits and challenges. *Curr. Fungal Infect. Rep.* **2014**, *8*, 203–210. [CrossRef]
12. Hagiwara, D.; Takahashi, H.; Watanabe, A.; Takahashi-Nakaguchi, A.; Kawamoto, S.; Kamei, K.; Gonoi, T. Whole-genome comparison of *Aspergillus fumigatus* strains serially isolated from patients with aspergillosis. *J. Clin. Microbiol.* **2014**, *52*, 4202–4209. [CrossRef] [PubMed]
13. Abdolrasouli, A.; Rhodes, J.; Beale, M.A.; Hagen, F.; Rogers, T.R.; Chowdhary, A.; Meis, J.F.; Armstrong-James, D.; Fisher, M.C. Genomic context of azole resistance mutations in *Aspergillus fumigatus* determined using whole-genome sequencing. *MBio* **2015**, *6*, e00536-15. [CrossRef] [PubMed]
14. Wei, X.; Chen, P.; Gao, R.; Li, Y.; Zhang, A.; Liu, F.; Lu, L. Screening and characterization of a non-cyp51A mutation in an *Aspergillus fumigatus* cox10 strain conferring azole resistance. *Antimicrob. Agents Chemother.* **2017**, *61*, e02101-16. [CrossRef] [PubMed]
15. Keller, N. Heterogeneity Confounds Establishment of "a" Model Microbial Strain. *mBio* **2017**, *8*, e00135-17. [CrossRef] [PubMed]
16. Nierman, W.C.; Pain, A.; Anderson, M.J.; Wortman, J.R.; Kim, H.S.; Arroyo, J.; Berriman, M.; Abe, K.; Archer, D.B.; Bermejo, C.; et al. Genomic sequence of the pathogenic and allergenic filamentous fungus *Aspergillus fumigatus*. *Nature* **2005**, *438*, 1151–1156. [CrossRef] [PubMed]
17. Fedorova, N.D.; Khaldi, N.; Joardar, V.S.; Maiti, R.; Amedeo, P.; Anderson, M.J.; Crabtree, J.; Silva, J.C.; Badger, J.H.; Albarraq, A.; et al. Genomic islands in the pathogenic filamentous fungus *Aspergillus fumigatus*. *PLoS Genet.* **2008**, *4*, e1000046. [CrossRef] [PubMed]
18. Rokas, A.; Payne, G.; Fedorova, N.D.; Baker, S.E.; Machida, M.; Yu, J.; Georgianna, D.R.; Dean, R.A.; Bhatnagar, D.; Cleveland, T.E.; et al. What can comparative genomics tell us about species concepts in the genus Aspergillus? *Stud. Mycol.* **2007**, *59*, 11–17. [CrossRef] [PubMed]
19. Garcia-Rubio, R.; Alcazar-Fuoli, L.; Monteiro, M.C.; Monzon, S.; Cuesta, I.; Pelaez, T.; Mellado, E. Insight into the significance of *Aspergillus fumigatus* cyp51A polymorphisms. *Antimicrob. Agents Chemother.* **2018**, *62*, e00241-18. [CrossRef] [PubMed]
20. Da Silva Ferreira, M.E.; Kress, M.R.V.Z.; Savoldi, M.; Goldman, M.H.S.; Härtl, A.; Heinekamp, T.; Brakhage, A.A.; Goldman, G.H. The akuBKU80 mutant deficient for nonhomologous end joining is a powerful tool for analyzing pathogenicity in *Aspergillus fumigatus*. *Eukaryot. Cell* **2006**, *5*, 207–211. [CrossRef] [PubMed]

21. Howard, S.J.; Cerar, D.; Anderson, M.J.; Albarrag, A.; Fisher, M.C.; Pasqualotto, A.C.; Laverdiere, M.; Arendrup, M.C.; Perlin, D.S.; Denning, D.W. Frequency and evolution of Azole resistance in *Aspergillus fumigatus* associated with treatment failure. *Emerg. Infect. Dis.* **2009**, *15*, 1068–1076. [CrossRef] [PubMed]

22. Takahashi-Nakaguchi, A.; Muraosa, Y.; Hagiwara, D.; Sakai, K.; Toyotome, T.; Watanabe, A.; Kawamoto, S.; Kamei, K.; Gonoi, T.; Takahashi, H. Genome sequence comparison of *Aspergillus fumigatus* strains isolated from patients with pulmonary aspergilloma and chronic necrotizing pulmonary aspergillosis. *Med. Mycol.* **2015**, *53*, 353–360. [CrossRef] [PubMed]

23. Tang, C.M.; Cohen, J.; Holden, D.W. An *Aspergillus fumigatus* alkaline protease mutant constructed by gene disruption is deficient in extracellular elastase activity. *Mol. Microbiol.* **1992**, *6*, 1663–1671. [CrossRef] [PubMed]

24. Alcazar-Fuoli, L.; Mellado, E.; Alastruey-Izquierdo, A.; Cuenca-Estrella, M.; Rodriguez-Tudela, J.L. Aspergillus section Fumigati: Antifungal susceptibility patterns and sequence-based identification. *Antimicrob. Agents Chemother.* **2008**, *52*, 1244–1251. [CrossRef] [PubMed]

25. Mellado, E.; Garcia-Effron, G.; Alcazar-Fuoli, L.; Melchers, W.J.; Verweij, P.E.; Cuenca-Estrella, M.; Rodriguez-Tudela, J.L. A new *Aspergillus fumigatus* resistance mechanism conferring in vitro cross-resistance to azole antifungals involves a combination of cyp51A alterations. *Antimicrob. Agents Chemother.* **2007**, *51*, 1897–1904. [CrossRef] [PubMed]

26. Bolger, A.M.; Lohse, M.; Usadel, B. Trimmomatic: A flexible trimmer for Illumina sequence data. *Bioinformatics* **2014**, *30*, 2114–2120. [CrossRef] [PubMed]

27. Langmead, B.; Salzberg, S.L. Fast gapped-read alignment with Bowtie2. *Nat. Methods* **2012**, *9*, 357–359. [CrossRef] [PubMed]

28. Quinlan, A.R.; Hall, I.M. BEDTools: A flexible suite of utilities for comparing genomic features. *Bioinformatics* **2010**, *26*, 841–842. [CrossRef] [PubMed]

29. McKenna, A.; Hanna, M.; Banks, E.; Sivachenko, A.; Cibulskis, K.; Kernytsky, A.; Garimella, K.; Altshuler, D.; Gabriel, S.; Daly, M.; et al. The Genome Analysis toolkit: A MapReduce framework for analyzing nextgeneration DNA sequencing data. *Genome Res.* **2010**, *20*, 1297–1303. [CrossRef] [PubMed]

30. Stamatakis, A. RAxML version 8: A tool for phylogenetic analysis and post-analysis of large phylogenies. *Bioinformatics* **2014**, *30*, 1312–1313. [CrossRef] [PubMed]

31. Yu, G.; Smith, D.K.; Zhu, H.; Guan, Y.; Lam, T.T.-Y. ggtree: An R package for visualization and annotation of phylogenetic trees with their covariates and other associated data. *Methods Ecol. Evol.* **2017**, *8*, 28–36. [CrossRef]

32. Kumar, S.; Stecher, G.; Tamura, K. MEGA7: Molecular Evolutionary Genetics Analysis version 7.0. *Mol. Biol. Evol.* **2016**, *33*, 1870–1874. [CrossRef] [PubMed]

33. Pearson, W.R.; Lipman, D.J. Improved tools for biological sequence comparison. *Proc. Natl. Acad. Sci. USA* **1988**, *85*, 2444–2448. [CrossRef] [PubMed]

34. Nei, M.; Kumar, S. *Molecular Evolution and Phylogenetics*; Oxford University Press: New York, NY, USA, 2000; ISBN 9780195135855.

35. Inouye, M.; Dashnow, H.; Raven, L.A.; Schultz, M.B.; Pope, B.J.; Tomita, T.; Zobel, J.; Holt, K.E. SRST2: Rapid genomic surveillance for public health and hospital microbiology labs. *Genome Med.* **2014**, *6*, 90. [CrossRef] [PubMed]

36. Krzywinski, M.; Schein, J.; Birol, I.; Connors, J.; Gascoyne, R.; Horsman, D.; Jones, S.J.; Marra, M.A. Circos: An information aesthetic for comparative genomics. *Genome Res.* **2009**, *19*, 1639–1645. [CrossRef] [PubMed]

37. Ashu, E.E.; Hagen, F.; Chowdhary, A.; Meis, J.F.; Xu, J. Global population genetic analysis of *Aspergillus fumigatus*. *mSphere* **2017**, *2*, e00019-17. [CrossRef] [PubMed]

38. Kowalski, C.H.; Beattie, S.R.; Fuller, K.K.; McGurk, E.A.; Tang, Y.-W.; Hohl, T.M.; Obar, J.J.; Cramer, R.A., Jr. Heterogeneity among isolates reveals that fitness in low oxygen correlates with *Aspergillus fumigatus* virulence. *mBio* **2016**, *7*, e01515-16. [CrossRef] [PubMed]

39. Popovich, K.J.; Snitkin, E.S. Whole Genome Sequencing—Implications for Infection Prevention and Outbreak Investigations. *Curr. Infect. Dis. Rep.* **2017**, *19*, 15. [CrossRef] [PubMed]

40. Wortman, J.R.; Fedorova, N.; Crabtree, J.; Joardar, V.; Maiti, R.; Haas, B.J.; Amedeo, P.; Lee, E.; Angiuoli, S.V.; Jiang, B.; et al. Whole genome comparison of the *A. fumigatus* family. *Med. Mycol.* **2006**, *44* (Suppl. 1), S3–S7. [CrossRef]

41. Fuller, K.K.; Cramer, R.A.; Zegans, M.E.; Dunlap, J.C.; Loros, J.J. *Aspergillus fumigatus* photobiology illuminates the marked heterogeneity between isolates. *mBio* **2016**, *7*, e01517-16. [CrossRef] [PubMed]

42. Rizzetto, L.; Giovannini, G.; Bromley, M.; Bowyer, P.; Romani, L.; Cavalieri, D. Strain dependent variation of immune responses to *A. fumigatus*: Definition of pathogenic species. *PLoS ONE* **2013**, *8*, e56651. [CrossRef] [PubMed]

43. Losada, L.; Sugui, J.A.; Eckhaus, M.A.; Chang, Y.C.; Mounaud, S.; Figat, A.; Joardar, V.; Pakala, S.B.; Pakala, S.; Venepally, P. Genetic Analysis Using an Isogenic Mating Pair of *Aspergillus fumigatus* Identifies Azole Resistance Genes and Lack of MAT Locus's Role in Virulence. *PLoS Pathog.* **2015**, *11*, e1004834. [CrossRef] [PubMed]

44. Dyer, P.S.; O'Gorman, C.M. Sexual development and cryptic sexuality in fungi: Insights from Aspergillus species. *FEMS Microbiol. Rev.* **2012**, *36*, 165–192. [CrossRef] [PubMed]

45. Dyer, P.S.; Inderbitzin, P.; Debuchy, R. Mating-type structure, function, regulation and evolution in the Pezizomycotina. In *Growth, Differentiation and Sexuality. The Mycota (A Comprehensive Treatise on Fungi as Experimental Systems for Basic and Applied Research)*; Wendland, J., Ed.; Springer International Publishing: Basel, Switzerland, 2016; Volume I, pp. 351–385. ISBN 978-3-319-25844-7.

46. Yu, Y.; Amich, J.; Will, C.; Eagle, C.E.; Dyer, P.S.; Krappmann, S. The novel *Aspergillus fumigatus* MAT1-2-4 mating-type gene is required for mating and cleistothecia formation. *Fungal Genet. Biol.* **2017**, *108*, 1–12. [CrossRef] [PubMed]

47. Diaz-Guerra, T.M.; Mellado, E.; Cuenca-Estrella, M.; Rodriguez-Tudela, J.L. A point mutation in the 14alpha-sterol demethylase gene cyp51A contributes to itraconazole resistance in *Aspergillus fumigatus*. *Antimicrob. Agents Chemother.* **2003**, *47*, 1120–1124. [CrossRef] [PubMed]

48. Mellado, E.; Garcia-Effron, G.; Alcazar-Fuoli, L.; Cuenca-Estrella, M.; Rodriguez-Tudela, J.L. Substitutions at methionine 220 in the 14-alpha sterol demethylase (Cyp51A) of *Aspergillus fumigatus* are responsible for resistance in vitro to azole antifungal drugs. *Antimicrob. Agents Chemother.* **2004**, *48*, 2747–2750. [CrossRef] [PubMed]

49. Howard, S.J.; Webster, I.; Moore, C.B.; Gardiner, R.E.; Park, S.; Perlin, D.S.; Denning, D.W. Multi-azole resistance in *Aspergillus fumigatus*. *Int. J. Antimicrob. Agents* **2006**, *28*, 450–453. [CrossRef] [PubMed]

50. Bellete, B.; Raberin, H.; Morel, J.; Flori, P.; Hafid, J.; Manhsung, R.T. Acquired resistance to voriconazole and itraconazole in a patient with pulmonary aspergilloma. *Med. Mycol.* **2010**, *48*, 197–200. [CrossRef] [PubMed]

51. Hodiamont, C.J.; Dolman, K.M.; Ten Berge, I.J.; Melchers, W.J.; Verweij, P.E.; Pajkrt, D. Multiple-azole-resistant *Aspergillus fumigatus* osteomyelitis in a patient with chronic granulomatous disease successfully treated with long-term oral posaconazole and surgery. *Med. Mycol.* **2009**, *47*, 217–220. [CrossRef] [PubMed]

52. Vermeulen, E.; Maertens, J.; Schoemans, H.; Lagrou, K. Azole-resistant *Aspergillus fumigatus* due to TR46/Y121F/T289A mutation emerging in Belgium, July 2012. *Eur. Surveill.* **2012**, *17*, 20326.

53. Chen, Y.; Li, Z.; Han, X.; Tian, S.; Zhao, J.; Chen, F.; Su, X.; Zhao, J.; Zou, Z.; Gong, Y.; et al. Elevated MIC values to imidazole drugs among *Aspergillus fumigatus* isolates with TR34/L98H/S297T/F495I mutation. *Antimicrob. Agents Chemother.* **2018**, *62*, e01549-17. [CrossRef] [PubMed]

54. Burgel, P.R.; Baixench, M.T.; Amsellem, M.; Audureau, E.; Chapron, J.; Kanaan, R.; Honoré, I.; Dupouy-Camet, J.; Dusser, D.; Klaassen, C.H.; et al. High prevalence of azole-resistant *Aspergillus fumigatus* in adults with cystic fibrosis exposed to itraconazole. *Antimicrob. Agents Chemother.* **2012**, *56*, 869–874. [CrossRef] [PubMed]

55. Chowdhary, A.; Kathuria, S.; Xu, J.; Sharma, C.; Sundar, G.; Singh, P.K.; Gaur, S.N.; Hagen, F.; Klaassen, C.H.; Meis, J.F. Clonal expansion and emergence of environmental multiple-triazole-resistant *Aspergillus fumigatus* strains carrying the TR34/L98H mutations in the *cyp51A* gene in India. *PLoS ONE* **2012**, *7*, e52871. [CrossRef] [PubMed]

56. Chang, H.; Ashu, E.; Sharma, C.; Kathuria, S.; Chowdhary, A.; Xu, J. Diversity and origins of Indian multi-triazole resistant strains of *Aspergillus fumigatus*. *Mycoses* **2016**, *59*, 450–466. [CrossRef] [PubMed]

57. Garcia-Rubio, R.; Cuenca-Estrella, M.; Mellado, E. Triazole Resistance in Aspergillus Species: An Emerging Problem. *Drugs* **2017**, *77*, 599–613. [CrossRef] [PubMed]

58. Wang, H.C.; Huang, J.C.; Lin, Y.H.; Chen, Y.H.; Hsieh, M.I.; Choi, P.C.; Lo, H.J.; Liu, W.L.; Hsu, C.S.; Shih, H.I.; et al. Prevalence, mechanisms and genetic relatedness of the human pathogenic fungus *Aspergillus fumigatus* exhibiting resistance to medical azoles in the environment of Taiwan. *Environ. Microbiol.* **2017**, *20*, 270–280. [CrossRef] [PubMed]

59. Snelders, E.; Rijs, A.J.; Kema, G.H.; Melchers, W.J.; Verweij, P.E. Possible environmental origin of resistance of *Aspergillus fumigatus* to medical triazoles. *Appl. Environ. Microbiol.* **2009**, *75*, 4053–4057. [CrossRef] [PubMed]

60. Camps, S.M.T.; Rijs, A.J.M.M.; Klaassen, C.H.W.; Meis, J.F.; O'Gorman, C.M.; Dyer, P.S.; Melchers, W.J.G.; Verweij, P.E. Molecular epidemiology of *Aspergillus fumigatus* isolates harboring the TR34/L98H azole resistance mechanism. *J. Clin. Microbiol.* **2012**, *50*, 2674–2680. [CrossRef] [PubMed]

61. Chowdhary, A.; Sharma, C.; Hagen, F.; Meis, J.F. Exploring azole antifungal drug resistance in *Aspergillus fumigatus* with special reference to resistance mechanisms. *Future Microbiol.* **2014**, *9*, 697–711. [CrossRef] [PubMed]

62. Moye-Rowley, W.S. Multiple mechanisms contribute to the development of clinically significant azole resistance in *Aspergillus fumigatus*. *Front. Microbiol.* **2015**, *6*, 70. [CrossRef] [PubMed]

63. Fraczek, M.G.; Bromley, M.; Buied, A.; Moore, C.B.; Rajendran, R.; Rautemaa, R.; Ramage, G.; Denning, D.W.; Bowyer, P. The cdr1B efflux transporter is associated with non-cyp51 a mediated itraconazole resistance in *Aspergillus fumigatus*. *J. Antimicrob. Chemother.* **2013**, *68*, 1486–1496. [CrossRef] [PubMed]

Article

Golgi Reassembly and Stacking Protein (GRASP) Participates in Vesicle-Mediated RNA Export in *Cryptococcus neoformans*

Roberta Peres da Silva [1,2,†], Sharon de Toledo Martins [3,†], Juliana Rizzo [4], Flavia C. G. dos Reis [3], Luna S. Joffe [5], Marilene Vainstein [6], Livia Kmetzsch [6], Débora L. Oliveira [5], Rosana Puccia [1], Samuel Goldenberg [3], Marcio L. Rodrigues [3,4] and Lysangela R. Alves [3,*]

1   Departamento de Microbiologia, Imunologia e Parasitologia da Escola Paulista de Medicina-UNIFESP, São Paulo, SP 04023-062, Brazil; roberta.peresdasilva@nottingham.ac.uk (R.P.d.S.); ropuccia@gmail.com (R.P.)
2   School of Life Sciences, University of Nottingham, Nottingham NG7 2RD, UK
3   Instituto Carlos Chagas, Fundação Oswaldo Cruz, Fiocruz-PR, Curitiba, PR 81310-020, Brazil; sdt.martins@gmail.com (S.d.T.M.); flaviar23@gmail.com (F.C.G.d.R.); sgoldenb@fiocruz.br (S.G.); marciolrodrig@gmail.com (M.L.R.)
4   Instituto de Microbiologia Professor Paulo de Góes, Universidade Federal do Rio de Janeiro, Rio de Janeiro, RJ 21941-901, Brazil; juju.rizzo@gmail.com
5   Centro de Desenvolvimento Tecnológico em Saúde (CDTS), Fundação Oswaldo Cruz, Rio de Janeiro, RJ 21040-900, Brazil; lujoffe@gmail.com (L.S.J.); debora_leite@yahoo.com.br (D.L.O.)
6   Centro de Biotecnologia e Departamento de Biologia Molecular e Biotecnologia, Universidade Federal do Rio Grande do Sul, Porto Alegre, RS 91501-970, Brazil; mhv@cbiot.ufrgs.br (M.V.); liviak@cbiot.ufrgs.br (L.K.)
*   Correspondence: lysangela.alves@fiocruz.br
†   These authors contributed equally to this work.

Received: 1 July 2018; Accepted: 31 July 2018; Published: 8 August 2018

Abstract: Golgi reassembly and stacking protein (GRASP) is required for polysaccharide secretion and virulence in *Cryptococcus neoformans*. In fungal species, extracellular vesicles (EVs) participate in the export of polysaccharides, proteins and RNA. In the present work, we investigated if EV-mediated RNA export is functionally connected with GRASP in *C. neoformans* using a *grasp*Δ mutant. Since GRASP-mediated unconventional secretion involves autophagosome formation in yeast, we included the *atg7*Δ mutant with defective autophagic mechanisms in our analysis. All fungal strains exported EVs but deletion of *GRASP* or *ATG7* profoundly affected vesicular dimensions. The mRNA content of the *grasp*Δ EVs differed substantially from that of the other two strains. The transcripts associated to the endoplasmic reticulum were highly abundant transcripts in *grasp*Δ EVs. Among non-coding RNAs (ncRNAs), tRNA fragments were the most abundant in both mutant EVs but *grasp*Δ EVs alone concentrated 22 exclusive sequences. In general, our results showed that the EV RNA content from *atg7*Δ and WT were more related than the RNA content of *grasp*Δ, suggesting that GRASP, but not the autophagy regulator Atg7, is involved in the EV export of RNA. This is a previously unknown function for a key regulator of unconventional secretion in eukaryotic cells.

Keywords: *Cryptococcus neoformans*; RNA; extracellular vesicles; GRASP; Atg7; unconventional secretory pathway

1. Introduction

Extracellular vesicle (EV) formation and release constitute a ubiquitous export mechanism of proteins, DNA and RNA [1,2]. EVs play key roles in processes of cell communication, homeostasis, immunopathogenesis and microbial virulence [1,2]. EV formation is a conserved mechanism in both prokaryotic and eukaryotic cells [3]. In fungi, EVs participate in the transport of macromolecules across the cell wall [4–6]. Fungal EVs transport a variety of macromolecules including proteins, lipids, glycans, pigments and, as more recently described, RNA [4,6–9].

EV biogenesis in fungi is still poorly understood. It has been hypothesized that EV biogenesis in eukaryotes is a complex process that is regulated at multiple levels [10,11]. EV formation is part of the unconventional secretion machinery in eukaryotes and general regulators of unconventional secretion have been identified. GRASP (Golgi reassembly and stacking protein) is a secretion regulator originally characterized in human cells as part of the Golgi cisternae stacking and ribbon formation [12,13]. During stress, GRASP is required for protein delivery to the plasma membrane or to the extracellular space by an unconventional pathway that involves autophagosome-like structures [14]. In mammalian cells, GRASP is also involved in the delivery of a mutant form of cystic fibrosis transmembrane conductance regulator to the plasma membrane in a Golgi-independent manner [15]. In *Drosophila melanogaster*, GRASP participates in the delivery of integrins from the ER directly to the plasma membrane, thus bypassing the Golgi [16]. In the amoeba *Dictyostelium discoideum*, a GRASP orthologue (GrpA) was necessary for acyl-coenzyme A-binding protein (AcbA) secretion during spore differentiation [17]. In the yeast species *Saccharomyces cerevisiae* and *Pichia pastoris*, another GRASP orthologous (Ghr1) was also required for starvation-induced secretion of AcbA [18,19].

In the yeast-like neuropathogen *Cryptococcus neoformans*, GRASP was required for polysaccharide export to the extracellular space. Polysaccharide secretion is fundamental for virulence in *C. neoformans* [20] and, in fact, a *grasp*Δ mutant was hypovirulent in mice [20]. Polysaccharide export in *C. neoformans* is mediated by EVs but connections between GRASP functions and EV cargo remain uncharacterized.

Autophagy is a self-degradative process conserved in eukaryotes, presenting a housekeeping role by degrading dysfunctional components such as organelles and misfolded proteins [21]. The Atg7 is an autophagy regulator protein member of the ubiquitin-activating enzyme (E1) family involved in this process [22]. The Atg proteins have non-canonical roles in distinct cellular pathways. For example, *Toxoplasma gondii* Atg8 localizes to the apicoplast and is essential for organelle homeostasis and survival of the tachyzoite stage of the parasite [23]. Atg7 non-autophagic roles include cathepsin K secretion in bone osteoclasts [24], IFNγ-mediated antiviral activity against virus replication [25], adipogenesis in mice [26] and cell cycle regulation via p53 interaction and expression of p21 in mouse embryonic fibroblasts [27].

Autophagy regulators play key roles in cryptococcal physiology and, in fact, we have recently demonstrated that the putative autophagy regulator Atg7 affects both physiological and pathogenic mechanisms in *C. neoformans* [28]. In *D. discoideum*, GRASP-mediated unconventional secretion is mediated by autophagosomes, showing that there is a connection between these processes [18,29].

The role of unconventional secretion regulators in vesicular export of RNA is unknown but the functional connections between GRASP and Atg7 led us to evaluate whether these proteins affected extracellular RNA export in *C. neoformans*. Our results suggest that GRASP, but not Atg7, is a key regulator of vesicular export of RNA in *C. neoformans*.

2. Material and Methods

2.1. Fungal Strains and Growth Conditions

The *C. neoformans* strains used in this study included the parental isolate H99 and the mutant strains *atg7*Δ and *grasp*Δ, which were generated in previous studies by our group [20,28]. Fungal cultures were maintained at 30 °C in Sabouraud dextrose plates (1% dextrose, 4% peptone).

Cells recovered from the stationary cultures were used to inoculate minimal medium composed of dextrose (15 mM), $MgSO_4$ (10 mM), KH_2PO_4 (29.4 mM), glycine (13 mM) and thiamine-HCl (3 μM) for further cultivation for three days at 30 °C, with shaking. All protocols adhered to the biosecurity demands of the Carlos Chagas Institute of Fiocruz (Curitiba, Brazil).

2.2. Extracellular Vesicle Isolation and Diameter Determination

EVs were isolated from fungal culture supernatants as previously described [4]. Briefly, cell-free culture supernatants were recovered by centrifugation at 4000× *g* for 15 min at 4 °C and the resulting supernatants were pelleted at 15,000× *g* for 30 min to remove small debris. The final supernatants were concentrated by a factor of 20 in an Amicon ultrafiltration system (100-kDa cutoff, Millipore, Burlington, VT, USA). Concentrated supernatants were centrifuged at 15,000× *g* for 30 min to ensure the removal of aggregates and the resulting supernatant was then ultracentrifuged at 100,000× *g* for 1 h to precipitate vesicles. Vesicle pellets were washed once in phosphate-buffered saline (PBS) and the final pellets were suspended in PBS. For analysis of EV dimensions, nanoparticle tracking analysis (NTA) was performed on a LM10 Nanoparticle Analysis System, coupled with a 488 nm laser and equipped with a sCMOS camera and a syringe pump (Malvern Panalytical, Malvern, UK). The data was acquired and analyzed using the NTA 3.0 Software (Malvern Panalytical). EVs from all samples were diluted 1:30 in filtered PBS (0.22 μM) and measured within the optimal dilution range previously described by Maas and colleagues (9×10^7–2.9×10^9 particles/mL) [30]. Polystyrene microspheres (100 nm) were used for equipment calibration. Samples were injected using a syringe pump speed of 50 and three videos of 60 s were captured per sample, with the camera level set to 15, gain set to 3 and viscosity set to water (0.954–0.955 cP). For data analysis, the gain was set to 10 and detection threshold was set to 5 for all samples. Levels of blur and max jump distance were automatically set. Particle detection values were normalized to the total number of cells in cultures from which each sample was obtained.

2.3. Small RNA Isolation

Small RNA (sRNA)-enriched fractions were isolated with the miRNeasy mini kit (Qiagen, Hilden, Germany) and then treated with the RNeasy MinElute Cleanup Kit (Qiagen), according to the manufacturer's protocol, to obtain small RNA-enriched fractions. The success of the sRNA extraction was assessed in representative EV preparations that were treated with 30 U DNase I (Qiagen) and characterized in an Agilent 2100 Bioanalyzer (Agilent Technologies, Santa Clara, CA, USA). To confirm that the RNA was confined within the EVs, vesicle samples were treated with 0.4 μg μl^{-1} RNase (Promega, Madison, WY, USA) for 10 min at 37 °C before RNA extraction, as previously described [9].

2.4. RNA Sequencing

One hundred ng of purified sRNA were used for RNA-seq analysis from two independent biological replicates. The RNA-seq was performed in a SOLiD 3 plus platform using the RNA-Seq kit (Life Technologies, Carlsbad, CA, USA) according to the manufacturer's recommendations.

2.5. Cellular RNA Isolation and Quantitative PCR

Yeast cells were grown in minimal medium for 72 h, pelleted by 1 min centrifugation at 14.000× *g*, washed in PBS, suspended in the lysis buffer provided in miRCURY™ RNA Isolation Kit–Cell & Plant (Exiqon, Vedbaek, Danmark) and vortexed 5 times in acid washed glass beads (425–600 micron, Sigma-Aldrich, St. Louis, MO, USA). The lysate was centrifuged for 2 min at 14.000× *g* and the supernatants were collected for RNA isolation with the mirCURY™ kit, following the manufacturer's instructions. The RNAs were eluted in ultrapure water and treated with RQ1 RNase-Free Dnase (Promega) following the manufacturer's instructions. Reverse transcription reactions with the DNAse-treated RNAs were performed with a random primer and the ImProm-II™ Reverse Transcription System (Promega), following the manufacturer's instructions. Real time PCR reactions were performed using SYBR® Select Master Mix and run and analyzed using the LightCycler® 96 System (Roche, Basel, Switzerland). The primers corresponded to CNAG_03103 Cullin3 Forward GCCATACGGGAGATACAGAAC, Reverse GAGGTGTTGGACGATGAGAG, CNAG_07590 V_typeH Forward TCATGCTCAACGAAGTCAGG, Reverse GGAAGCAGTGGTTGTGAATG, CNAG_03337 hypothetic Forward CGGTCTTTATCGCTGCTGTAT, Reverse ATTGAAGAGTGGATGTCGTGG and CNAG_00483 Actin Forward CCACACTGTCCCCATTTACGA, Reverse CAGCAAGATCGATACGG AGGAT Each reaction was performed using 10 ng of cDNA. The experiment was performed in triplicates and the expression levels relative to actin were calculated according to Pfaffl's method using *t*-test for the statistical analysis [31].

2.6. In Silico Data Analysis

The sequencing data were analyzed using the version 9.1 of CLC Genomics Workbench©. The reads were trimmed on the basis of quality, with a threshold Phred score of 15. The reference genomes used for mapping were obtained from the NCBI database (*C. neoformans*-GCA_000149245.3). The alignment was performed as follows: additional 100-base upstream and downstream sequences; 10 minimum number of reads; 2 maximum number of mismatches; −2 nonspecific match limit and minimum fraction length of 0.9 for the genome mapping or 1.0 for the RNA mapping. The minimum reads similarity mapped on the reference genome was 80%. Only uniquely mapped reads were considered in the analysis. The libraries were normalized per million and the expression values for the transcripts were recorded in RPKM (reads per kilobase per million), we also analyzed the other expression values-TPM (transcripts per million) and CPM (counts per million).

2.7. Data Access

The data is deposited to the Sequence Read Archive (SRA) database of NCBI (Bethesda, MA, USA) under study accession number (SRA: SRX2793565 to 67).

3. Results

3.1. Lack of GRASP Results in Changes in the RNA Content of Cryptococcus neoformans Extracellular Vesicles

Our experimental model included wild type (WT) and two mutant strains of *C. neoformans*. WT cells corresponded to strain H99, a standard and widely investigated clinical isolate. Knockout mutant strains (KO) lacked expression of two regulators of cryptococcal pathogenicity, *GRASP* and *ATG7* [20,28].

We first asked whether the lack of either *GRASP* (*grasp*Δ) or *ATG7* (*atg7*Δ) expression would affect the EVs composition. The analysis of diameter distribution of wild type EVs by nanoparticle tracking analysis (Figure 1) revealed a major population of cryptococcal vesicles in the 50–250 nm range. Peaks of EVs corresponding to approximately 300, 410, 500 and 630 nm were also observed. Although the dimensions of cryptococcal EVs have been traditionally determined by dynamic light scattering and/or electron microscopy, the results obtained by nanoparticle tracking analysis were consistent with the previous literature [32]. Deletion of *GRASP* or *ATG7* produced a clear impact on the size distribution of cryptococcal EVs. In comparison to WT cells, peaks corresponding to sizes higher than 300 nm were no longer observed. A minor peak at 225 nm and major, sharp peaks at 100 and 140 nm were observed in EVs produced by both mutants. Complementation of mutant cells resulted in EV fractions enriched in the 100–300 nm range, but the minor peaks at 415, 500 and 600 nm observed in WT cells were still not detectable. Although deletion of *GRASP* or *ATG7* resulted in modified EV detection, no statistical differences were observed between the different samples. In summary, the nanoparticle tracking analysis revealed that deletion of *GRASP* and *ATG7* affected EV properties in *C. neoformans*. We then asked whether the differences in EV diameters correlated with the RNA content in *C. neoformans* EVs.

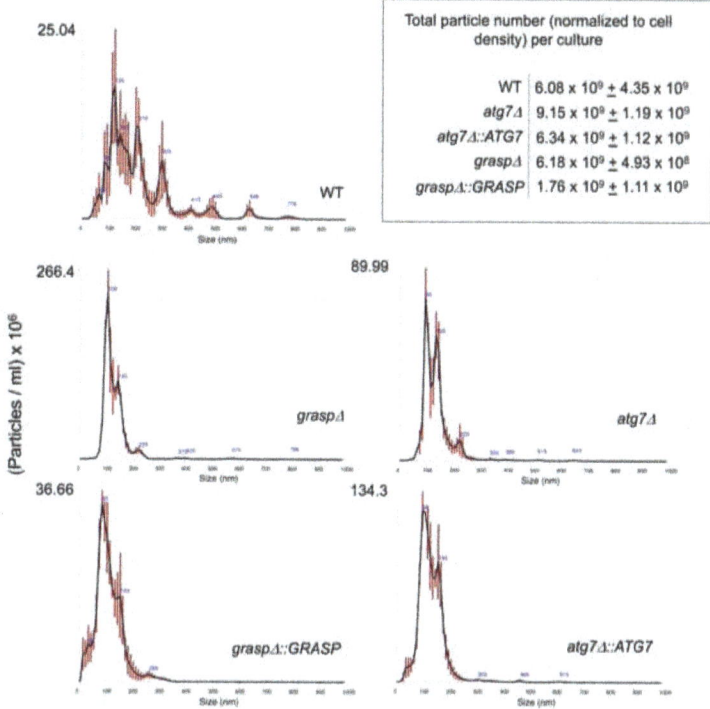

Figure 1. Nanoparticle tracking analysis of *Cryptococcus neoformans* extracellular vesicles (EVs) comparing wild type (WT), mutant (*graspΔ* and *atg7Δ*) and complemented (*graspΔ::GRASP* and *atg7Δ::ATG7*) cells. Results are representative of two independent biological replicates producing similar profiles. Particles were quantified in EV samples suspended in 150 mL phosphate-buffered saline (PBS). Particle detection values shown in the upper, right panel were normalized to the total number of cells in the cultures from which each sample was obtained.

Total RNA was isolated from fungal EVs and two independent biological replicates were subjected to RNA-seq (Figure S1). In order to compare the EV-RNA composition between the knockout (*atg7Δ* and *graspΔ*) and the WT strains we first aligned the RNA-seq reads with the *C. neoformans* H99 genome (GCA_000149245.3) sequences. We used the raw data available for isolate H99 from our previous work [9] and compared them with the *atg7Δ* and *graspΔ* EV RNA (Table 1). For all *C. neoformans* strains about 85% of the EV-RNA reads mapped to intronic regions, while less than 10% mapped to exons. A similar profile was observed for the *C. neoformans* WT strain (H99) in our previous work [9].

Analysis of EV-mRNAs showed that the correlation between WT and *atg7Δ* (r 0.71) sequences was greater than that for WT and *graspΔ* (r 0.22) (Figure 2A,B). This result indicates that the mRNA content in WT EVs was closer to that of *atg7Δ* vesicles than to the content of *graspΔ* EVs.

Table 1. RNA-seq mapping statistics. The values refer to the average of the replicates.

	C. neoformans					
	WT		*atg7Δ*		*graspΔ*	
	Uniquely Mapped	% of Total Mapped	Uniquely Mapped	% of Total Mapped	Uniquely Mapped	% of Total Mapped
Exon	5030	0.4	60,683	9.2	59,425	7.5
Exon-exon	10,664	0.6	1458	0.2	2350	0.3
Total exon	113,655	9.7	62,141	9.4	61,774	7.8
Total intron	1,003,971	90.3	568,003	84.9	758,109	86.9
Total gene	1,117,625	100	667,288	100.0	861,092	100.0

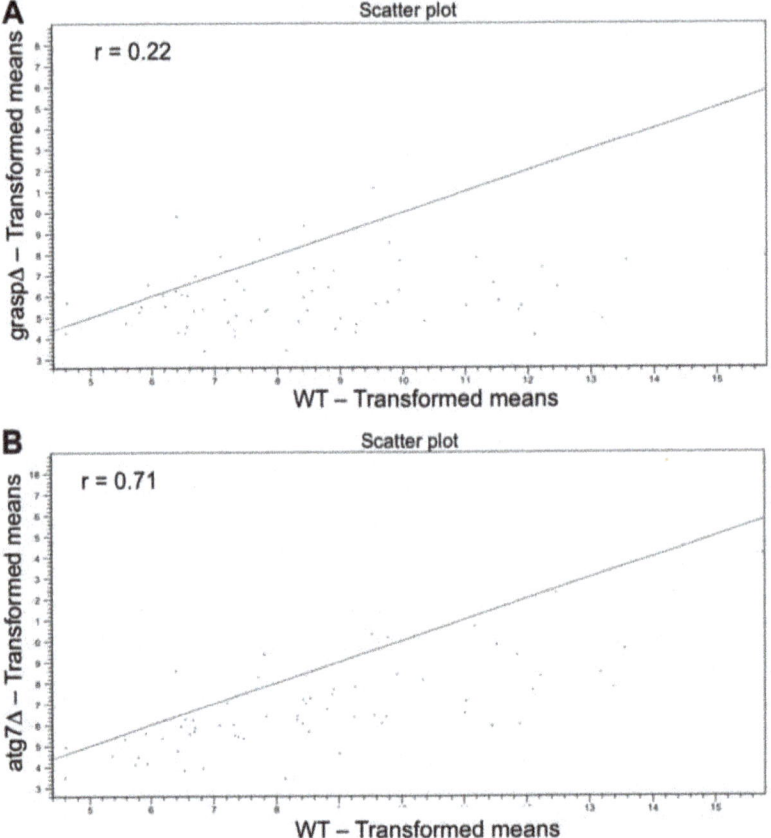

Figure 2. Correlation between the EV-mRNA sequences of *graspΔ* vs. WT samples (**A**) and *atg7Δ* vs. WT preparations (**B**). The transformed mean read values for WT EVs are in the X-axis, while those obtained from mutant vesicles are in the y-axis.

We next performed paired comparisons (WT versus *graspΔ* and WT versus *atg7Δ*) and applied the statistical negative binomial test [33] and the filters RPKM $\geq$ 50, log2 $\geq$ 2 and false discovery rate (FDR) $\leq$ 0.01. From the WT versus *graspΔ* analysis, 266 mRNAs were identified as enriched in the EVs from the *graspΔ* mutant (Table S1). From these transcripts, we observed enrichment in cellular components ($p \leq$ 0.03) such as membrane and endoplasmic reticulum (Figure 3). For biological processes ($p \leq$ 0.03), the enriched terms included organelle organization, cell cycle and gene expression (Figure 3). For the WT versus *atg7*Δ analysis, 74 mRNAs were found enriched in the *atg7*Δ compared to the WT strain (Table S2). The most abundant cellular components mRNAs (*n* = 75) in *atg7*Δ EVs were the nucleus and the mitochondrion (Figure 4). Biological processes were associated to transcription, transcription regulation and RNA processing (Figure 4). However, the score values for some terms did not meet the statistics criteria (*p* ≤ 0.03).

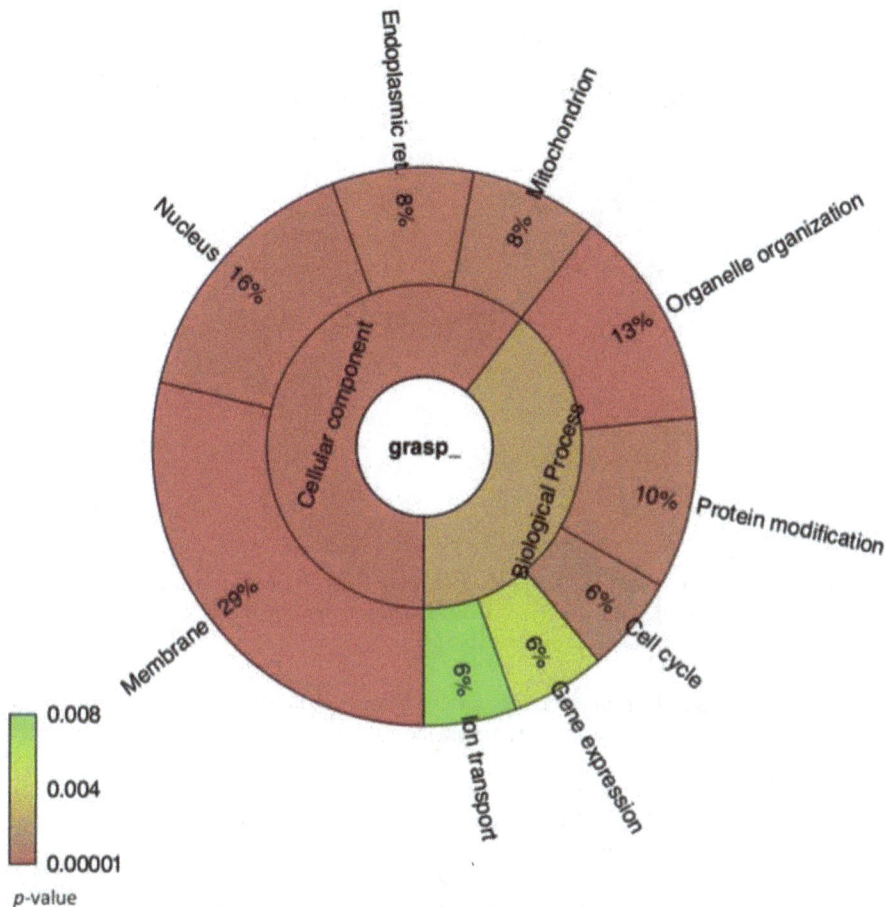

Figure 3. Krona chart representing the gene ontology of mRNA sequences enriched in EVs isolated from the *C. neoformans grasp*Δ mutant. The percentage refers to the relative enrichment for the Gene Ontology (GO) terms. The colors represent the *p*-value for each term plotted in the chart.

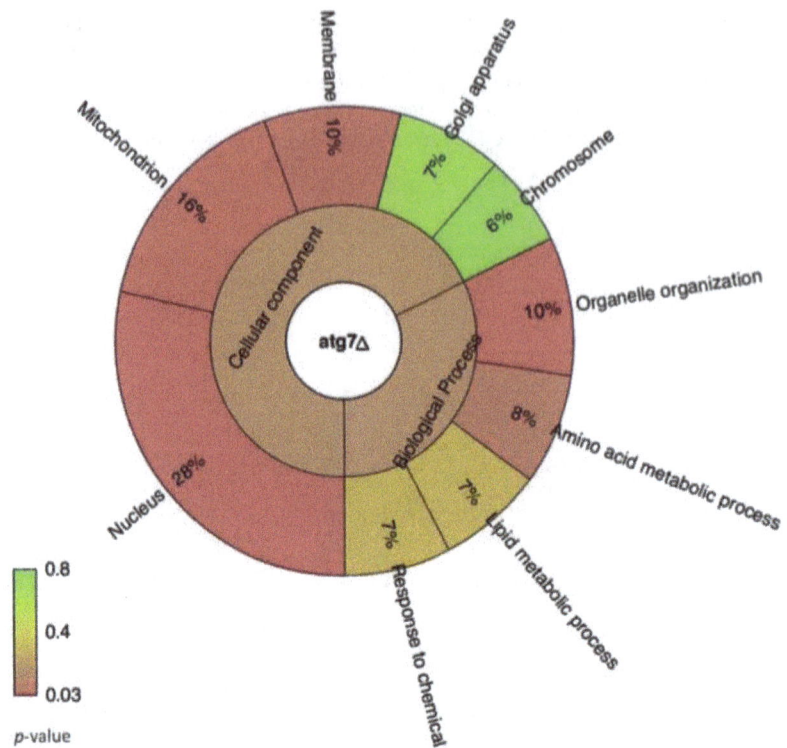

Figure 4. Krona chart representing the gene ontology of mRNA sequences enriched in EVs isolated from the *C. neoformans atg7Δ* mutant. The percentage refers to the relative enrichment for the GO terms. The colors represent the *p*-value for each term plotted in the chart.

Noteworthy, the second and third most abundant transcripts exclusively identified in *graspΔ* EVs were those for the ER lumen protein retaining receptor and the regulator of vesicle transport through interaction with t-SNAREs 1 (Table S1). The former determines specificity of the luminal ER protein retention system and is required for normal vesicular traffic through the Golgi. The latter is involved in multiple transport pathways [34,35]. In addition, most of the transcripts were associated to organelles, such as the nucleus, the mitochondrion and the endoplasmic reticulum, suggesting that somehow the *GRASP* knockout resulted in altered population of transcripts composing the EVs. This enrichment profile was not observed in the *ATG7* knockout (Table S2), thus validating the differences observed in the *graspΔ* mutant EVs.

As we observed this alteration in the EV-RNA composition for the *graspΔ* mutant we asked if this difference was due to a general alteration in the cell transcriptome caused by the *GRASP* knockout. We then selected three of the most enriched transcripts found in the *graspΔ* mutant EVs and assessed their expression value by qPCR in WT, mutant (*graspΔ* and *atg7Δ*) and complemented (*graspΔ::GRASP* and *atg7Δ::ATG7*) strains (Figure 5). The expression values of cullin 3, hypothetical protein CNAG_03337 and the V-type H transporting ATPase subunit C transcripts were similar in WT and *graspΔ* mutant strains, despite the mRNA alteration in the EVs obtained from these two strains. The *atg7Δ* mutant showed the highest expression levels of these mRNAs when compared to the WT and *graspΔ* strains (Figure 5). In addition, these transcripts had very low identification or were not detected in EVs from the *atg7Δ* strain (Table S2). Therefore, despite of the fact that *atg7Δ* mutant showed high expression levels, this variation did not correlate with the presence of these transcripts in EV fractions. Analysis of the complemented strains demonstrated a partial restoration of the wild-type phenotype in the *atg7Δ* system (Figure 5). Altogether, these results reinforce the notion that the *GRASP* deletion lead to a shift in the RNA composition of cryptococcal EVs.

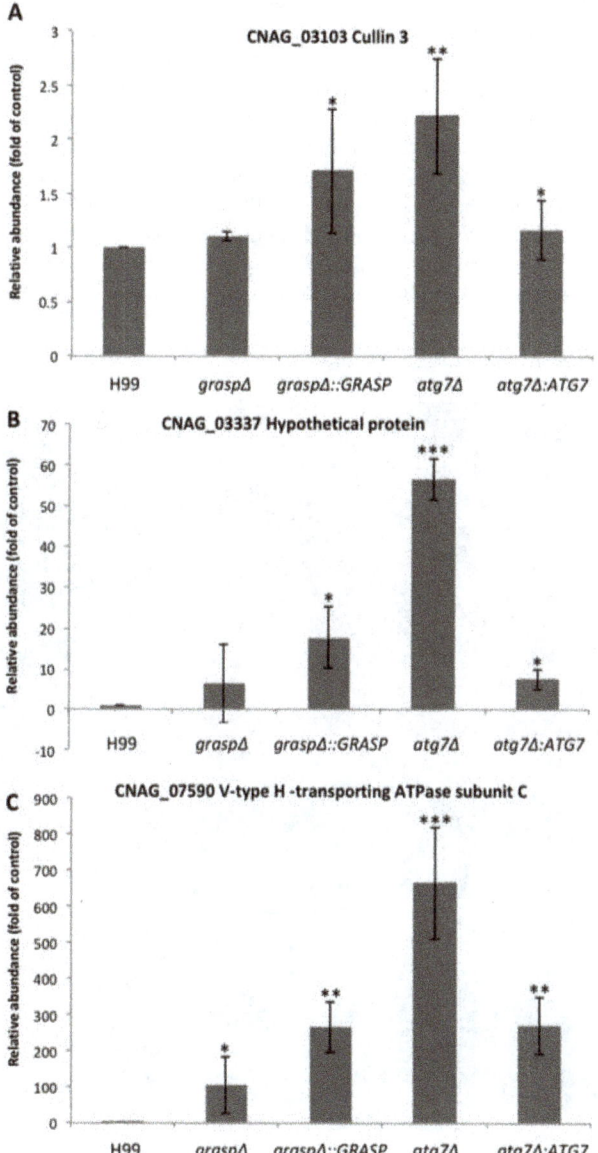

Figure 5. Analysis of the cellular transcription levels of three vesicular RNA sequences. Transcript levels for (**A**) Cullin 3; (**B**) hypothetical protein CNAG_03337 and (**C**) V-type H⁺-transporting ATPase subunit C were normalized to the levels of actin transcripts. The X-axis corresponds to each strain analyzed (WT, *graspΔ*, *graspΔ::GRASP*, *atg7Δ* and *atg7Δ::ATG7*). The y-axis corresponds to the relative expression level of the mRNAs in the cell. Each bar represents the mean and standard error of triplicate samples. * $p < 0.05$; ** $p < 0.01$; *** $p < 0.001$.

3.2. Comparison of Cellular RNA Versus Extracellular Vesicle RNA Composition

The differences in the RNA composition of EVs produced by the *grasp*Δ strain led us to question whether the mRNAs in the EVs correspond to those highly expressed in the cell, likely resulting from random incorporation into vesicular carriers [36]. To address this hypothesis, we compared the *C. neoformans* transcriptome (H99 strain) with the vesicular RNA sequences [9,37] (Table S3). After applying the differential gene expression analysis (DGE) we observed that, for several transcripts, there was an inversion between the expression patterns in the cell and the RNA abundance in the EVs (Figure 6 and Table 2). For example, one of the most enriched transcripts in the EVs presented low levels of expression in the cell (CNAG_06651 amidohydrolase). On the other hand, CNAG_03012 (encoding a quorum sensing-like molecule) had an RPKM value greater than 20,000 in the cell but showed low abundance in the EVs (average RPKM value of 36; Table 3 and Table S3). This observation indicates a lack of correlation between the most expressed cellular mRNAs and EV cargo, therefore reinforcing the supposition that RNA loading into WT or mutant EVs is not random.

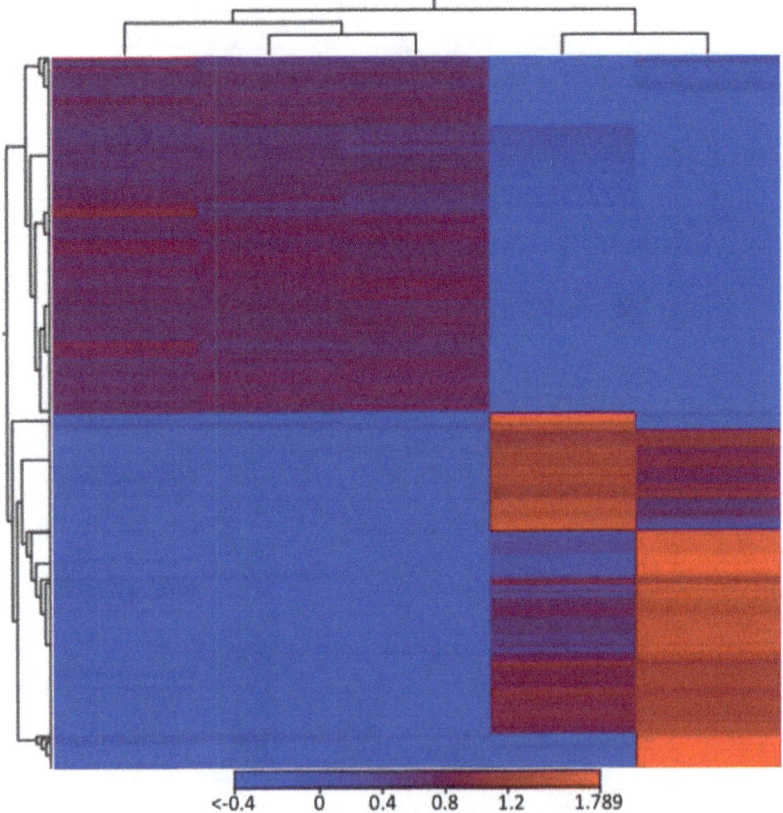

Figure 6. Heat map illustration of the comparison between cellular and EV RNAs. The expression levels are visualized using a gradient color scheme, where the red color is used for high expression levels and the blue color is used for low expression levels. Each line corresponds to a gene of the *C. neoformans* H99 strain.

Table 2. Comparison between cellular and vesicular RNA in *C. neoformans* (H99 strain). The top ten most expressed transcripts in the cell are shown in light blue. The most represented RNAs in the EVs are illustrated in light red.

Name	Product	EV vs. Cell-Log Fold Change	EV vs. Cell-FDR *p*-Value	SRR3199612 Cell 1 RPKM	SRR3199613 Cell 2-RPKM	SRR3199614 Cell 3-RPKM	EV RNA 1-RPKM	EV RNA 2-RPKM
CNAG_03012	quorum sensing-like molecule	−5.06	0.00%	20,332.13	18,844.93	20,155.19	48.40	24.28
CNAG_06207	hypothetical protein	−6.93	0.00%	16,304.37	14,037.19	13,815.62	8.35	6.02
CNAG_04105	hypothetical protein	−2.31	2.29%	16,003.01	10,010.74	16,467.89	254.98	73.62
CNAG_03143	hypothetical protein	−2.09	3.87%	13,070.91	8338.53	12,401.56	231.90	78.39
CNAG_01735	hypothetical protein	−3.60	0.03%	9034.37	6373.02	7970.58	56.97	16.11
CNAG_06075	hypothetical protein	−2.98	0.39%	6021.45	5119.83	6051.19	66.75	13.02
CNAG_03007	hypothetical protein	−6.58	0.00%	5861.25	5356.39	4321.23	3.72	2.64
CNAG_06298	hypothetical protein	−7.02	0.00%	5319.83	5499.72	6635.36	2.65	3.96
CNAG_06101	ADP, ATP carrier protein	−3.04	0.11%	4475.82	5415.80	4535.39	44.21	32.06
CNAG_07466	U3 small nucleolar RNA-associated protein 7, U3 small nucleolar RNA-associated protein 7, variant 1, U3 small nucleolar RNA-associated protein 7, variant 2	10.05	0.00%	392.25	765.29	487.61	39,204.58	43,212.28
CNAG_01093	hypothetical protein	8.04	0.00%	45.01	41.14	33.25	831.32	480.72
CNAG_06651	amidohydrolase	12.70	0.00%	3.80	4.64	3.38	777.33	4354.12
CNAG_00311	3-hydroxyisobutyryl-CoA hydrolase	6.97	0.00%	62.61	78.16	54.48	650.07	373.07
CNAG_02129	hypothetical protein	2.12	3.76%	423.90	380.09	587.72	178.91	56.55
CNAG_05774	hypothetical protein, hypothetical protein, variant	4.48	0.00%	87.07	87.32	82.46	146.53	104.07
CNAG_05651	hypothetical protein	7.83	0.00%	5.28	8.41	7.53	138.73	51.57
CNAG_07515	hypothetical protein	4.79	0.00%	57.55	52.26	77.35	118.53	127.65
CNAG_04124	hypothetical protein	7.66	0.00%	6.60	6.26	5.59	113.90	21.83
CNAG_07028	26S proteasome regulatory subunit N11	4.03	0.00%	103.45	112.54	86.34	112.59	118.66

Table 3. Intron retention in EV RNAs.

ID	Data obtained from Gonzalez-Hilarion et al., 2016 [38]				WT			ΔAtg7			ΔGRASP			Product
	intron	Type	RPKM	Exons	RPKM	Unique Exon Reads	Unique Intron Reads	RPKM	Unique Exon Reads	Unique Intron Reads	RPKM	Unique Exon Reads	Unique Intron Reads	
CNAG_03602	Ic2-554 Ic2-555	in5UTR in5UTR	6.66 204.80	5	23.5	4	228	40.59	3	87.5	28.57	1.5	81	U3 small nucleolar RNA-associated protein 5
CNAG_03645	Ic2-787 Ic2-788	inCDS inCDS	8.01 12.93	8	6.3	1.5	73.5	53.43	5.5	9.5	58.98	6.5	14.5	NET1-associated nuclear protein 1 (U3 small nucleolar RNA-associated protein 17)
CNAG_04068	Ic2-3155	inCDS	3.18	4	71.2	3.5	10	270.93	6	11	522.07	12	4.5	large subunit ribosomal protein L28e
CNAG_07982	Ic4-247 Ic4-248	inCDS inCDS	6.96 17.47	5	1061.8	60.5	100.5	81.42	5	3	60.25	3.5	7	hypothetical protein
CNAG_00930	Ic4-351 Ic4-349 Ic4-350	inCDS in5UTR in5UTR	2.61 1066.15 81.59	7	50.1	5.5	328	81.47	4	303.5	106.30	6	397	argininosuccinate synthase
CNAG_07884	Ic8-1359	inCDS	10.84	3	7.4	0.5	66	18.73	0.5	21.5	85.27	2.5	21	hypothetical protein
CNAG_07813	Ic12-778 Ic12-776	inCDS in5UTR	8.46 7.38	5	75.4	6.5	231.5	14.71	0.5	141.5	37.66	1	208.5	hypothetical protein
CNAG_06167	Ic13-990	in5UTR	26.36	5	131.1	11	770	103.14	6.5	152.5	146.05	10.5	295	metal homeostatis protein bsd2
CNAG_01820	Ic3-1947	in3UTR	28.14	12	342.4	47	180.5	236.94	17	15.5	344.85	27.5	15	pyruvate kinase, pyruvate kinase, variant
CNAG_06033	Ic13-230	inCDS	10.93	7	99.4	15	22	47.14	3.5	5	33.92	3.5	63.5	pfkB family carbohydrate kinase superfamily
CNAG_03730	Ic1-1335	in5UTR	35.87	4	41.7	2	479.5	102.24	2	23.5	0.00	0	175.5	DNA-directed RNA polymerase II subunit RPB11
CNAG_06401	Ic14-772	in5UTR	14.61	11	6.1	0.5	23	52.25	4.5	162.5	28.12	2.5	27.5	hypothetical protein

3.3. Intronic Reads

We have previously observed that a great number of *C. neoformans* EV-RNA reads mapped to intronic regions of the genome [9], which is in agreement with our current findings with the knockout strains. To analyze intronic reads in mRNAs and exclude non-coding RNAs (ncRNAs), such as ribosomal RNA (rRNA) or transfer RNA (tRNAs), we used the presence of exons as a criterion to ensure ncRNAs were excluded (Table 3). We observed two types of patterns, including reads mapping to both exons and introns in variable proportions and those that mapped only to introns in the messenger RNAs. The intronic mapping shared by the EV RNAs from the WT and mutant strains of *C. neoformans* were associated to translation and also to transmembrane proteins (Table 3). From the 32 mRNAs with intronic reads found in the EV samples, 12 have previously been described as transcripts with intron retention [38]. It has already been reported that 59% of the genes from *C. neoformans* use alternative splicing (AS) that varies depending on the growth conditions. The intron retention (IR) is the prevalent AS mechanism in this fungus [38]. We also observed differences in abundance between the cell mRNAs compared to those in the EVs (Table 3). For example, the mRNA CNAG_07982 that codes for a hypothetical protein is 10 times more abundant in the EV than in the cell. A similar profile was observed for sequence CNAG_01820, which encodes a pyruvate kinase (Table 3). It has been speculated that the mRNAs that present IR are not the most expressed in cells based on a negative correlation between the highly expressed transcripts and the presence of IR [38]. However, our present data show that most of the reads that were considered as aligned in introns, are in fact the rRNAs 25S, 18S and 5.8S (data not shown). Nevertheless, we obtained highly abundant transcripts that are likely to be intron-retaining mRNAs (Table 3), suggesting that somehow these IR mRNAs might be directed to the EVs. The function of these transcripts needs to be further investigated.

3.4. Non-Coding RNAs

The EV-RNA sequences obtained in this work also mapped to ncRNAs. The most abundant molecules were the 25S, 18S and 5.8S rRNAs, accounting for more than 90% of the ncRNA and intronic reads (data not shown). As described for the mRNA analysis, we performed paired comparisons (WT versus *grasp*Δ and WT vs. *atg7*Δ) and applied the statistical negative binomial test [33] and the filters RPKM $\geq$ 50, log2 $\geq$ 2 and FDR $\leq$ 0.01. For the WT versus *grasp*Δ we observed 43 ncRNAs enriched in *grasp*Δ (Table S4). For WT versus *atg7*Δ 30 ncRNAs were enriched in the *atg7*Δ strain (Table S5). From these results, it was possible to observe that the tRNA-derived fragments (tRFs) were enriched in both knockouts (Table S4 and S5). tRFs have been identified in EVs from organisms in all kingdoms, including archaea, bacteria and eukaryotes, where they play different biological roles [39].

4. Discussion

Fungal extracellular vesicles might correspond to structures that randomly incorporate cytosolic molecules that are released extracellularly or in the cell wall [11]. Our current results, however, suggest that EV RNA cargo can be finely regulated. Our model consisted of an investigation of the role of *C. neoformans* proteins GRASP and Atg

7 in the vesicular export of RNA. Although these proteins are functionally connected in other systems [12,17,26,27], our findings suggest that GRASP, but not Atg7, has a fundamental role in addressing RNA to cryptococcal EVs. The Atg proteins, which are primarily linked to autophagy processes, have non-canonical roles in distinct cellular pathways. It seems clear, however, that despite the variety of functions played by Atg7 and the significant alterations that its gene deletion causes in *C. neoformans*, the RNA populations transported by EVs were not greatly affected by the *atg7* knockout in *C. neoformans*. The phenotypic characteristics of this mutant included more efficient melanization, larger cell size, autophagic bodies formation and virulence attenuation [28].

Remarkably, phenotypic traits including EV dimensions were only partially recovered in complemented strains. This observation is likely related to methodological particularities intrinsic to the genetic manipulation of *C. neoformans*. For instance, biolistic transformation usually results in large chromosomic alterations but most importantly, gene complementation results in random insertion of *ATG7-* or *GRASP*-containing cassettes in multiple chromosome loci. Under these conditions, many phenotypic traits can be unpredictably affected and complemented genes can have their expression altered. In the specific case of *GRASP*, complementation of the *graspD* strain used in this study resulted in *GRASP* overexpression [20,28], which might be related to the unique phenotypic properties of the complemented strain.

Sequencing analysis of vesicular RNA obtained from mutant strains suggested that important biological functions are associated with nucleic acid-containing fungal vesicles. For example, the tRF-3′end derived (or CCA) uses the canonical miRNA machinery to downregulate replication of protein A1 mRNA and other transcripts in B cell lymphoma [40]. Regulation of translation is also a potential process where tRFs participate. It was demonstrated that tRF derived from tRNA-Val in the archaebacteria *Haloferax volcanii* binds to the small ribosomal subunit, consequently repressing translation by preventing a peptidyl transferase activity [41]. tRFs are also associated to the regulation of cell viability, RNA turnover and RNA stability [42–45]. The roles of GRASP and Atg7 in these processes have not been established but the enrichment of specific classes of RNA in mutant EVs suggests the existence of robust connections between EV traffic and tRFs. In *Trypanosoma cruzi*, the causing agent of Chagas disease, tRF-containing EVs can be transferred to other parasites and/or to host cells to modulate gene expression or facilitate infection [46,47]. In EVs from dendritic and T cells there are different populations of tRFs indicating selective loading of these molecules into the vesicles [48]. Human semen EVs are enriched with tRFs that hypothetically act as translational repressors [49]. It is unknown whether fungal vesicles can be transferred to other cells and consequently regulate metabolism and gene expression but it is tempting to speculate this hypothesis based on the findings mentioned above.

The mRNA population from *graspΔ* EVs had low correlation with WT vesicles. In addition, ncRNA populations were also clearly distinct in EVs from WT and *graspΔ* cells, where snoRNA predominated in the WT and tRNA/tRFs in the KOs.

The distinct RNA cargo in the mutants analyzed in this study is in agreement with a key and general role of GRASP in unconventional secretion in *C. neoformans* and a minor participation of Atg7. Polysaccharides, which lack secretory tags, require GRASP for efficient secretion in *C. neoformans* [20]. Deletion of *ATG7*, however, did not affect polysaccharide export in this fungus [28]. Multivesicular body formation and consequent exosome release involve a number of cellular regulators whose functions directly affect EVs [50,51]. In fungi, a number of regulators affect biogenesis of exosome-like EVs, including the ESCRT machinery, flippases and GRASP [52]. It has been hypothesized that GRASP (Grh1) could participate in this process by acting as a chaperone and directly influencing the cargo of EVs [53]. This GRASP chaperone function could be linked to our current results since RNA cargo was deeply affected in the *graspΔ* mutant. Altogether, these results strongly indicate a novel function for the GRASP family in eukaryotes that could directly affect cell communication, gene expression and host-pathogen interactions.

Supplementary Materials: The following are available online at http://www.mdpi.com/2073-4425/9/8/400/s1, Table S1: List of transcripts enriched in *graspΔ* compared to the wild type strain (H99), Table S2: List of transcripts enriched in *atg7Δ* compared to the wild type strain (H99), Table S3: List of transcripts enriched in the EVs compared to the cell in the wild type strain H99, Table S4: List of ncRNAs enriched in *graspΔ* compared to the wild type strain (H99), Table S5: List of ncRNAs enriched in *atg7Δ* compared to the wild type strain (H99), Figure S1: Electropherograms of the small-RNA content of EVs from the WT, *graspΔ* and *atg7Δ* strains of *C. neoformans*. The size in nucleotides (nt) and the fluorescence intensity (FU) are indicated on the corresponding axes of the graphs generated from the profiles shown on the left.

Author Contributions: R.P.d.S.: obtained the EVs, isolated the RNA, performed the analysis; S.d.T.M.: isolated the EVs, performed NTA analysis and the qPCR validation; J.R.: isolated the EVs and interpreted NTA results; F.C.G.d.R. isolated the EVs and interpreted NTA results; L.S.J. performed the EVs characterization; M.V. produced the mutant strains; L.K. obtained the EVs; D.L.O. obtained the EVs; R.P. analyzed the data, discussed the results, wrote the manuscript; S.G. discussed the results, wrote the manuscript; M.L.R. analyzed the data, wrote the manuscript; L.R.A. performed the RNA-seq, analyzed the data, wrote the manuscript. All authors discussed the results, wrote and approved the final manuscript.

Funding: This research received no external funding.

Acknowledgments: Rosana Puccia was supported by grants from the Brazilian agencies FAPESP, CNPq and CAPES. Marcio Lourenço Rodrigues was supported by grants from the Brazilian agencies FAPERJ and CNPq and by the Instituto Nacional de Ciência e Tecnologia de Inovação em Populações de Doenças Negligenciadas (INCT-IDPN). Samuel Goldenberg was supported by grants from the Brazilian agencies Fundação Araucária–PRONEX and CNPq.

Conflicts of Interest: The authors declare no conflict of interest.

References

1. Coakley, G.; Maizels, R.M.; Buck, A.H. Exosomes and other extracellular vesicles: The new communicators in parasite infections. *Trends Parasitol.* **2015**, *31*, 477–489. [CrossRef] [PubMed]
2. Tkach, M.; Théry, C. Communication by extracellular vesicles: Where we are and where we need to go. *Cell* **2016**, *164*, 1226–1232. [CrossRef] [PubMed]
3. Deatheragea, B.L.; Cooksona, B.T. Membrane vesicle release in bacteria, eukaryotes Eukaryotes and Archaea: A conserved yet underappreciated aspect of microbial life. *Infect. Immun.* **2012**, *80*, 1948–1957. [CrossRef] [PubMed]
4. Rodrigues, M.L.; Nimrichter, L.; Oliveira, D.L.; Frases, S.; Miranda, K.; Zaragoza, O.; Alvarez, M.; Nakouzi, A.; Feldmesser, M.; Casadevall, A. Vesicular polysaccharide export in *Cryptococcus neoformans* is a eukaryotic solution to the problem of fungal trans-cell wall transport. *Eukaryot. Cell* **2007**, *6*, 48–59. [CrossRef] [PubMed]
5. Albuquerque, P.C.; Nakayasu, E.S.; Rodrigues, M.L.; Frases, S.; Casadevall, A.; Zancope-Oliveira, R.M.; Almeida, I.C.; Nosanchuk, J.D. Vesicular transport in *Histoplasma capsulatum*: An effective mechanism for trans-cell wall transfer of proteins and lipids in ascomycetes. *Cell Microbiol.* **2008**, *10*, 1695–1710. [CrossRef] [PubMed]
6. Rodrigues, M.L.; Nakayasu, E.S.; Oliveira, D.L.; Nimrichter, L.; Nosanchuk, J.D.; Almeida, I.C.; Casadevall, A. Extracellular vesicles produced by *Cryptococcus neoformans* contain protein components associated with virulence. *Eukaryot. Cell* **2008**, *7*, 58–67. [CrossRef] [PubMed]
7. Eisenman, H.C.; Frases, S.; Nicola, A.M.; Rodrigues, M.L.; Casadevall, A. Vesicle-associated melanization in *Cryptococcus neoformans*. *Microbiology* **2009**, *155*, 3860–3867. [CrossRef] [PubMed]
8. Rizzo, J.; Oliveira, D.L.; Joffe, L.S.; Hu, G.; Gazos-Lopes, F.; Fonseca, F.L.; Almeida, I.C.; Frases, S.; Kronstad, J.W.; Rodrigues, M.L. Role of the Apt1 protein in polysaccharide secretion by *Cryptococcus neoformans*. *Eukaryot. Cell* **2014**, *13*, 715–726. [CrossRef] [PubMed]
9. Peres da Silva, R.; Puccia, R.; Rodrigues, M.L.; Oliveira, D.L.; Joffe, L.S.; César, G.V.; Nimrichter, L.; Goldenberg, S.; Alves, L.R. Extracellular vesicle-mediated export of fungal RNA. *Sci. Rep.* **2015**, *5*, 7763. [CrossRef] [PubMed]
10. Oliveira, D.L.; Nakayasu, E.S.; Joffe, L.S.; Guimarães, A.J.; Sobreira, T.J.; Nosanchuk, J.D.; Cordero, R.J.; Frases, S.; Casadevall, A.; Almeida, I.C.; et al. Biogenesis of extracellular vesicles in yeast: Many questions with few answers. *Commun. Integr. Biol.* **2010**, *3*, 533–535. [CrossRef] [PubMed]
11. Rodrigues, M.L.; Franzen, A.J.; Nimrichter, L.; Miranda, K. Vesicular mechanisms of traffic of fungal molecules to the extracellular space. *Curr. Opin. Microbiol.* **2013**, *16*, 414–420. [CrossRef] [PubMed]
12. Barr, F.A.; Puype, M.; Vandekerckhove, J.; Warren, G. GRASP65, a protein involved in the stacking of Golgi cisternae. *Cell* **1997**, *91*, 253–262. [CrossRef]
13. Shorter, J.; Watson, R.; Giannakou, M.E.; Clarke, M.; Warren, G.; Barr, F.A. GRASP55, a second mammalian GRASP protein involved in the stacking of Golgi cisternae in a cell-free system. *EMBO J.* **1999**, *18*, 4949–4960. [CrossRef] [PubMed]
14. Rabouille, C.; Malhotra, V.; Nickel, W. Diversity in unconventional protein secretion. *J. Cell Sci.* **2012**, *125*, 5251–5255. [CrossRef] [PubMed]
15. Gee, H.Y.; Noh, S.H.; Tang, B.L.; Kim, K.H.; Lee, M.G. Rescue of ΔF508-CFTR trafficking via a GRASP-dependent unconventional secretion pathway. *Cell* **2011**, *146*, 746–760. [CrossRef] [PubMed]
16. Grieve, A.G.; Rabouille, C. Extracellular cleavage of E-cadherin promotes epithelial cell extrusion. *J. Cell Sci.* **2014**, *127*, 3331–3346. [CrossRef] [PubMed]

17. Kinseth, M.A.; Anjard, C.; Fuller, D.; Guizzunti, G.; Loomis, W.F.; Malhotra, V. The Golgi-associated protein GRASP is required for unconventional protein secretion during development. *Cell* **2007**, *130*, 524–534. [CrossRef] [PubMed]

18. Duran, J.M.; Anjard, C.; Stefan, C.; Loomis, W.F.; Malhotra, V. Unconventional secretion of Acb1 is mediated by autophagosomes. *J. Cell Biol.* **2010**, *188*, 527–536. [CrossRef] [PubMed]

19. Manjithaya, R.; Anjard, C.; Loomis, W.F.; Subramani, S. Unconventional secretion of *Pichia pastoris* Acb1 is dependent on GRASP protein, peroxisomal functions and autophagosome formation. *J. Cell Biol.* **2010**, *188*, 537–546. [CrossRef] [PubMed]

20. Kmetzsch, L.; Joffe, L.S.; Staats, C.C.; de Oliveira, D.L.; Fonseca, F.L.; Cordero, R.J.; Casadevall, A.; Nimrichter, L.; Schrank, A.; Vainstein, M.H.; et al. Role for Golgi reassembly and stacking protein (GRASP) in polysaccharide secretion and fungal virulence. *Mol. Microbiol.* **2011**, *81*, 206–218. [CrossRef] [PubMed]

21. Glick, D.; Barth, S.; Macleod, K.F. Autophagy: Cellular and molecular mechanisms. *J. Pathol.* **2010**, *221*, 3–12. [CrossRef] [PubMed]

22. Nakatogawa, H.; Suzuki, K.; Kamada, Y.; Ohsumi, Y. Dynamics and diversity in autophagy mechanisms: Lessons from yeast. *Nat. Rev. Mol. Cell Biol.* **2009**, *10*, 458–467. [CrossRef] [PubMed]

23. Lévêque, M.F.; Berry, L.; Cipriano, M.J.; Nguyen, H.M.; Striepen, B.; Besteiro, S. Autophagy-related protein ATG8 has a noncanonical function for apicoplast inheritance in *Toxoplasma gondii*. *MBio* **2015**, *6*, e01446-15. [CrossRef] [PubMed]

24. DeSelm, C.J.; Miller, B.C.; Zou, W.; Beatty, W.L.; van Meel, E.; Takahata, Y.; Klumperman, J.; Tooze, S.A.; Teitelbaum, S.L.; Virgin, H.W. Autophagy proteins regulate the secretory component of osteoclastic bone resorption. *Dev. Cell* **2011**, *21*, 966–974. [CrossRef] [PubMed]

25. Dreux, M.; Chisari, F.V. Impact of the autophagy machinery on hepatitis Hepatitis C virus infection. *Viruses* **2011**, *3*, 1342–1357. [CrossRef] [PubMed]

26. Zhang, Y.; Goldman, S.; Baerga, R.; Zhao, Y.; Komatsu, M.; Jin, S. Adipose-specific deletion of autophagy-related gene 7 (atg7) in mice reveals a role in adipogenesis. *Proc. Natl. Acad. Sci. USA* **2009**, *106*, 19860–19865. [CrossRef] [PubMed]

27. Lee, I.H.; Kawai, Y.; Fergusson, M.M.; Rovira, I.I.; Bishop, A.J.; Motoyama, N.; Cao, L.; Finkel, T. Atg7 modulates p53 activity to regulate cell cycle and survival during metabolic stress. *Science* **2012**, *336*, 225–228. [CrossRef] [PubMed]

28. Oliveira, D.L.; Fonseca, F.L.; Zamith-Miranda, D.; Nimrichter, L.; Rodrigues, J.; Pereira, M.D.; Reuwsaat, J.C.; Schrank, A.; Staats, C.; Kmetzsch, L.; et al. The putative autophagy regulator Atg7 affects the physiology and pathogenic mechanisms of *Cryptococcus neoformans*. *Future Microbiol.* **2016**, *11*, 1405–1419. [CrossRef] [PubMed]

29. Bruns, C.; McCaffery, J.M.; Curwin, A.J.; Duran, J.M.; Malhotra, V. Biogenesis of a novel compartment for autophagosome-mediated unconventional protein secretion. *J. Cell Biol.* **2011**, *195*, 979–992. [CrossRef] [PubMed]

30. Maas, S.L.N.; De Vrij, J.; Van Der Vlist, E.J.; Geragousian, B.; Van Bloois, L.; Mastrobattista, E.; Schiffelers, R.M.; Wauben, M.H.M.; Broekman, M.L.D.; Nolte-'t Hoen, E.N. Possibilities and limitations of current technologies for quantification of biological extracellular vesicles and synthetic mimics. *J. Control. Release* **2015**, *200*, 87–96. [CrossRef] [PubMed]

31. Pfaffl, M.W. A new mathematical model for relative quantification in real-time RT-PCR. *Nucleic Acids Res.* **2001**, *29*, e45. [CrossRef] [PubMed]

32. Rodrigues, M.L.; Oliveira, D.L.; Vargas, G.; Girard-Dias, W.; Franzen, A.J.; Frasés, S.; Miranda, K.; Nimrichter, L. Analysis of yeast extracellular vesicles. *Methods Mol. Biol.* **2016**, *1459*, 175–190. [CrossRef] [PubMed]

33. Baggerly, K.A.; Deng, L.; Morris, J.S.; Aldaz, C.M. Differential expression in SAGE: Accounting for normal between-library variation. *Bioinformatics* **2003**, *19*, 1477–1483. [CrossRef] [PubMed]

34. Semenza, J.C.; Hardwick, K.G.; Dean, N.; Pelham, H.R. ERD2, a yeast gene required for the receptor-mediated retrieval of luminal ER proteins from the secretory pathway. *Cell* **1990**, *61*, 1349–1357. [CrossRef]

35. Von Mollard, G.; Stevens, T.H. The *Saccharomyces cerevisiae* v-SNARE Vti1p is required for multiple membrane transport pathways to the vacuole. *Mol. Biol. Cell* **1999**, *10*, 1719–1732. [CrossRef] [PubMed]

36. Abels, E.R.; Breakefield, X.O. Introduction to extracellular vesicles: Biogenesis, RNA cargo selection, content, release and uptake. *Cell. Mol. Neurobiol.* **2016**, *36*, 301–312. [CrossRef] [PubMed]

37. Li, C.; Lev, S.; Saiardi, A.; Desmarini, D.; Sorrell, T.C.; Djordjevic, J.T. Identification of a major IP5 kinase in *Cryptococcus neoformans* confirms that PP-IP5/IP7, not IP6, is essential for virulence. *Sci. Rep.* **2016**, *6*, 23927. [CrossRef] [PubMed]

38. Gonzalez-Hilarion, S.; Paulet, D.; Lee, K.T.; Hon, C.C.; Lechat, P.; Mogensen, E.; Moyrand, F.; Proux, C.; Barboux, R.; Bussotti, G.; et al. Intron retention-dependent gene regulation in *Cryptococcus neoformans*. *Sci. Rep.* **2016**, *6*, 32252. [CrossRef] [PubMed]

39. Keam, S.P.; Hutvagner, G. tRNA-Derived Fragments (tRFs): Emerging new roles for an ancient RNA in the regulation of gene expression. *Life* **2015**, *5*, 1638–1651. [CrossRef] [PubMed]

40. Maute, R.L.; Schneider, C.; Sumazin, P.; Holmes, A.; Califano, A.; Basso, K.; Dalla-Favera, R. tRNA-derived microRNA modulates proliferation and the DNA damage response and is down-regulated in B cell lymphoma. *Proc. Natl. Acad. Sci. USA* **2013**, *110*, 1404–1409. [CrossRef] [PubMed]

41. Gebetsberger, J.; Zywicki, M.; Künzi, A.; Polacek, N. tRNA-derived fragments target the ribosome and function as regulatory non-coding RNA in *Haloferax volcanii*. *Archaea* **2012**, *2012*, 260909. [CrossRef] [PubMed]

42. Lee, Y.S.; Shibata, Y.; Malhotra, A.; Dutta, A. A novel class of small RNAs: tRNA-derived RNA fragments (tRFs). *Genes Dev.* **2009**, *23*, 2639–2649. [CrossRef] [PubMed]

43. Haussecker, D.; Huang, Y.; Lau, A.; Parameswaran, P.; Fire, A.Z.; Kay, M.A. Human tRNA-derived small RNAs in the global regulation of RNA silencing. *RNA* **2010**, *16*, 673–695. [CrossRef] [PubMed]

44. Couvillion, M.T.; Bounova, G.; Purdom, E.; Speed, T.P.; Collins, K. A *Tetrahymena Piwi* bound to mature tRNA 3′ fragments activates the exonuclease Xrn2 for RNA processing in the nucleus. *Mol. Cell* **2012**, *48*, 509–520. [CrossRef] [PubMed]

45. Goodarzi, H.; Liu, X.; Nguyen, H.C.; Zhang, S.; Fish, L.; Tavazoie, S.F. Endogenous tRNA-derived fragments suppress breast cancer progression via YBX1 displacement. *Cell* **2015**, *161*, 790–802. [CrossRef] [PubMed]

46. Garcia-Silva, M.R.; Cabrera-Cabrera, F.; das Neves, R.F.; Souto-Padrón, T.; de Souza, W.; Cayota, A. Gene expression changes induced by *Trypanosoma cruzi* shed microvesicles in mammalian host cells: Relevance of tRNA-derived halves. *BioMed Res. Int.* **2014**, *2014*, 305239. [CrossRef] [PubMed]

47. Garcia-Silva, M.R.; das Neves, R.F.; Cabrera-Cabrera, F.; Sanguinetti, J.; Medeiros, L.C.; Robello, C.; Naya, H.; Fernandez-Calero, T.; Souto-Padron, T.; de Souza, W.; et al. Extracellular vesicles shed by *Trypanosoma cruzi* are linked to small RNA pathways, life cycle regulation and susceptibility to infection of mammalian cells. *Parasitol. Res.* **2014**, *113*, 285–304. [CrossRef] [PubMed]

48. Nolte-'t Hoen, E.N.; Buermans, H.P.; Waasdorp, M.; Stoorvogel, W.; Wauben, M.H.; 't Hoen, P.A. Deep sequencing of RNA from immune cell-derived vesicles uncovers the selective incorporation of small non-coding RNA biotypes with potential regulatory functions. *Nucleic Acids Res.* **2012**, *40*, 9272–9285. [CrossRef] [PubMed]

49. Vojtech, L.; Woo, S.; Hughes, S.; Levy, C.; Ballweber, L.; Sauteraud, R.P.; Strobl, J.; Westerberg, K.; Gottardo, R.; Tewari, M.; et al. Exosomes in human semen carry a distinctive repertoire of small non-coding RNAs with potential regulatory functions. *Nucleic Acids Res.* **2014**, *42*, 7290–7304. [CrossRef] [PubMed]

50. Huotari, J.; Helenius, A. Endosome maturation. *EMBO J.* **2011**, *30*, 3481–3500. [CrossRef] [PubMed]

51. Hanson, P.I.; Cashikar, A. Multivesicular body morphogenesis. *Annu. Rev. Cell Dev. Biol.* **2012**, *28*, 337–362. [CrossRef] [PubMed]

52. Oliveira, D.L.; Rizzo, J.; Joffe, L.S.; Godinho, R.M.; Rodrigues, M.L. Where do they come from and where do they go: Candidates for regulating extracellular vesicle formation in fungi. *Int. J. Mol. Sci.* **2013**, *14*, 9581–9603. [CrossRef] [PubMed]

53. Malhotra, V. Unconventional protein secretion: An evolving mechanism. *EMBO J.* **2013**, *32*, 1660–1664. [CrossRef] [PubMed]

Article

Pathogenesis of the *Candida parapsilosis* Complex in the Model Host *Caenorhabditis elegans*

Ana Carolina Remondi Souza [1,2] ⓘ, Beth Burgwyn Fuchs [2,*] ⓘ, Viviane de Souza Alves [3] ⓘ, Elamparithi Jayamani [2], Arnaldo Lopes Colombo [1] and Eleftherios Mylonakis [2,*]

[1] Special Mycology Laboratory, Division of Infectious Diseases, Federal University of São Paulo-UNIFESP, 04039-032 São Paulo, SP, Brazil; carolina.remondi@yahoo.com.br (A.C.R.S.); arnaldolcolombo@gmail.com (A.L.C.)

[2] Division of Infectious Diseases, Rhode Island Hospital, Alpert Medical School of Brown University, Providence, RI 02903, USA; ejayamani@partners.org

[3] Microorganisms Cell Biology Laboratory, Microbiology Department, Biological Sciences Institute, Federal University of Minas Gerais, Belo Horizonte 31270-901, MG, Brazil; gouveiava@ufmg.br

* Correspondence: helen_fuchs@brown.edu (B.B.F.); emylonakis@lifespan.org (E.M.); Tel.: +1-401-444-7309 (B.B.F.); +1-401-444-7856 (E.M.)

Received: 18 June 2018; Accepted: 25 July 2018; Published: 8 August 2018

Abstract: *Caenorhabditis elegans* is a valuable tool as an infection model toward the study of *Candida* species. In this work, we endeavored to develop a *C. elegans*-*Candida parapsilosis* infection model by using the fungi as a food source. Three species of the *C. parapsilosis* complex (*C. parapsilosis* (*sensu stricto*), *Candida orthopsilosis* and *Candida metapsilosis*) caused infection resulting in *C. elegans* killing. All three strains that comprised the complex significantly diminished the nematode lifespan, indicating the virulence of the pathogens against the host. The infection process included invasion of the intestine and vulva which resulted in organ protrusion and hyphae formation. Importantly, hyphae formation at the vulva opening was not previously reported in *C. elegans*-*Candida* infections. Fungal infected worms in the liquid assay were susceptible to fluconazole and caspofungin and could be found to mount an immune response mediated through increased expression of *cnc-4*, *cnc-7*, and *fipr-22/23*. Overall, the *C. elegans*-*C. parapsilosis* infection model can be used to model *C. parapsilosis* host-pathogen interactions.

Keywords: *Candida parapsilosis*; *Caenorhabditis elegans*; hyphae; invertebrate infection model; host-pathogen interaction

1. Introduction

Candida parapsilosis is a common human opportunistic pathogen able to cause superficial and invasive diseases. Most notably, it causes bloodstream infections (BSIs) in very low birth weight neonates and in patients with catheter-associated candidemia and/or intravenous hyperalimentation [1–3]. In 2005, the genetically heterogeneous taxon *C. parapsilosis* was reclassified into three species: *C. parapsilosis* (*sensu stricto*), *Candida orthopsilosis*, and *Candida metapsilosis* [4]. As of yet, it is unclear if there are putative differences between virulence traits among species within the *C. parapsilosis* complex [5–7]. *C. parapsilosis* (*sensu lato*) is the most common non-*albicans Candida* species (NAC) isolated from BSIs in Spain, Italy, many countries in Latin America, while being described as prevalent in U.S. medical centers [8–11].

Over the past decade, invertebrate models have become increasingly valuable to facilitate the study of fungal pathogenesis [12]. Several factors triggered the development of these models, including ethical issues, costs, and physiological simplicity. Moreover, the innate immune mechanisms between invertebrates and mammals share evolutionary conservation, which provides insight into common virulence factors involved in fungal pathogenesis of different types of hosts [13–17].

In particular, *Caenorhabditis elegans* has been used successfully as a candidiasis infection model [18], and its utility has been demonstrated in the assessment of fungal virulence traits and identification of new anti-fungal compounds [18–20]. Nematodes consume fungal pathogens, substituting for the normal laboratory diet, *Escherichia coli*. The ingested fungi establish an infection within the worm gut that can be characterized by the accumulation of yeast and distention of the intestine. The infected nematodes can be followed with either solid or liquid media assay conditions. In liquid medium assays, yeast form hyphae that protrude through the worm cuticle [18,21]. Both *Candida albicans* and non-*albicans* species have been found to cause lethal infections in *C. elegans* [18].

Although *C. elegans* has proven to be a valuable host to study *C. albicans* and a limited number of non-*albicans* species, there are still limited evaluations applied to the study of *C. parapsilosis* species complex infections [22]. In this study, we developed a *C. parapsilosis* (*sensu lato*)-*C. elegans* infection model and demonstrated the utility of this model to study virulence traits of this pathogenic yeast. Furthermore, our endeavors provide insight into the host's defense mechanisms involved against *C. parapsilosis* infection. We describe the reduced lifespan of worms that ingest *C. parapsilosis* and the host symptoms that follow, which differ from those involved in *C. albicans*-*C. elegans* infection.

2. Materials and Methods

2.1. Strains and Media

The *Candida* strains used in these experiments were obtained from the American Type Culture Collection (ATCC) and included: *C. parapsilosis* (*sensu stricto*) ATCC 22019, *C. orthopsilosis* ATCC 96141, and *C. metapsilosis* ATCC 96143. The *C. elegans* strains described in this study were: N2 bristol [23], *glp-4*; *sek-1* [24] and *pmk-1* [24] (Table 1). The *C. elegans* strains were maintained at 25 °C (N2 and *pmk-1*) and 15 °C (*glp-4*; *sek-1*) and propagated on Nematode Growth Medium (NGM) agar plates seeded with the *E. coli* strain HB101 using established procedures [23]. Yeast cultures were grown in yeast extract, peptone, dextrose (YPD) medium at 30 °C. *E. coli* (HB101) was grown in Luria Broth (LB, Sigma Aldrich, Saint Louis, MO, USA) at 37 °C.

Table 1. *Candida parapslosis* complex and *Caenorhabditis elegans* strains used in this study.

Strain	Description	Purpose	Reference
Candida species			
C. parapsilosis (sensu stricto) ATCC 22019	WT [a]	All experiments	ATCC
Candida orthopsilosis ATCC 96141	WT	All experiments	ATCC
Candida metapsilosis ATCC 93143	WT	All experiments	ATCC
C. elegans			
N2	WT	Immunity response	[23]
glp-4; sek-1	glp-4(bn2) I; sek-1(km4)	Killing assay, treatment with antifungal drugs, microscopic studies	[24]
pmk-1	pmk-1(km25)	Immunity response	[24]

[a] Wild-type.

2.2. Caenorhabditis elegans Liquid Medium Killing Assays

The infection assay was performed as previously described [14]. In brief, worms grown on NGM plates were washed with M9 buffer and placed on 24 h old *C. parapsilosis* species complex lawns (on brain heart infusion (BHI) agar plates) for 4 h. After this, the worms were washed off the plates and transferred to wells (*n* = 30 per well) in a twelve-well plate that contained 2 mL of liquid medium

(20% BHI, 80% M9, 45 µg/mL Kan). The plates were incubated at 25 °C and nematode survival was examined at 24 h intervals for the subsequent 144 h.

2.3. Antifungal Drug Treatments

To study the efficacy of antifungal agents against the *C. parapsilosis* complex in this model, fluconazole (Sigma Aldrich) and caspofungin (Merck, Kenilworth, NJ, USA) were dissolved in dimethyl sulphoxide (DMSO) and added to the liquid assay. The exposure was relative to the minimum inhibitory concentration (MIC) (Table 2): 1×MIC, 2×MIC, and 0.5×MIC. Therefore, for fluconazole, the following concentrations were tested: 1.0 µg/mL (1×), 2.0 µg/mL (2×) and 0.5 µg/mL (0.5×). For caspofungin, the concentrations tested were: 0.5 µg/mL (1×), 1 µg/mL (2×) and 0.25 µg/mL (0.5×) (Table 2). Worms were incubated at 25 °C and survival was monitored daily [18]. Worms were considered dead when they failed to respond to the touch of a platinum wire pick [14].

Table 2. In vitro activity against *C. parapsilosis* species complex reference strains.

Strain	MIC (µg/mL)	
	Fluconazole	Caspofungin
ATCC 22019	1.0	0.5
ATCC 96141	1.0	0.5
ATCC 96143	1.0	0.5

MIC: minimum inhibitory concentration.

2.4. Microscopic Studies

To study *C. parapsilosis* colonization in *C. elegans*, *glp-4*, *sek-1* nematodes were pre-infected with *C. parapsilosis* reference strains for 4 h at 25 °C [14]. Then, the worms were washed three times in M9 buffer and transferred to fresh BHI:M9 medium and incubated at 25 °C for 20 and 48 h. The worms were paralyzed with 1 mM sodium azide solution and placed on 2% agarose pads to capture images at 20 and 48 h post infection [25]. A confocal laser microscope was used for observation (Carl Zeiss M1, Oberkochen, Germany).

2.5. Quantitative RT-PCR Analyses of Candida parapsilosis Infected Nematodes

Following infection, N2 worms were treated and RNA was extracted as previously described [26]. The sample quality was assessed through RNA concentration and the 260/280 or 260/230 ratios using a Nanovue spectrophotometer (GE LifeSciences, Piscataway, NJ, USA).

RNA was reverse transcribed to cDNA using the Verso cDNA Synthesis Kit (Thermo Scientific, Waltham, MA, USA). cDNA was analyzed by quantitative real-time (qRT-PCR) using iTaq Universal SYBR Green Supermix® (Bio-Rad, Hercules, CA, USA) at CFX1000 machine (Bio-Rad) and specific primers to the following targets: *Fipr22/23*, *abf-1*, *abf-2*, *cnc-4*, and *cnc-7* (Table 3). All values were normalized against the reference gene *act-1* [19,27–30]. The thermal cycling conditions were comprised of an initial step at 95 °C for 30 s, followed by thirty-five cycles involving denaturation at 95 °C for 5 s, annealing at 58 °C for 15 s and extension at 72 °C for 1 min. The $2^{-\Delta\Delta Ct}$ was calculated for relative quantification of gene expression.

Table 3. Oligonucleotide sequences used in this study.

Oligonucleotide [a]	Sequence 5′ to 3′	Reference
ABF-1/Fw	CTGCCTTCTCCTTGTTCTCCTACT	[19]
ABF-1/Rv	CCTCTGCATTACCGGAACATC	[19]
ABF-2/Fw	TTTCCTTGCACTTCTCCTGG	This study [b]
ABF-2/Rv	CGGTTCCACAGTTTTGCATAC	This study
CNC-4/Fw	ACAATGGGGCTACGGTCCATAT	This study
CNC-4/Rv	ACTTTCCAATGAGCATTCCGAGGA	This study
CNC-7/Fw	CAGGTTCAATGCAGTATGGCTATGG	This study
CNC-7/Rv	GGACGGTACATTCCCATACC	This study
FIPR-22/23 Fw	GCTGAAGCTCCACACATCC	[19]
FIPR-22/23 Rv	TATCCCATTCCTCCGTATCC	[19]

[a] The letters Fw and Rv in the primers names describe the orientation of the primers 5′ to 3′: F for forward (sense) and R for reverse (antisense); [b] the efficiency of primers was evaluated based on the slope of the standard curve constructed by a 10 fold-dilution series using the cDNA.

2.6. Statistics

Killing curves were plotted and the estimation of differences in survival (log-rank and Wilcoxon tests) was performed by the Kaplan-Meier method using GraphPad Prism 5 (GraphPad Software, La Jolla, CA, USA). A p-value of <0.05 was considered significant. Relative gene expression was compared using Bonferroni's Method with GraphPad Prism 5 (GraphPad Software). Each experiment was repeated at least three times, and each independent experiment gave similar results.

3. Results

3.1. Killing Caenorhabditis elegans by Candida parapsilosis Species Complex

First, we assessed the ability of different species within the *C. parapsilosis* complex to cause infection. The results showed that all three species (*C. parapsilosis* (*sensu stricto*), *C. orthopsilosis* and *C. metapsilosis*) were able to kill *C. elegans*. More specifically, in triplicate experiments, we found that *C. parapsilosis* (*sensu stricto*) 22019, *C. orthopsilosis* 96141 and *C. metapsilosis* 96143 killed *C. elegans* with the time to 50% mortality ranging from four to six days for *C. parapsilosis* (*sensu stricto*), *C. orthopsilosis* and *C. metapsilosis*. In all cases, mortality was higher than the *E. coli* (HB101) control group (*p*-value = 0.003 for *C. parapsilosis* (*sensu stricto*); *p*-value = 0.009 for *C. orthopsilosis* and m for *C. metapsilosis*). Interestingly, there was no significant difference between the *C. parapsilosis* species complex infection groups.

3.2. Hyphal Formation of Candida parapsilosis Species Complex within Caenorhabditis elegans

Candida albicans directed killing of *C. elegans* host is characterized by the formation of filaments that pierce the worm cuticle in a liquid media assay [18]. We investigated whether filaments could be observed in the *C. parapsilosis*-*C. elegans* infection model. Nematode morphology of the infected worms was observed at 20 h and 48 h post-infection with the reference strains ATCC22019, ATCC96141, and ATCC96143. Worms that consumed *E. coli* appeared in good health (Figure 1A). As shown in Figure 1, the progress of infection and death of *C. elegans* was similar when they were infected by *C. parapsilosis* (*sensu stricto*) (Figure 1B), *C. orthopsilosis* (Figure 1C) or *C. metapsilosis* (Figure 1D). The intestine was distended after ingesting fungal cells (Figure 1B–D), and hyphae start to accumulate within the intestines of live animals 20 h after infection. Filaments were observed breaching the worm at the vulva by 48 h and were observed fully protruding only in lethally infected nematodes.

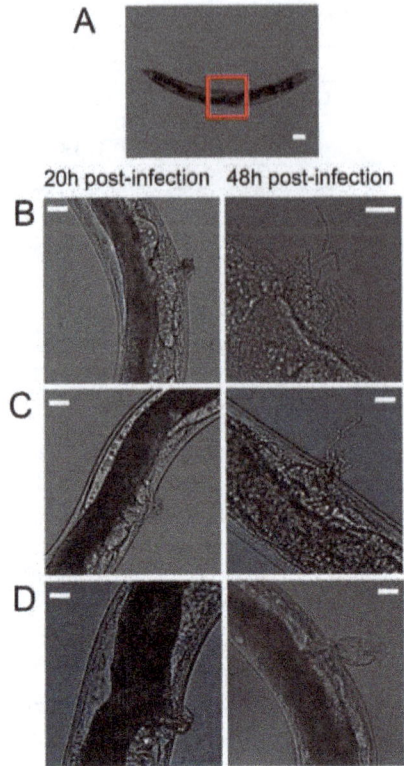

Figure 1. *C. elegans* physiological effects after *C. parapsilosis* exposure. (**A**) *Escherichia coli* exposure shows no adverse effects. The vulva region is highlighted in red; (**B**) *C. parapsilosis* (*sensu stricto*); (**C**) *C. orthopsilosis*; (**D**) *C. metapsilosis*. Scale Bar-20 µm.

3.3. Treatment with Antifungal Drugs

To investigate if compounds could inhibit the fungal infection, *C. elegans* were challenged with *C. parapsilosis* (*sensu stricto*), *C. orthopsilosis*, or *C. metapsilosis* by ingesting the three investigational strains individually on solid media and were then transferred to liquid media, where they were treated with fluconazole and caspofungin. As demonstrated in Figure 2, we observed a dose dependent prolonged survival in response to caspofungin or fluconazole, so that the wells containing antifungal at a concentration below the effective dose (0.5×MIC) resulted in dead worms, similar to those observed in wells that contained no antifungal. On the other hand, a statistically significant increase in survival ($p < 0.001$) was detected when the worms were treated with caspofungin and fluconazole in concentrations at 1×MIC and 2×MIC. In fact, on average, after administration for 144 days, 1×MIC and 2×MIC of fluconazole allowed for, at least, 57% of the nematodes to survive, while 0.5×MIC of this azole resulted in only 38% of the worms being alive. Similarly, caspofungin treatment at 0.5×MIC resulted in 31% nematode survival, whereas at a dose of 1×MIC and 2×MIC, this increased to 69% and 74%, respectively.

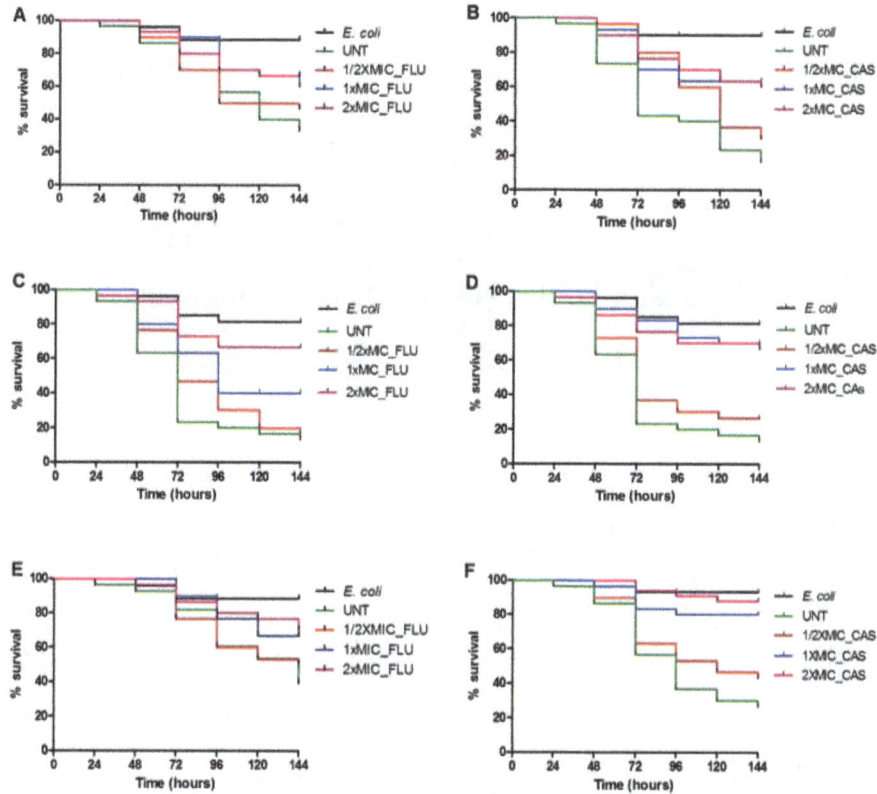

Figure 2. Efficacy of fluconazole (FLU) and caspofungin (CAS) during *C. elegans* infection with *C. parapsilosis* (*sensu stricto*), *C. orthopsilosis* and *C. metapsilosis* reference strains. (**A**) *C. parapsilosis* ATCC 22019, FLU treatment. $p < 0.05$ to FLU_1×MIC. (**B**) *C. parapsilosis* ATCC 22019, CAS treatment. $p = 0.0004$ to CAS_1×MIC. (**C**) *C. orthopsilosis* ATCC 96141, FLU treatment. $p < 0.007$ to FLU_1×MIC. (**D**) *C. orthopsilosis* ATCC 96141, CAS treatment. $p = 0.0001$ to CAS_1×MIC. (**E**) *C. metapsilosis* ATCC 96143, FLU treatment. $p < 0.05$ to FLU_1×MIC. (**D**) *C. metapsilosis* ATCC 96143, CAS treatment. $p < 0.0001$ to CAS_1×MIC.

3.4. Caenorhabditis elegans Immune Response to Candida parapsilosis Complex Infection

In order to understand the defense mechanisms involved against *C. parapsilosis* infection, we focused our attention on five antimicrobial peptides (AMP), which are postulated to have antifungal activity in vivo [19,31]. As shown in Figure 3, 4 h after exposure to *C. parapsilosis* (*sensu stricto*), *C. orthopsilosis*, and *C. metapsilosis*, the expression of the AMP's *cnc-4*, *cnc-7*, and *fipr22/23* increased significantly in response to the presence of the fungal pathogens compared to an *E. coli* control group. The expression of *abf-1* increased only when *C. elegans* was challenged with *C. orthopsilosis* and the *abf-2* expression was unchanged for any of the species involved. Corroborating these findings, we found that *C. elegans pmk-1* (*km25*) mutants were hyper-susceptible to infection with *C. parapsilosis* complex ($p < 0.05$). The PMK-1 mitogen-activated protein (MAP) kinase, orthologous to the mammalian p38 MAPK, in *C. elegans* immunity, is a central regulator of nematode defenses and is required for the basal and pathogen-induced expression of three antifungal immune effectors (*cnc-4*, *cnc-7* and *fipr22/23*), but not *abf-2* (Figure 4).

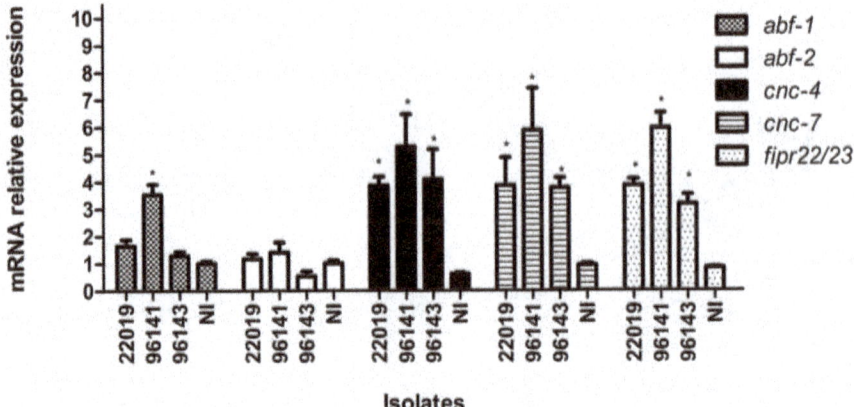

Figure 3. Relative expression of antimicrobial peptides after infection with *C. parapsilosis* (*sensu stricto*) (ATCC 22019), *C. orthopsilosis* (ATCC 96141), *C. metapsilosis* (ATCC 96143), and in non-infected worms (NI). Data are presented as the average of three biological replicates each normalized to a control gene. The error bars represent the standard errors of the mean for three independent biological replicates. * $p < 0.005$.

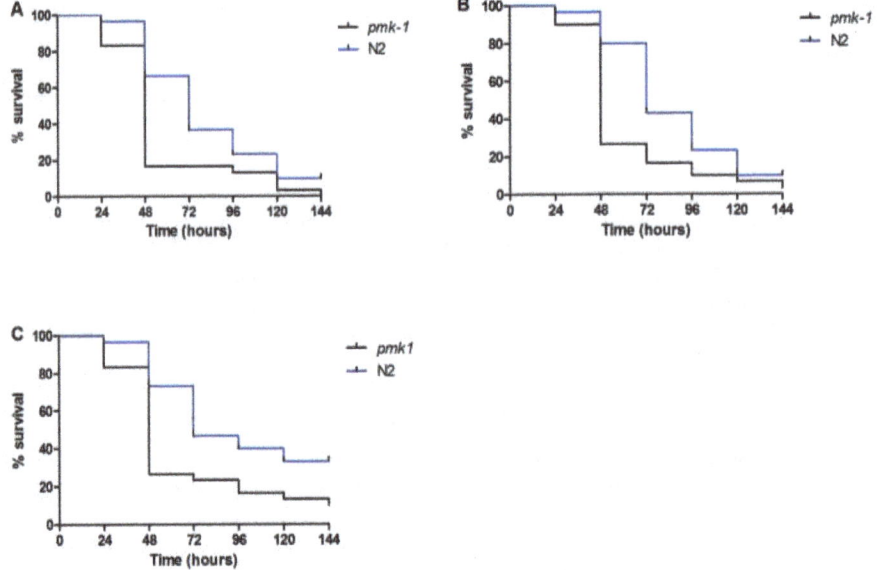

Figure 4. Infection assay with *C. elegans* wild-type (N2) and *pmk-1* animals shows that *pmk-1* was more susceptible to (**A**) *C. parapsilosis* (*sensu stricto*) (ATCC 22019), (**B**) *C. orthopsilosis* (ATCC 96141) and (**C**) *C. metapsilosis* (ATCC 96143) infection ($p < 0.05$).

4. Discussion

In this study, the *C. elegans* invertebrate infection model emerges as a valuable tool for the study of *C. parapsilosis*. Using reference strains, we revealed that *C. parapsilosis* species complex cells, ingested by *C. elegans*, infect and kill the nematode. These data corroborate previous studies showing that

C. orthopsilosis and *C. metapsilosis* are capable of causing invasive infection in mouse models and in humans [8,32,33].

Although the pathogenic potential of the three *C. parapsilosis* complex species is characterized, little is known about the putative differences of their virulence. Previous studies have reported that *C. metapsilosis* seems to be the least virulent member of the *C. parapsilosis* complex in both in vitro and in vivo assays [17,34–36]. However, there is still some controversy within the topic. In a study by Treviño-Rangel et al. (2014), the authors suggest that the three species of the *C. parapsilosis* group possess a similar pathogenic potential in disseminated candidiasis [32]. Corroborating these data, we found that, in *C. elegans* infection model, *C. metapsilosis* had similar mortality rates to that of *C. parapsilosis* (*sensu stricto*) and *C. orthopsilosis*.

Candida hyphal formation is a key virulence determinant that allows cells to invade host tissues and escape phagocytic destruction [37,38]. Pukkila-Worley et al. (2009) demonstrated that the switching from budding yeast to a filamentous (hyphal) form results in aggressive tissue destruction and death of the nematode [21]. In this context, we investigated if this change also took place during infection by *C. parapsilosis*. As expected, all three *C. parapsilosis* strains produced filaments and this phenomenon seems to be associated with *C. elegans* killing. The *C. elegans*-*C. albicans* infection model revealed that *C. albicans* infections within this host begin to accumulate hyphae at the upper part of the gut at the initial stages of the infection process that then spread to consume the entire worm [21]. By contrast, in *C. elegans* infected with *C. parapsilosis* species complex filaments initiated at the vulva rather than the gut, indicating differences in the *C. elegans*-*C. albicans* versus *C. elegans*-*C. parapsilosis* infection processes and host-pathogen interactions.

Although the pathogenicity process may be altered between the two host-pathogen infection models, a response to drug treatment remains conserved. Different studies have demonstrated the utility of invertebrate infection models, including *C. elegans*, as a screening method for potential antifungal compounds [18,25]. In this study, we found a correlation between the in vivo efficacy of antifungals during *C. parapsilosis* (*sensu stricto*), *C. orthopsilosis*, and *C. metapsilosis* infection and their in vitro susceptibility profiles for the standard care therapeutic agents, fluconazole and caspofungin, in a dose dependent manner. By demonstrating that both therapeutic agents have a protective effect during infection in the *C. elegans* model, we gave evidence that drug discovery assays applied to other *Candida* spp.-*C. elegans* models are potentially applicable.

As a whole organism, the model also yields the ability to investigate host immune responses to the pathogen. The nematode *C. elegans* is able to specifically recognize and defend itself against bacterial and fungal pathogens due the presence of complex, inducible, antimicrobial, innate immune responses, which involve the activation of antifungal effectors and core immune genes [19,31,39–42]. Pukkila-Worley et al. (2011) showed that exposure to *C. albicans* stimulated a rapid host response involving approximately 1.6% of the genome, with the majority of the genes encoding antimicrobial, secreted, or detoxification proteins [19]. We therefore used qPCR to check the expression of five AMP genes (*abf-1*, *abf-2*, *cnc-4*, *cnc-7*, and *fipr22/23*) and two transcriptional factors (*zip-2* and *atf-7*). The *abf-1* and *abf*-2 genes belong to a family of six genes encoding antibacterial factors (ABFs) in *C. elegans* and their antimicrobial action were previously described [19,31,43,44]. Expression of these two ABFs seems to be species-specific [41,45]. We found induction of *abf-1* only after *C. orthopsilosis* infection. Regarding *abf-2*, no expression was observed for any of the species.

The second family of AMP we evaluated was the caenacins (CNCs), which are expressed in the *C. elegans* epidermis and, therefore, play a direct role against pathogens that infect worms via the intestinal lumen or cuticle [41,46–48]. Induction of *cnc-4* and *cnc-7* after *C. albicans* has been previously described [41]. Accordingly, we found that the expression of both genes increases significantly in worms fed with any of the three *C. parapsilosis* species, when compared to an *E. coli* control.

In 2008, Pujol et al. described a group of uncharacterized genes that seem to be specifically induced upon fungal infection and could potentially encode AMPs [47]. These genes were called fungus-induced peptide related (*fipr*). During infection with *C. albicans*, the expression of *fipr22/23* is up-regulated [19]. In our study, we also observed induction of the *fipr22/23* gene by the three species belonging to the *C. parapsilosis* complex, suggesting that the FIPRs are involved in the defense mechanisms against *C. parapsilosis* infection. Taken together, our data demonstrated that *C. elegans* mounts a specific defense response against the three different species of *C. parapsilosis* complex.

5. Conclusions

In summary, we demonstrated that *C. elegans* can be used as an appropriate infection model to study the pathogenicity of *C. parapsilosis* (*sensu stricto*), *C. orthopsilosis* and *C. metapsilosis*, not only for evaluating the virulence traits of these species, but also to screen antifungal agents, and study mechanisms of innate immune response against these yeasts. Future studies should expand the model described here to yield more insights about the pathogenicity of these species, especially *C. orthopsilosis* and *C. metapsilosis*.

Author Contributions: B.B.F., A.C.R.S., A.L.C., E.J. and E.M. conceived and designed the experiments. A.C.R.S., V.d.S.A., and E.J. performed the experiments and contributed to analysis. A.L.C. contributed strains and materials. A.C.R.S. and B.B.F wrote the first draft of the article and all other authors contributed to the final version of the text.

Funding: This study was supported by the Fundação de Amparo a Pesquisa do Estado de São Paulo (FAPESP), (2017/02203-7) and Conselho Nacional de Desenvolvimento Científico e Tecnológico (CNPq), Brazil, (308011/2010-4). A.C.R.S. received a doctoral fellowship from FAPESP (2012/04769-4, 2013/07405-6) and a post-doc fellowship from Comissão de Aperfeiçoamento de Pessoal do Nível Superior (CAPES). V.d.S.A. received a post-doctoral fellowship from CNPq, A.L.C. received grants from FAPESP and CNPq. Beth Fuchs was supported by a grant from the Brown-Brazil Initiative.

Conflicts of Interest: The authors declare no conflict of interest.

References

1. Pammi, M.; Holland, L.; Butler, G.; Gacser, A.; Bliss, J.M. *Candida parapsilosis* is a significant neonatal pathogen: A systematic review and meta-analysis. *Pediatr. Infect. Dis. J.* **2013**, *32*, e206–e216. [CrossRef] [PubMed]

2. Trofa, D.; Gacser, A.; Nosanchuk, J.D. *Candida parapsilosis*, an emerging fungal pathogen. *Clin. Microbiol. Rev.* **2008**, *21*, 606–625. [CrossRef] [PubMed]

3. Quindos, G. Epidemiology of candidaemia and invasive candidiasis. A changing face. *Rev. Iberoam. Micol.* **2014**, *31*, 42–48. [CrossRef] [PubMed]

4. Tavanti, A.; Davidson, A.D.; Gow, N.A.; Maiden, M.C.; Odds, F.C. *Candida orthopsilosis* and *Candida metapsilosis* spp. Nov. to replace *Candida parapsilosis* groups II and III. *J. Clin. Microbiol.* **2005**, *43*, 284–292. [CrossRef] [PubMed]

5. Gomez-Lopez, A.; Alastruey-Izquierdo, A.; Rodriguez, D.; Almirante, B.; Pahissa, A.; Rodriguez-Tudela, J.L.; Cuenca-Estrella, M.; Barcelona Candidemia Project Study Group. Prevalence and susceptibility profile of *Candida metapsilosis* and *Candida orthopsilosis*: Results from population-based surveillance of candidemia in Spain. *Antimicrob. Agents Chemother.* **2008**, *52*, 1506–1509. [CrossRef] [PubMed]

6. Lockhart, S.R.; Messer, S.A.; Pfaller, M.A.; Diekema, D.J. Geographic distribution and antifungal susceptibility of the newly described species *Candida orthopsilosis* and *Candida metapsilosis* in comparison to the closely related species *Candida parapsilosis*. *J. Clin. Microbiol.* **2008**, *46*, 2659–2664. [CrossRef] [PubMed]

7. Goncalves, S.S.; Amorim, C.S.; Nucci, M.; Padovan, A.C.; Briones, M.R.; Melo, A.S.; Colombo, A.L. Prevalence rates and antifungal susceptibility profiles of the *Candida parapsilosis* species complex: Results from a nationwide surveillance of candidaemia in Brazil. *Clin. Microbiol. Infect.* **2010**, *16*, 885–887. [CrossRef] [PubMed]

8. Nucci, M.; Queiroz-Telles, F.; Alvarado-Matute, T.; Tiraboschi, I.N.; Cortes, J.; Zurita, J.; Guzman-Blanco, M.; Santolaya, M.E.; Thompson, L.; Sifuentes-Osornio, J.; et al. Epidemiology of candidemia in Latin America: A laboratory-based survey. *PLoS ONE* **2013**, *8*, e59373. [CrossRef] [PubMed]

9. Marcos-Zambrano, L.J.; Escribano, P.; Sanchez, C.; Munoz, P.; Bouza, E.; Guinea, J. Antifungal resistance to fluconazole and echinocandins is not emerging in yeast isolates causing fungemia in a Spanish tertiary care center. *Antimicrob. Agents Chemother.* **2014**, *58*, 4565–4572. [CrossRef] [PubMed]

10. Colombo, A.L.; Guimaraes, T.; Sukienik, T.; Pasqualotto, A.C.; Andreotti, R.; Queiroz-Telles, F.; Nouer, S.A.; Nucci, M. Prognostic factors and historical trends in the epidemiology of candidemia in critically ill patients: An analysis of five multicenter studies sequentially conducted over a 9-year period. *Intens. Care Med.* **2014**, *40*, 1489–1498. [CrossRef] [PubMed]

11. Pfaller, M.A.; Jones, R.N.; Castanheira, M. Regional data analysis of *Candida non-albicans* strains collected in United States medical sites over a 6-year period, 2006–2011. *Mycoses* **2014**, *57*, 602–611. [CrossRef] [PubMed]

12. Arvanitis, M.; Glavis-Bloom, J.; Mylonakis, E. Invertebrate models of fungal infection. *Biochim. Biophys. Acta* **2013**, *1832*, 1378–1383. [CrossRef] [PubMed]

13. Desalermos, A.; Fuchs, B.B.; Mylonakis, E. Selecting an invertebrate model host for the study of fungal pathogenesis. *PLoS Pathog.* **2012**, *8*, e1002451. [CrossRef] [PubMed]

14. Muhammed, M.; Coleman, J.J.; Mylonakis, E. *Caenorhabditis elegans*: A nematode infection model for pathogenic fungi. *Methods Mol. Biol.* **2012**, *845*, 447–454. [PubMed]

15. Junqueira, J.C. *Galleria mellonella* as a model host for human pathogens: Recent studies and new perspectives. *Virulence* **2012**, *3*, 474–476. [CrossRef] [PubMed]

16. Fallon, J.P.; Reeves, E.P.; Kavanagh, K. The *Aspergillus fumigatus* toxin fumagillin suppresses the immune response of *Galleria mellonella* larvae by inhibiting the action of haemocytes. *Microbiology* **2011**, *157*, 1481–1488. [CrossRef] [PubMed]

17. Gago, S.; Garcia-Rodas, R.; Cuesta, I.; Mellado, E.; Alastruey-Izquierdo, A. *Candida parapsilosis*, *Candida orthopsilosis*, and *Candida metapsilosis* virulence in the non-conventional host *Galleria mellonella*. *Virulence* **2014**, *5*, 278–285. [CrossRef] [PubMed]

18. Breger, J.; Fuchs, B.B.; Aperis, G.; Moy, T.I.; Ausubel, F.M.; Mylonakis, E. Antifungal chemical compounds identified using a *C. elegans* pathogenicity assay. *PLoS Pathog.* **2007**, *3*, e18. [CrossRef] [PubMed]

19. Pukkila-Worley, R.; Ausubel, F.M.; Mylonakis, E. *Candida albicans* infection of *Caenorhabditis elegans* induces antifungal immune defenses. *PLoS Pathog.* **2011**, *7*, e1002074. [CrossRef] [PubMed]

20. Mylonakis, E.; Moreno, R.; El Khoury, J.B.; Idnurm, A.; Heitman, J.; Calderwood, S.B.; Ausubel, F.M.; Diener, A. *Galleria mellonella* as a model system to study *Cryptococcus neoformans* pathogenesis. *Infect. Immun.* **2005**, *73*, 3842–3850. [CrossRef] [PubMed]

21. Pukkila-Worley, R.; Peleg, A.Y.; Tampakakis, E.; Mylonakis, E. *Candida albicans* hyphal formation and virulence assessed using a *Caenorhabditis elegans* infection model. *Eukaryot. Cell* **2009**, *8*, 1750–1758. [CrossRef] [PubMed]

22. Ortega-Riveros, M.; De-la-Pinta, I.; Marcos-Arias, C.; Ezpeleta, G.; Quindos, G.; Eraso, E. Usefulness of the non-conventional *Caenorhabditis elegans* model to assess *Candida* virulence. *Mycopathologia* **2017**, *182*, 785–795. [CrossRef] [PubMed]

23. Brenner, S. The genetics of *Caenorhabditis elegans*. *Genetics* **1974**, *77*, 71–94. [PubMed]

24. Kim, D.H.; Feinbaum, R.; Alloing, G.; Emerson, F.E.; Garsin, D.A.; Inoue, H.; Tanaka-Hino, M.; Hisamoto, N.; Matsumoto, K.; Tan, M.W.; et al. A conserved p38 map kinase pathway in *Caenorhabditis elegans* innate immunity. *Science* **2002**, *297*, 623–626. [CrossRef] [PubMed]

25. Huang, X.; Li, D.; Xi, L.; Mylonakis, E. *Caenorhabditis elegans*: A simple nematode infection model for *Penicillium marneffei*. *PLoS ONE* **2014**, *9*, e108764. [CrossRef] [PubMed]

26. Rohlfing, A.K.; Miteva, Y.; Hannenhalli, S.; Lamitina, T. Genetic and physiological activation of osmosensitive gene expression mimics transcriptional signatures of pathogen infection in *C. elegans*. *PLoS ONE* **2010**, *5*, e9010. [CrossRef] [PubMed]

27. Hoogewijs, D.; Houthoofd, K.; Matthijssens, F.; Vandesompele, J.; Vanfleteren, J.R. Selection and validation of a set of reliable reference genes for quantitative *sod* gene expression analysis in *C. elegans*. *BMC Mol. Biol.* **2008**, *9*, 9. [CrossRef] [PubMed]

28. Li, J.; Ebata, A.; Dong, Y.; Rizki, G.; Iwata, T.; Lee, S.S. *Caenorhabditis elegans* HCF-1 functions in longevity maintenance as a DAF-16 regulator. *PLoS Biol.* **2008**, *6*, e233. [CrossRef] [PubMed]

29. Pujol, N.; Cypowyj, S.; Ziegler, K.; Millet, A.; Astrain, A.; Goncharov, A.; Jin, Y.; Chisholm, A.D.; Ewbank, J.J. Distinct innate immune responses to infection and wounding in the *C. elegans* epidermis. *Curr. Biol.* **2008**, *18*, 481–489. [CrossRef] [PubMed]

30. Zhang, Y.; Chen, D.; Smith, M.A.; Zhang, B.; Pan, X. Selection of reliable reference genes in *Caenorhabditis elegans* for analysis of nanotoxicity. *PLoS ONE* **2012**, *7*, e31849. [CrossRef] [PubMed]

31. Kato, Y.; Aizawa, T.; Hoshino, H.; Kawano, K.; Nitta, K.; Zhang, H. abf-1 and abf-2, ASABF-type antimicrobial peptide genes in *Caenorhabditis elegans*. *Biochem. J.* **2002**, *361*, 221–230. [CrossRef] [PubMed]

32. Treviño-Rangel Rde, J.; Rodriguez-Sanchez, I.P.; Elizondo-Zertuche, M.; Martinez-Fierro, M.L.; Garza-Veloz, I.; Romero-Diaz, V.J.; Gonzalez, J.G.; Gonzalez, G.M. Evaluation of in vivo pathogenicity of *Candida parapsilosis*, *Candida orthopsilosis*, and *Candida metapsilosis* with different enzymatic profiles in a murine model of disseminated candidiasis. *Med. Mycol.* **2014**, *52*, 240–245. [CrossRef] [PubMed]

33. Blanco-Blanco, M.T.; Gomez-Garcia, A.C.; Hurtado, C.; Galan-Ladero, M.A.; Lozano Mdel, C.; Garcia-Tapias, A.; Blanco, M.T. *Candida orthopsilosis* fungemias in a Spanish tertiary care hospital: Incidence, epidemiology and antifungal susceptibility. *Rev. Iberoam. Micol.* **2014**, *31*, 145–148. [CrossRef] [PubMed]

34. Gacser, A.; Trofa, D.; Schafer, W.; Nosanchuk, J.D. Targeted gene deletion in *Candida parapsilosis* demonstrates the role of secreted lipase in virulence. *J. Clin. Investig.* **2007**, *117*, 3049–3058. [CrossRef] [PubMed]

35. Nemeth, T.; Toth, A.; Szenzenstein, J.; Horvath, P.; Nosanchuk, J.D.; Grozer, Z.; Toth, R.; Papp, C.; Hamari, Z.; Vagvolgyi, C.; et al. Characterization of virulence properties in the *C. parapsilosis sensu lato* species. *PLoS ONE* **2013**, *8*, e68704. [CrossRef] [PubMed]

36. Bertini, A.; De Bernardis, F.; Hensgens, L.A.; Sandini, S.; Senesi, S.; Tavanti, A. Comparison of *Candida parapsilosis*, *Candida orthopsilosis*, and *Candida metapsilosis* adhesive properties and pathogenicity. *Int. J. Med. Microbiol.* **2013**, *303*, 98–103. [CrossRef] [PubMed]

37. Liu, H. Transcriptional control of dimorphism in *Candida albicans*. *Curr. Opin. Microbiol.* **2001**, *4*, 728–735. [CrossRef]

38. Navarro-Garcia, F.; Sanchez, M.; Nombela, C.; Pla, J. Virulence genes in the pathogenic yeast *Candida albicans*. *FEMS Microbiol. Rev.* **2001**, *25*, 245–268. [CrossRef] [PubMed]

39. Engelmann, I.; Pujol, N. Innate immunity in *C. elegans*. *Adv. Exp. Med. Biol.* **2010**, *708*, 105–121. [PubMed]

40. Ermolaeva, M.A.; Schumacher, B. Insights from the worm: The *C. elegans* model for innate immunity. *Semin. Immunol.* **2014**, *26*, 303–309. [CrossRef] [PubMed]

41. Engelmann, I.; Griffon, A.; Tichit, L.; Montanana-Sanchis, F.; Wang, G.; Reinke, V.; Waterston, R.H.; Hillier, L.W.; Ewbank, J.J. A comprehensive analysis of gene expression changes provoked by bacterial and fungal infection in *C. elegans*. *PLoS ONE* **2011**, *6*, e19055. [CrossRef] [PubMed]

42. Zugasti, O.; Bose, N.; Squiban, B.; Belougne, J.; Kurz, C.L.; Schroeder, F.C.; Pujol, N.; Ewbank, J.J. Activation of a G protein-coupled receptor by its endogenous ligand triggers the innate immune response of *Caenorhabditis elegans*. *Nat. Immunol.* **2014**, *15*, 833–838. [CrossRef] [PubMed]

43. Kato, Y.; Komatsu, S. ASABF, a novel cysteine-rich antibacterial peptide isolated from the nematode *Ascaris suum*. Purification, primary structure, and molecular cloning of cDNA. *J. Biol. Chem.* **1996**, *271*, 30493–30498. [CrossRef] [PubMed]

44. Zhang, H.; Kato, Y. Common structural properties specifically found in the CSαβ-type antimicrobial peptides in nematodes and mollusks: Evidence for the same evolutionary origin? *Dev. Comp. Immunol.* **2003**, *27*, 499–503. [CrossRef]

45. Means, T.K.; Mylonakis, E.; Tampakakis, E.; Colvin, R.A.; Seung, E.; Puckett, L.; Tai, M.F.; Stewart, C.R.; Pukkila-Worley, R.; Hickman, S.E.; et al. Evolutionarily conserved recognition and innate immunity to fungal pathogens by the scavenger receptors SCARF1 and CD36. *J. Exp. Med.* **2009**, *206*, 637–653. [CrossRef] [PubMed]

46. Couillault, C.; Pujol, N.; Reboul, J.; Sabatier, L.; Guichou, J.F.; Kohara, Y.; Ewbank, J.J. TLR-independent control of innate immunity in *Caenorhabditis elegans* by the TIR domain adaptor protein TIR-1, an ortholog of human SARM. *Nat. Immunol.* **2004**, *5*, 488–494. [CrossRef] [PubMed]

47. Pujol, N.; Zugasti, O.; Wong, D.; Couillault, C.; Kurz, C.L.; Schulenburg, H.; Ewbank, J.J. Anti-fungal innate immunity in *C. elegans* is enhanced by evolutionary diversification of antimicrobial peptides. *PLoS Pathog.* **2008**, *4*, e1000105. [CrossRef] [PubMed]

48. Zugasti, O.; Ewbank, J.J. Neuroimmune regulation of antimicrobial peptide expression by a noncanonical TGF-β signaling pathway in *Caenorhabditis elegans* epidermis. *Nat. Immunol.* **2009**, *10*, 249–256. [CrossRef] [PubMed]

Article

Role of Homologous Recombination Genes in Repair of Alkylation Base Damage by *Candida albicans*

Toni Ciudad [†] , Alberto Bellido [†], Encarnación Andaluz, Belén Hermosa and Germán Larriba *

Departamento de Microbiología, Facultad de Ciencias, Universidad de Extremadura, 06071 Badajoz, Spain;
aciudad@unex.es (T.C.); abdiaz@unex.es (A.B.); eandaluz@unex.es (E.A.); belenh@unex.es (B.H.)
* Correspondence: glarriba@unex.es
† These authors contributed equally to this work.

Received: 14 August 2018; Accepted: 27 August 2018; Published: 7 September 2018

Abstract: *Candida albicans* mutants deficient in homologous recombination (HR) are extremely sensitive to the alkylating agent methyl-methane-sulfonate (MMS). Here, we have investigated the role of HR genes in the protection and repair of *C. albicans* chromosomes by taking advantage of the heat-labile property (55 °C) of MMS-induced base damage. Acute MMS treatments of cycling cells caused chromosome fragmentation in vitro (55 °C) due to the generation of heat-dependent breaks (HDBs), but not in vivo (30 °C). Following removal of MMS wild type, cells regained the chromosome ladder regardless of whether they were transferred to yeast extract/peptone/dextrose (YPD) or to phosphate buffer saline (PBS); however, repair of HDB/chromosome restitution was faster in YPD, suggesting that it was accelerated by metabolic energy and further fueled by the subsequent overgrowth of survivors. Compared to wild type CAI4, chromosome restitution in YPD was not altered in a C*arad59* isogenic derivative, whereas it was significantly delayed in C*arad51* and C*arad52* counterparts. However, when post-MMS incubation took place in PBS, chromosome restitution in wild type and HR mutants occurred with similar kinetics, suggesting that the exquisite sensitivity of C*arad51* and C*arad52* mutants to MMS is due to defective fork restart. Overall, our results demonstrate that repair of HDBs by resting cells of *C. albicans* is rather independent of CaRad51, CaRad52, and CaRad59, suggesting that it occurs mainly by base excision repair (BER).

Keywords: homologous recombination; *Candida albicans*; alkylation damage; repair

---

1. Introduction

Methyl-methane-sulfonate (MMS) is used for the analysis of pathways involved in repair/tolerance to methylation [1]. Methyl-methane-sulfonate generates methylated bases on dsDNA whose repair can cause nicks, gaps, and, indirectly, double-strand breaks (DSBs) that can engage in homologous recombination (HR) directly or cause fork stalling during the next replication round [2–4]. Although methylated bases can be directly removed by DNA-methyl-transferases, the major repair pathway consists of a step-wise process known as base excision repair (BER) [1,5]. In *Saccharomyces cerevisiae*, BER is initiated by specific DNA-*N*-glycosylases that remove the damaged bases. The apurinic/apyrimidic (AP) sites generated are removed by redundant AP endonucleases Apn1 and Apn2, which cleave 5′ the AP-site to form nicks with a 5′ desoxyribose phosphate (5′dRP). Removal of 5′dRP is carried out by the coordinated action of DNA polymerase (δ or ε) and the flap endonuclease Rad27/Fen1, followed by ligation. Alternatively, AP-sites can be processed by unspecific Ntg1, Ntg2, or Ogg1 lyases to generate 3′AP sites (3′-dRP), which are then removed by the 3′-diesterase activity of Apn1/Apn2 or as part of an oligonucleotide generated by the endonuclease Rad1–Rad10. Finally, the gap is filled by DNA Pol and the backbone sealed by DNA ligase [1,4,6]. It is likely that similar enzymes and reactions account for BER in other fungi. However, only one APN endonuclease

(*APN1*) and one NTG lyase (*NTG1*), in addition to an *OGG1* homolog, have been so far identified and partially characterized in *Candida albicans* [7]. Importantly, null mutants in each of these *C. albicans* genes as well as the triple mutant exhibited wild type sensitivities to MMS suggesting the presence of redundant enzymes involved in repair of the methylated bases [7].

Importantly, in methylated DNA, AP-sites can arise from elimination of methylated bases by the action of glycosylase Mag1 [6] or, non-enzymatically, through the spontaneous depurination of methylated N3 and N7 purines, a process that is accelerated by heat [8]. It is well known that AP-sites are heat labile [9,10] and can be converted into single-strand breaks (SSBs) at 55 °C (referred to as HDB, for heat dependent breaks), a temperature generally used to prepare plugs for pulse-filed gel electrophoresis (PFGE) [11]. In fact, in *S. cerevisiae* stationary, or G1-arrested cells, DSBs and chromosome degradation was observed when methylated DNA or DNA carrying AP-sites was incubated at 55 °C [4,6,11]. Methyl-methane-sulfonate treatment can also cause opposed closely-spaced nicks in vivo (referred to as heat independent breaks (HIB) because they are detected when the same DNA samples are incubated at 30 °C), which can also result in secondary DSBs, and therefore in chromosome fragmentation [4,6,11]. For *S. cerevisiae* wild type, typical MMS treatments generate a large amount of HDBs and a low amount of HIB [12]. However, the latter are substantially increased in some BER-deficient cells as in *apn1/apn2* double mutants [4,6,12].

Rad51 and Rad52 are evolutionary conserved proteins that play crucial roles in HR. In yeast, recombinase Rad51 is required for recombination pathways involving strand invasion. These pathways also require Rad52, which is thought to mediate the Rad51-ssDNA nucleofilament assembly. In addition, Rad52 participates in all recombination processes that require single strand annealing. For this reason, the *rad52* mutation is epistatic to deletion of any other gene of the *RAD52* epistasis group. Rad59 is a yeast paralog of Rad52 that exhibits strand annealing activity but lacks the ability of loading Rad51 onto ssDNA (for reviews see References [3,13]). Several studies have indicated that, although *rad51* and *rad52* mutants are extremely sensitive to MMS, HR does not play any role in the repair of MMS-born HIB of haploid *S. cerevisiae* G1 stationary cells which do not have a partner for engaging in HR, and the same was true for diploid G2-arrested cells which do [4,6,11,14]. However, Rad52 was crucial for that repair of HIB in G2-arrested cells in the absence of Anp1 and Anp2 endonucleases, suggesting that HR acts as a backup to repair lesions produced by AP-lyases Ntg1/2 and Ogg1 in the absence of Anp1,2 [6]. Besides, Rad51 and Rad52 are required for replication of methylated DNA [15–17] as well as for repair of the gaps generated in the process when cells reach the G2 phase [17]. Methyl-methane-sulfonate lesions and BER intermediates that are not repaired before they encounter the replication fork may cause replication fork stalling and collapse, unless stalled forks are bypassed by translation synthesis (stalled forks) causing increased mutagenesis or faithfully repaired by HR (both stalled and collapsed forks) [6,16,17]. The importance of HR in repair of methylated DNA extends to *C. albicans* where mutants affected in either HR (C*arad52*) or resection of DSBs (C*arad50*, C*amre11*) were significantly more sensitive to MMS than single mutants in BER genes [7,18].

In addition to MMS, a number of both endogenous and environmental agents including anticancer drugs (i.e., 4-methyl-5-oxo-2,3,4,6,8-pentazabicyclo[4.3.0]nona-2,7,9-triene-9-carboxamide temozolomide) can cause methylation damage [8,19,20], and therefore may affect viability and virulence of commensal opportunistic pathogens such as *C. albicans*. We previously reported that C*arad52*-ΔΔ, and to a lesser extent C*arad51*-ΔΔ cells from *C. albicans*, exhibit increased sensitivity to MMS [21,22]. In the current study, we have determined methylation base damage and recovery in *C. albicans*, taking advantage of the secondary DSBs and subsequent chromosome fragmentation generated during preparation of samples for pulse-field gel electrophoresis (PFGE) at 55 °C. We also show that resting cells of *C. albicans* can repair HDBs in the absence of HR, whereas repair of HDBs (and other BER intermediates) by cycling cell was mostly dependent on efficient HR.

2. Materials and Methods

2.1. Strains

The *C. albicans* and *S. cerevisiae* strains used in this study are shown in Table 1. Strain CAF2-1 derives from reference strain SC5314 by deletion of one copy of *URA3* whereas CAI4 is the Uri⁻ auxotrophic derivative of CAF2-1. All three strains are wild type for DNA recombination and repair. They were routinely grown in either yeast extract/peptone/dextrose (YPD) medium or synthetic complete (SC) medium supplemented with 33 mM uridine when necessary (i.e., Uri⁻ strains) [23]. The haploid *S. cerevisiae* strain LSY0695-7D (Table 1) is a W303 derivative kindly provided by Lorraine Symington, from Columbia University [24]. Diploid W303 was obtained by standard genetic crosses [25].

Table 1. Strains used in this study.

Strains (Old Name)	Genotype	Parental	Reference
Candida albicans			
SC5314	Wild type		Gillum et al., 1984 [26]
CAF2-1	Δ*ura3::imm434/URA3*	SC5314	Fonzi and Irwin, 1993 [27]
CAI4	Δ*ura3::imm434/*Δ*ura3::imm434*	CAF2-1	Fonzi and Irwin, 1993 [27]
CAGL4 (TCR2.1)	Δ*ura3::imm434/*Δ*ura3::imm434* *rad52::hisG/*Δ*rad52::hisG-URA3-hisG*	CAI4	Ciudad et al., 2004 [28]
CAGL4.1 (TCR2.1.1)	Δ*ura3::imm434/*Δ*ura3::imm434* Δ*rad52::hisG/*Δ*rad52::hisG*	CAGL4	Ciudad et al., 2004 [28]
CAGL17 (BNC1.1)	Δ*ura3::imm434/*Δ*ura3::imm434* Δ*rad59::hisG/*Δ*rad59::hisG-URA3-hisG*	CAI4	García-Prieto et al., 2010 [21]
CAGL17.1 (BNC23.1)	Δ*ura3::imm434/*Δ*ura3::imm434* Δ*rad59::hisG/*Δ*rad59::hisG*	CAGL17	Bellido et al., 2015 [22]
CAGL19 (JGR5)	Δ*ura3::imm434/*Δ*ura3::imm434* Δ*rad51::hisG/*Δ*rad51::hisG-URA3-hisG*	CAI4	García-Prieto et al., 2010 [21]
CAGL19.1 (JGR5A)	Δ*ura3::imm434/*Δ*ura3::imm434* Δ*rad51::hisG/*Δ*rad51::hisG*	CAGL19	García-Prieto et al., 2010 [21]
S. cerevisiae			
LSY0695-7D W303 haploid	*MATa; ADE2 RAD5 met17-s ade2-1 trp1-1 his3-11,15 can1-100 ura3-1 leu2-3,112*		Bärtsch et al., 2000 [24]
W303 diploid	*MATa/α; ADE2 RAD5 met17-s/met17-s ade2-1/ade2-1 trp1-1/trp1-1 his3-11,15/his3-11,15 can1-100/can1-100 ura3-1/ura3-1 leu2-3,112/leu2-3,112*		Bellido et al., 2015 [22]

2.2. DNA Extraction and Analysis and Cell Transformation

The DNA preparation for PFGE, as well as resolution of the samples, was carried out as reported before [29,30] using a Bio-Rad Chef Dry III. Gels were stained with ethidium bromide (0.5 μg/mL) for 2–4 h and imaged using a Molecular Imager (Bio-Rad Laboratories, Madrid, Spain). *C. albicans* cells were transformed using the lithium acetate method [31]. *C. albicans* chromosome fragments generated during incubation of methylated DNA with proteinase K at 55 °C were sized by using *S. cerevisiae* chromosomes as molecular weight (MW) markers (316–1091 kb) (see below).

A functional copy of the *C. albicans* URA3 marker was obtained by digestion of pLUBP plasmid with Pst1-BglII [32]. The resulting 4.9 kb fragment containing the *URA3* gene was used to transform the Uri⁻ strains CAI4 and its derivatives *rad59*-ΔΔ and *rad51*-ΔΔ. Uri⁺ transformants were selected on minimal SC medium minus uridine and correct integration was verified by PCR using oligonucleotide URA3 left flank and URA3 right flank [32].

2.3. Sensitivity to DNA-Damaging Agents

For determination of survival following an acute short-term MMS treatment, about 5×10^5 exponentially growing cells suspended in 1 mL YPD were incubated with MMS (0.05%, final concentration) for 30 min. Incubation mixtures were diluted 10^3-fold with PBS, and 50 µL (containing 300–400 colony forming units (CFU) before treatment) were plated on YPD plates for 48 h to determine the number of colonies. All the assays were done in duplicate and repeated four or more times.

2.4. Generation of MMS-Induced Heat Labile Breaks

Alkylation was induced as described by Lundin et al. [11]. Briefly, the indicated yeast strain was grown overnight in YPD until $OD_{600} \approx 5$–7. MMS was added to a final concentration of 0.05% and cell suspensions were shaken at 30 °C for 15–30 min. The MMS was neutralized with 5% sodium thiosulfate by mixing 1:1 (v/v) ratio with 10% $Na_2S_2O_3$ and washed twice with phosphate buffer saline (PBS). An aliquot was processed for PFGE ($t = 0$) and the rest was incubated in MMS-free YPD medium at 30 °C. At regular intervals (1, 2, 4, 8 and 24 h) samples were taken for determination of OD_{600}, CFU, morphology, and repair of DNA damage (PFGE). Plugs for PFGE were prepared as described above and subsequently treated for 24 h with proteinase K (1 mg/mL) at 55 °C for analysis of HDBs or at 30 °C for analysis of HIBs as reported [4,6]. When indicated, thiosulfate neutralized cells were allowed to stand in PBS (resting cells) and samples were taken at the indicated times [4,6,11].

2.5. Calculations of Chromosome Fragments Sizes from Closely Spaced Single-Strand Breaks

Our calculations are based on the assumptions that SSBs (or heat-labile sites, in our assay) are distributed evenly between two DNA strands of a chromosome and that for a given SSB to form a DSB, a second SSB must appear on the opposite strand within an interval $\leq S$ [6]. For further calculations of the number of DSBs, Ma et al. [6] used a circularized ChrIII, which only enters the gel following induction of a single DSB, to quantitatively determine the ratio ChrII to ChrIII using Southern blotting with a probe that hybridizes to both chromosomes. Since *C. albicans* does not maintain circular plasmids [28], we determined the range of sizes of the smear produced by the MMS treatment (0.05% MMS, 30 min) using the *S. cerevisiae* chromosomes as size markers. For *S. cerevisiae* diploid W303 strain (24 Mb), chromosome fragment sizes generated by MMS treatment ranged between 250 and 666 Kb, whose mean is 455 kb. Generation of uniform fragments of this size would require an average of 52 DSB per diploid genome or 26 DSB per haploid genome. For *C. albicans* (32 Mb), chromosome fragments sizes ranged between 316 and 1091 kb. By analogy, for a mean of 700kb we calculated 46 DSB per diploid genome or 23 DSB per haploid genome. Importantly, these values are not far from the range of 30 and 40 DSB per *S. cerevisiae* haploid genome previously reported [6] for acute (30 min) 0.1% MMS treatments taking into account that we have used half MMS concentration (0.05% MMS) (Figure S1).

To quantify the extent of chromosome restitution during the post-MMS incubation, we estimated the intensities associated to Chr2, Chr5 bands (the same area for each chromosome) and smear from pulsed-field gels stained with ethidium bromide using the ImageJ software. Then, the ratio of the intensities Chr/smear at each time point was graphed versus the post-incubation time using Microsoft Excel. Data obtained with this approach agreed well with visual inspections of the gels. Chr2 and Chr5 were chosen as indicators of chromosome restitution because both are well resolved in PFGE and differ widely in size (2.23 Mb for Chr2 and 1.19 Mb for Chr5).

3. Results

3.1. The Role of HR Genes in Growth Polarization in Response to MMS

Genotoxic stress, including hydroxyurea and MMS treatments, triggers growth polarization of wild type *C. albicans* SC5314 and derivatives CAF2-1 (*URA3/ura3*) and CAI4 (*ura3/ura3*) (Table 1) generating elongated cells [33,34]. We have recently shown that the Uri⁻ strain CAI4 and its derivatives carrying additional auxotrophies (RM10, RM1000, BWP37 and SN strains) exhibit an enhanced growth polarization and susceptibility to 0.02% MMS compared to parental CAF2-1 [35]. This differential behavior is due to a spontaneous loss of heterozygosity (LOH) event during the generation of CAI4 on the right arm of chromosome 3 (Chr3R) that homozygosed *MBP1a*, which regulates expression of DNA repair genes at G1/S phase of the cell cycle [35]. To investigate if MMS-induced filamentation was further affected by HR mutations, we subjected wild type and mutant strains to 0.02% MMS for 16 h at 30 °C in liquid YPD (Figure 1). As expected, CAF2-1 displayed chains of elongated cells but no filaments, whereas its Uri$^-$ derivative CAI4 showed long filaments. Reintegration of one copy of *URA3* in its own locus improved growth rate but did not affect filamentation of the CAI4 strain [35]. All three C*arad59*-ΔΔ strains filamented as CAI4, regardless of whether they were Uri$^+$ or Uri$^-$, or if one copy of *URA3* had been reintegrated into its own locus in the Uri$^-$ version of the mutant (Figure 1). The same was true for C*arad51*-ΔΔ strains in its Uri$^-$, Uri$^+$, or *URA3*-reintegrated versions (Figure 1). For C*arad52*-ΔΔ, Uri$^+$ and Uri$^-$ versions behaved also similarly (Figure 1); however, as described [28], it was not possible to reintegrate *URA3* into its own locus in *rad52*-ΔΔ strains. Importantly, in addition to the typical filamentous cells, C*arad51*-ΔΔ and C*arad52*-ΔΔ cultures also contain yeast cells [23]. In response to MMS, yeast cells also formed "germinative tubes" whose form and length were similar to their counterparts from C*arad59*-ΔΔ or CAI4 strains. We conclude that recombination mutants retain the ability to filament in response to MMS. It should be, finally, noted that cell elongation was a specific trait of *C. albicans* since it was not observed when diploid *S. cerevisiae* W303 or its S*crad52* derivative were incubated in MMS or during their post-MMS incubation in either PBS or YPD (not shown).

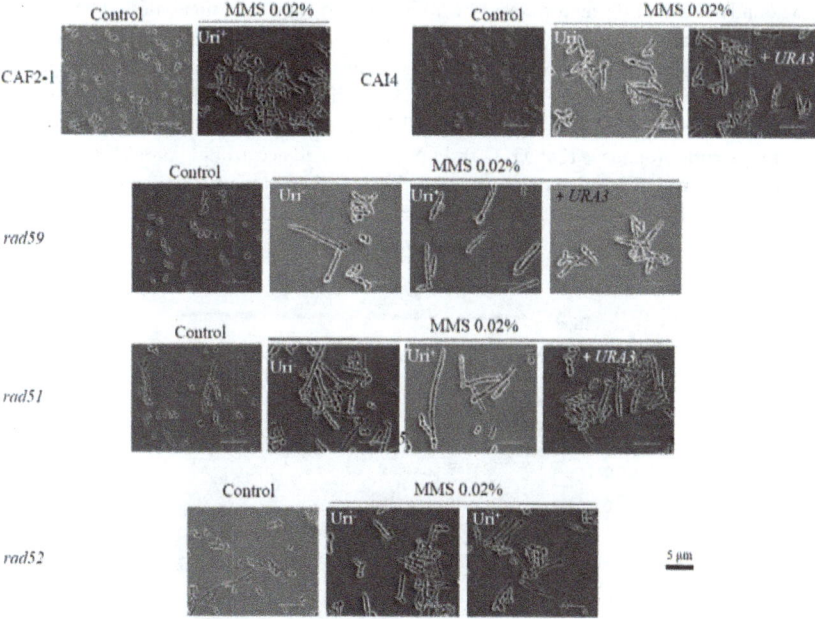

Figure 1. Methyl-methane-sulfonate (MMS) induces constitutive filamentous growth in wild type and HR mutants of *Candida albicans*. A yeast extract/peptone/dextrose (YPD) overnight culture of exponentially growing cells from the indicated strains was refreshed and adjusted to OD$_{600}$ = 1. Following a further incubation for 2 h at 30 °C with shaking, one half was suspended in YPD supplemented or not (control) with 0.02% MMS. After 12 h at 30 °C, with gentle shaking, samples were photographed using a Nikon Eclipse 600 microscope with a 60× DIC objective. A CC-12 digital camera interfaced with Soft Imaging System software was used for imaging (Izasa Scientific, Alcobendas, Madrid, Spain). Each bar corresponds to 5 μm.

3.2. Generation and Repair of MMS-Induced Heat Labile Lesions on Growing C. albicans Wild Type Cells

Although MMS treatment of *S. cerevisiae* and human cells in vivo does not cause DSBs, methylated bases and AP-sites have been shown to be heat labile and converted into single strand breaks (SSBs) during the 55 °C proteinase K treatment used for preparation of PFGE plugs [6,9,11]. Furthermore, closely spaced SSBs located in opposite DNA strands may result in DSBs, and therefore chromosome fragmentation [6,9,11]. With these premises in mind, we investigated the extent to which MMS causes HDBs and HIBs in DNA from *C. albicans*.

Parental CAF2-1 strain exhibits the standard *C. albicans* karyotype. However, when medium-to-late exponentially growing cells (OD$_{600}$ = 9; of note, when grown in YPD, *C. albicans* reaches OD$_{600}$ up to 18) were subjected to an acute MMS treatment (0.05%, 15–30 min in YPD) and chromosome preparation for PFGE (which includes incubation with protease K) was conducted at 55 °C there was complete loss of chromosomal bands, which migrated now below Chr7 as a smear, indicating the presence of DSBs (Figure 2). Furthermore, the average size of the pool of degraded chromosomes obtained at 55 °C decreased with the incubation time in MMS, suggesting progressive chromosome fragmentation (Figure S2, lanes 1–5). By contrast, when following the acute MMS treatment (0.05%, 15–30 min in YPD) proteinase K incubation was conducted at 30 °C, strain CAF2-1 displayed the standard PFGE karyotype, and chromosome degradation was negligible (Figure S3, CAF2-1, lanes 1 and 2). Under these conditions, extension of the acute MMS treatment up to 120 min did not lead to a significant increase in chromosomal degradation (Figure S3, lanes 3–5). Some smear was likely due to the generation of a few closely opposed SSBs during manipulation of the samples, since it also was occasionally shown by preparations of untreated wild type cells incubated at 55 °C; however, the induction of small amounts of HIBs by MMS cannot be ruled out. As expected from previous reports [11,12], MMS-treated late-exponential phase cells of *S. cerevisiae* W303 subjected to the same process (i.e., preparation of samples for PFGE at 55 °C) displayed also chromosome fragmentation (Figure S1). These results indicate that almost all DSBs were generated during the incubation of *C. albicans* plugs with proteinase K at 55 °C and not in vivo, and accordingly provide an assessment of the overall number of HDBs [6]. Consistent with this interpretation, incubation of plugs at a lower temperature (50 °C) resulted in a reduced migration of the PFGE smear accompanied by the presence of vestiges of intact chromosomal bands (Figure S4, compare panels A and B, CAF2-1 lanes).

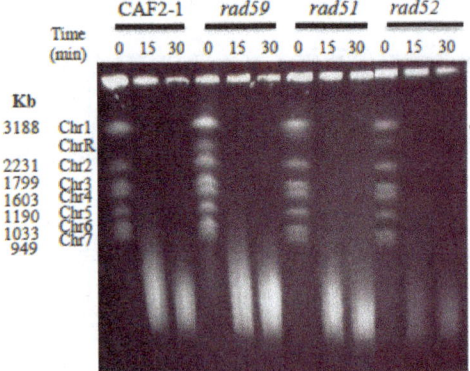

Figure 2. Determination of chromosome breaks (HDBs) following acute MMS treatments. Cells from wild type CAF2-1 and the several HR mutant derivatives were treated with 0.05% MMS for 15 and 30 min, followed by immediate DNA purification and pulse-field gel electrophoresis (PFGE). Proteinase K digestion was carried out at 55 °C to detect heat dependent breaks (HDBs). Chromosomes were visualized with ethidium bromide staining (see also Figure S2).

In *S. cerevisiae*, in vivo repair of the MMS-induced HDB prevented chromosome degradation in the subsequent 55 °C incubation with proteinase K [6]. In order to determine the time window required for repair of HDB in *C. albicans*, MMS-treated wild type CAF2-1 cells were post-incubated in either YPD or PBS lacking MMS and analyzed for chromosome restoration in time course experiments. Following a standard MMS treatment (0.05% MMS for 30 min in YPD and 55 °C incubations), no vestige of the chromosome ladder was apparent (Figure 3A, lanes 1 and 2). Traces of chromosomal bands were first seen after 2 h of recovery in YPD (Figure 3A, lane 4) and full restoration was accomplished by 4 h (Figure 3A, lane 5), with no significant changes being detected at later times (8–24 h) (Figure 3A, lanes 6 and 7). A similar conclusion was reached by quantification of Chr2 and Chr5 repair kinetics (Figure 3B). As expected from a true repair process, restoration of the chromosome ladder was paralleled by a significant reduction in the intensity of the smear (Figure 3A, compare lanes 2 to 7). It is worthy to mention that when PFGE samples were incubated at 50 °C "restitution" of the chromosomal ladder took only one hour (Figure S4A,B, lanes 1 to 3), as one could expect from the lower amount of SSBs generated at that temperature. Importantly, restitution of chromosomes was accelerated by the supply of metabolic energy (YPD) since it was significantly delayed when post-MMS incubation of CAF2-1 cells was carried out in PBS instead YPD (Figure 3A, lanes 8–12 and Figure 3B,C).

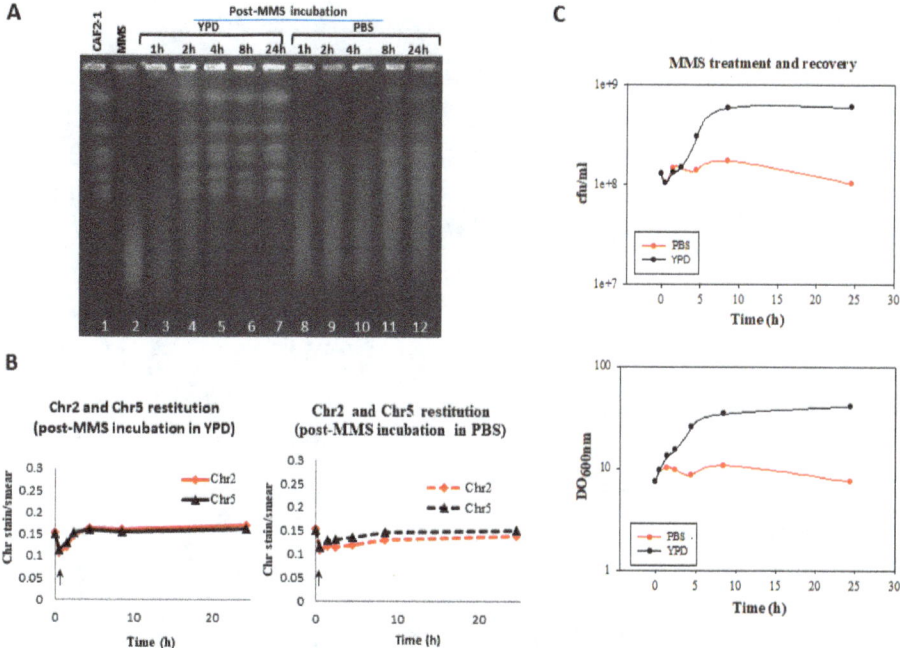

Figure 3. Induction and repair of HDBs by *C. albicans* cycling and resting wild type cells. *C. albicans* wild type (CAF2-1) was subjected to an acute MMS treatment (0.05% MMS for 30 min). Then cells were treated with 5% sodium thiosulphate, by mixing 1:1 (*v*/*v*) ratio with 10% Na$_2$S$_2$O$_3$, washed with PBS, and transferred to MMS-free YPD (cycling cells) or PBS (resting cells) at 30 °C. Samples were taken at the indicated times to determine PFGE profiles (**A**), to quantify Chr2 and Chr5 (**B**) and to calculate CFU and OD$_{600}$ (**C**). For PFGE profiles, DNA plugs were incubated with proteinase K at 55 °C. Of note, in panels **B** and **C**, initial time (0) corresponds to samples before MMS treatment. Arrows in panel B shows the MMS-treated sample.

The CFU number of strain CAF2-1, that had been slightly reduced (4%) by the MMS-treatment, increased significantly throughout the post-MMS incubation in YPD (2–8 h) (Figure 3C). Importantly, by 4 h elongated cells carried apical and lateral buds further suggesting that they had undergone mitosis; later on, elongated branched cells steadily returned to the typical yeast form which became predominant by 24 h (Figure 4). In agreement with CFUs, the OD_{600} value steadily increased throughout the post-MMS incubation, including the first hour (Figure 3C), when few 55 °C -HDB had been repaired as indicated by the absence of chromosome "restitution" (see Figure 3A and Figure S4). It is likely that the initial increase in OD_{600} was mostly due to the formation of elongated cells (a retarded effect of the acute MMS treatment (Figure 1)), which reached maximal length by 2–4 h (Figure 4) whereas the late increase was due to cell division. Importantly, in contrast to the significant increase in cell number in YPD, neither cell division (Figure 3C) nor cell elongation (Figure 4) were detected when MMS-treated populations were transferred to PBS. Considering that under these conditions repair of HDB was significantly delayed, we conclude that overgrowth of survivors contributed significantly to the fast restitution of chromosomes during the post-MMS incubation in YPD (see Section 4, Discussion).

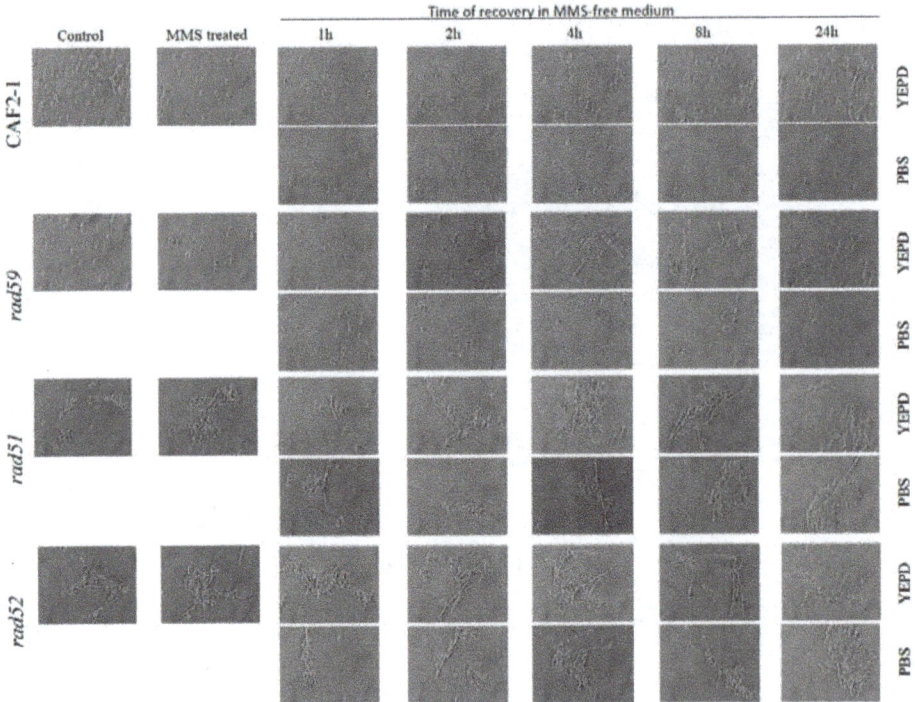

Figure 4. Cell morphology of MMS-treated exponentially growing cells from wild type and the indicated HR mutants during recovery in MMS-free YPD medium or PBS. For experimental conditions, see legend of Figure 3.

Because generation of CAI-4 from CAF2-1 resulted in increased sensitivity to low doses of MMS [35] (see above) and HR mutants (see below) were derived from CAI4, we compared chromosome recovery in MMS-treated CAF2-1 and CAI4 strains. As expected, CAI4 exhibited a small delay compared to CAF2-1 since no traces of chromosome bands could be detected in the former after 2 h of incubation in YPD. However, after 4 h both strains showed a full chromosome ladder (Figure S5).

3.3. Role of HR on the Repair of HDB

We have previously reported that *C. albicans* single mutants Ca*rad51-ΔΔ*, Ca*rad52-ΔΔ*, and Ca*rad59-ΔΔ* retained wild type PFGE karyotype profiles [21,22] (see also Figure 2 and Figure S3). When subjected to a standard acute MMS treatment and PFGE samples were incubated at 55 °C, all the mutants generated a smear similar to that shown by CAF2-1 cells (Figure 2). The average size of the pool of degraded chromosomes also decreased with the incubation time in MMS, suggesting a progressive accumulation of HDBs (Figure 2 and Figure S2). Besides, for each time-point (0–120 min in 0.05% MMS), the extent of chromosomal fragmentation was slightly higher for Ca*rad51-ΔΔ* and, to a larger extent, Ca*rad52-ΔΔ* strains compared to wild type suggesting that during the acute MMS treatment some repair in wild type is slowed down or blocked in Ca*rad51-ΔΔ* and Ca*rad52-ΔΔ* mutants (Figure S2). Finally, similarly to wild type, chromosomal fragmentation required in vitro incubations at high temperature since MMS-treated HR mutants also displayed standard chromosomal ladder when PFGE samples were incubated at 30 °C with proteinase K. Besides, prolongation of the MMS treatment up to 120 min did not result a significant increase in chromosomal degradation (Figure S3). We conclude that, as shown for wild type strain CAF2-1, treatment of HR mutants with MMS caused little or no induction of direct DSBs (HIBs).

Next, we examined the role of HR proteins in repair of HDBs during the post-MMS incubation at 30 °C in either YPD or PBS. In YPD, the Ca*rad59-ΔΔ* mutant showed a delay (4 h) in the repair of HDB/"restitution" of chromosomal bands (incubations of plugs conducted at 55 °C), compared to the wild type CAF2-1 control (2 h) (Figure S4). This delay was also detected in a parallel experiment in which samples were incubated at 50 °C but, under these conditions, repair of HDBs by Ca*rad59-ΔΔ* and wild type took only two and one hours respectively (Figure S4). However, no noticeable differences between CAI4-URA3 and Ca*rad59-ΔΔ* strains were detected regardless the post-MMS incubation took place in YPD or PBS (Figure 5A,B), suggesting that *MBP1a* homozygosis in Ca*rad59-ΔΔ* and not the absence of Rad59 itself could be responsible for the delay in chromosome restitution in YPD when compared to CAF2-1 (*MBP1a/MBP1b*). Importantly, as shown above for CAF2-1 (Figure 3), chromosome restitution in CAI4-URA3 and Ca*rad59-ΔΔ* was accelerated in YPD (Figures 5 and 6). In order to circumvent unwanted effects derived from the zygotic status of *MBP1*, CAI4-URA3 instead CAF2-1 was also used as a wild type control to investigate chromosome restitution in Ca*rad51-ΔΔ* and Ca*rad52-ΔΔ* mutants.

When post-incubated in YPD, MMS-treated Ca*rad51-ΔΔ* cultures exhibited a significant delay in chromosome restitution compared to CAI4-URA3 counterparts. As shown in Figures 5C and 6C, Ca*rad51-ΔΔ* exhibited a progressive increase in the average size of the smear followed by the appearance of some faint bands by 8 h and distinct clear chromosomal bands by 24 h (Figure 5C, lanes 3–7 and Figure 6C). A similar restitution pattern was shown by the Ca*rad52-ΔΔ* mutant (Figure 5D, lanes 3–7 and Figure 6D). However, when the post-MMS incubation took place in PBS chromosome restitution in both mutants occurred with kinetics similar to those of wild type and Ca*rad59-ΔΔ* (Figures 5 and 6). It is likely that replication of the few Ca*rad51-ΔΔ* and Ca*rad52-ΔΔ* survivors left by the MMS-treatment (12% and 7%, respectively) when transferred to YPD accounts by these differences.

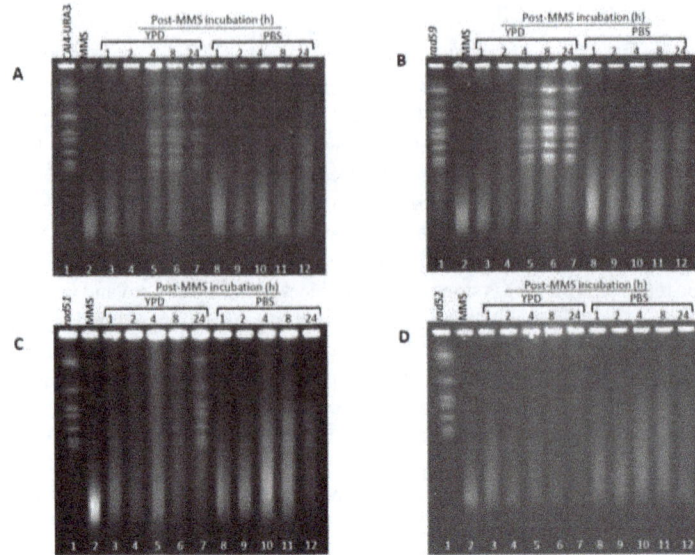

Figure 5. Induction and repair of HDBs by cycling and resting wild type and HR mutants cells of *C. albicans*. *C. albicans* wild type (CAI4-URA3) (**A**) and the indicated HR mutants (*rad59* –**B**-, *rad51* –**C**- and *rad52* –**D**-) were subjected to an acute MMS treatment (0.05% MMS for 30 min). Then cells were treated with 5% sodium thiosulphate, by mixing 1:1 (*v/v*) ratio with 10% Na$_2$S$_2$O$_3$, washed with PBS, and resuspended in MMS-free YPD (cycling cells) or PBS (resting cells) at 30 °C. Samples were taken at the indicated times to calculate OD$_{600}$, CFU (see Figure 7), and to determine PFGE profiles. For PFGE profiles, DNA plugs were incubated with proteinase K at 55 °C.

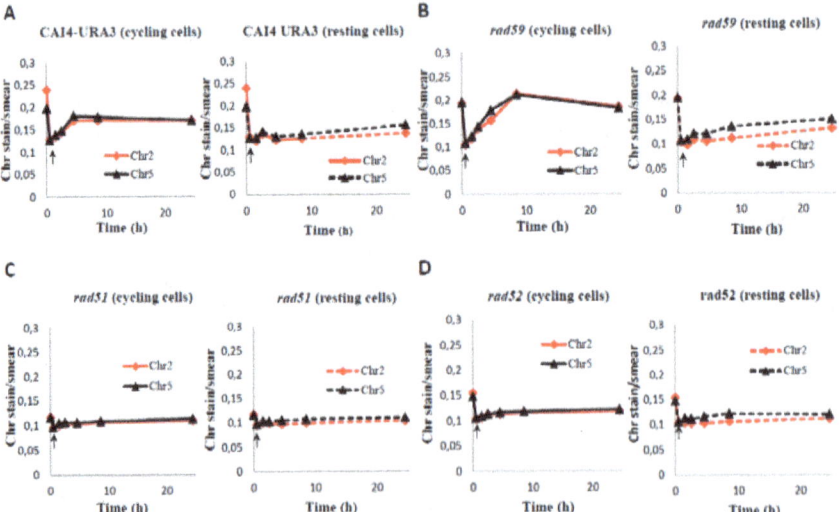

Figure 6. Quantification of Chr2 and Chr5 intensities during the MMS-treatment and the post-MMS incubation of wild type CAI4-URA3 (**A**) and HR Uri+ mutants C*arad59* (**B**), C*arad51* (**C**) and C*arad52* (**D**). Post-MMS incubation was carried out in YPD (cycling cells) or PBS (resting cells). PFGE gels shown in Figure 5 were quantified as described in Materials and Methods. Initial time (0) corresponds to samples before MMS treatment. Arrows within each panel shows the MMS-treated sample.

3.4. Effect of the MMS Treatment on Cell Number and Morphology during the Post-Incubation Recovery of Mutants Populations

Null Ca*rad59*-ΔΔ showed CFU and OD$_{600}$ values similar to those of wild type CAF2-1 throughout the MMS-treatment and post-MMS incubation (Figure 7), but, as shown for CAI4 (Figure 1), mutant cells displayed a more elongated morphology due to the *MBP1a* homozygosis (Figure 4). As expected, the MMS treatment strongly reduced Ca*rad51*-ΔΔ and Ca*rad52*-ΔΔ survival to 12% and 7%, respectively, in terms of CFUs (Figure 7A). For both mutants, CFUs and OD$_{600}$ continuously increased during the post-MMS recovery in YPD indicating replication of survivors (Figure 7B,C). It is worthy to notice the exacerbated elongation of Ca*rad51*-ΔΔ and Ca*rad52*-ΔΔ cells throughout the first 8 h of post-recovery as well as their stickiness and subsequent tendency to form aggregates (Figure 4). Importantly, as shown for wild type strains, no changes in morphology and a gentle decrease in OD$_{600}$ were observed during the post-MMS incubation of CAF2-1, Ca*rad59*-ΔΔ, Ca*rad51*-ΔΔ and Ca*rad52*-ΔΔ cells in PBS (Figures 4 and 7E). However, in contrast to wild type and Ca*rad59*-ΔΔ, Ca*rad51*-ΔΔ and Ca*rad52*-ΔΔ CFUs increased significantly throughout post-MMS incubation in PBS despite the absence of cell replication (Figure 7D). It is likely that in the absence of nutrients damaged cells do not advance in the cell cycle and may repair HDBs by BER before being plated in YPD. Therefore, an increasing fraction of Ca*rad51*-ΔΔ and Ca*rad52*-ΔΔ cells can enter the S-phase with a repaired genome and give rise to healthy colonies on solid YPD.

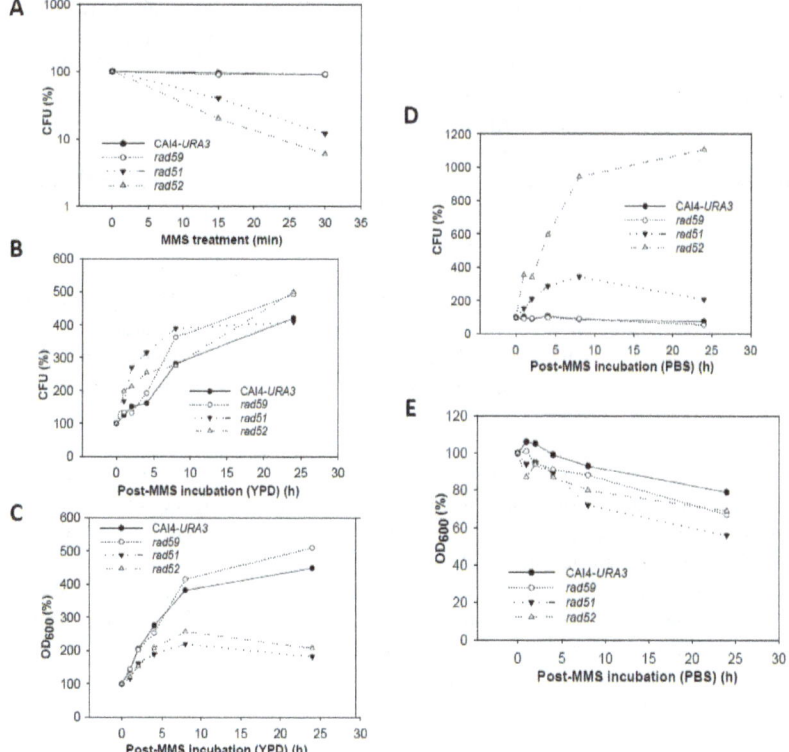

Figure 7. Changes in CFU and OD$_{600}$ from wild type and HR mutants during MMS treatment and recovery. (**A**) Cell survival after the acute MMS-treatment. Survivability in terms of CFU was 95% for wild type (CAI4-URA3) and Ca*rad59*-ΔΔ, 12% for Ca*rad51*-ΔΔ, and 5% for Ca*rad52*-ΔΔ. (**B,C**) Variation of CFUs and OD$_{600}$, respectively, during recovery of MMS-treated cells in YPD. (**D,E**) Variation in CFU and OD$_{600}$, respectively during recovery in PBS.

4. Discussion

4.1. Role of HR in Repair of HDB by Proliferating C. Albicans Cells

In this study, we have analyzed the generation of HDBs by MMS in the genome of *C. albicans* as well as the role of HR genes Ca*RAD51*, Ca*RAD52*, and Ca*RAD59* in their repair. Importantly, we have used MMS concentrations reported not to cause direct DSBs in *S. cerevisiae* [9,12,36,37]. This was further confirmed by the absence of significant chromosome fragmentation in PFGE gels of MMS-treated *C. albicans* cells when incubation of plugs was conducted at 30 °C to detect exclusively HIB (Figure S3). As previously described [6], chromosome fragmentation was due to transformation of HDBs into SSBs at high temperature (55 °C). When close enough in opposite strands, SSBs are converted into secondary DSBs, which are manifested as genome shattering in PFGE gels. Our results are consistent with the observation that depurination of methylated DNA is a function of temperature [8]. When temperature dropped from 55 °C to 50 °C less HDBs/DSB were generated in vitro; this was reflected by the lower degree of chromosome fragmentation.

In *S. cerevisiae* repair of methylated bases by BER is constant and requires the action of the Mag1 glycosylase, which removes methylated bases leaving apurinic heat-sensitive sites throughout all the stages of the cell cycle [4,12,15]. However, Rad51 and Rad52 were also required for DNA replication in the presence of MMS [11,15,16] (see below). We found that for replicating cells of *C. albicans*, survival of an acute MMS-treatment was strongly dependent on HR proteins CaRad51 and, to a larger extent, CaRad52. In *S. cerevisiae*, the requirement of ScRad51 and ScRad52 for survival and further growth of survivors could be attributed to the role of both proteins in facilitating fork bypass through methylated DNA and subsequent repair of the resulting ssDNA gaps in G2/M [17]. It is likely that both functions are conserved in *C. albicans* where they are responsible for the rescue of >90% of damaged cells. In this scenario, the severe drop in survivability observed for cycling Ca*rad51*-ΔΔ and Ca*rad52*-ΔΔ cells is likely due to defective fork restart of damaged DNA. This S-phase block can potentially be bypassed by translesion synthesis which causes error-prone repair [38] and generates highly mutagenized survivors. In contrast to the high vulnerability of cells that transit S-phase, G1- or G2-cells present in our asynchronic cultures can potentially repair HDBs faithfully using BER and NER enzymes or other pathways before entering the S phase, thus providing additional non-mutagenized survivors [14,39]. In fact, the increase in CFU shown by Ca*rad51*-ΔΔ and Ca*rad52*-ΔΔ MMS-treated populations when incubated in YPD suggests that survivors have undergone at least one replication round (Figure 7), contributing in this way to restitution of the chromosome ladder. Consistent with this possibility, the relative amount of smear detected in PFGE gels by 8 and 24 h was significantly reduced.

4.2. Role of HR in Repair of HDB by Resting C. albicans Cells

Importantly, not only chromosomes were restituted, albeit with a lower kinetic, during the post-MMS incubation in PBS, but noticeable differences in repair kinetics between wild type and HR mutants Ca*rad51*-ΔΔ, Ca*rad52*-ΔΔ and Ca*rad59*-ΔΔ were not observed. Under these conditions, no cell proliferation was detected in wild type and Ca*rad59*-ΔΔ, but CFU increased in Ca*rad51*-ΔΔ and Ca*rad52*-ΔΔ mutants, indicating that HDBs are being repaired using pathways other than HR. Consistent with the absence of cell proliferation, OD$_{600}$ did not increased (in fact some decrease was observed for all strains, likely as a consequence of cell lysis) and reduction of the smear throughout the post-MMS incubations was negligible compared to that observed in YPD. We conclude that in the absence of cell proliferation repair of HDBs caused by an acute MMS-treatment is independent of CaRad51, CaRad52, and CaRad59.

In *S. cerevisiae*, HR was shown to be crucial for the repair of alkylation damage by G2/M cells in the absence of Anp1 and Anp2 endonucleases, but not in its presence, suggesting that HR acts as a backup to repair lesions caused by unspecific Ntg1, Ntg2, and Ogg1 lyases [4]. Nothing is known on BER regulation in *C. albicans*. In this organism, only one *ANP* endonuclease (*APN1*) and one *NTG* lyase (*NTG1*), in addition to one *OGG1* lyase, have been reported. Furthermore, single (*anp1*, *ntg1*, *ogg1*) and double (*anp1 ntg1* and *ogg1 ntg1*) BER deletion mutants exhibited wild type sensitivity to MMS, while HR mutants were exquisitely sensitive, suggesting the existence of additional BER activities or that BER may be less important for repair of MMS damage in *C. albicans* compared to *S. cerevisiae* [7,18]. According to the present results, the single ANP endonuclease present in *C. albicans* seems to be able to repair methylation damage by resting cells whereas HR is mainly, if not exclusively, needed to allow replication of methylated DNA and further repair of the subsequent lesions when cells reach the G2/M phase. Work is in progress to determine if HR is needed for repair of methylation damage by G2/M-arrested *C. albicans* cells.

Supplementary Materials: The following are available online at http://www.mdpi.com/2073-4425/9/9/447/s1. Figure S1. Determination of chromosome breaks (HDBs) following a MMS treatment. Figure S2. Kinetics of chromosome breaks (HDBs) throughout a prolonged MMS treatment. Figure S3. Determination of heat independent breaks (HIBs) following a prolonged MMS treatment. Figure S4. Effect of temperature on the induction and repair of HDBs by *C. albicans* cycling cells. *C. albicans* wild type (CAF2-1) and Carad59-mutant were subjected to an acute MMS treatment (0.05% MMS for 30 min). Figure S5. Influence of the MBP1 homozygosis on the repair of HDBs by *C. albicans* cycling cells.

Author Contributions: Conceived and designed the experiments: G.L. Performed the experiments: T.C., A.B., E.A. and B.H. Edited the manuscript: G.L. and T.C.

Funding: This research received no external funding.

Acknowledgments: We thank to Andrés Aguilera (CABIMER and Universidad de Sevilla) for allowing us to use his laboratory for several assays and Wenjian Ma for critical reading of an early version of this manuscript.

Conflicts of Interest: The authors declare no conflicts of interest.

References

1. Boiteux, S.; Jinks-Robertson, S. DNA repair mechanisms and the bypass of DNA damage in *Saccharomyces cerevisiae*. *Genetics* **2013**, *193*, 1025–1064. [CrossRef] [PubMed]

2. Budzowska, M.; Kanaar, R. Mechanisms of dealing with DNA damage-induced replication problems. *Cell Biochem. Biophys.* **2009**, *53*, 17–31. [CrossRef] [PubMed]

3. Heyer, W.D.; Ehmsen, K.T.; Liu, J. Regulation of homologous recombination in eukaryotes. *Annu. Rev. Genet.* **2010**, *44*, 113–139. [CrossRef] [PubMed]

4. Ma, W.; Westmoreland, J.W.; Gordenin, D.A.; Resnick, M.A. Alkylation base damage is converted into repairable double-strand breaks and complex intermediates in G2 cells lacking AP endonuclease. *PLoS Genet.* **2011**, *7*, e1002059. [CrossRef] [PubMed]

5. Sedgwick, B.; Bates, P.A.; Paik, J.; Jacobs, S.C.; Lindahl, T. Repair of alkylated DNA: Recent advances. *DNA Repair* **2007**, *6*, 429–442. [CrossRef] [PubMed]

6. Ma, W.; Resnick, M.A.; Gordenin, D.A. Apn1 and Apn2 endonucleases prevent accumulation of repair-associated DNA breaks in budding yeast as revealed by direct chromosomal analysis. *Nucleic Acids Res.* **2008**, *36*, 1836–1846. [CrossRef] [PubMed]

7. Legrand, M.; Chan, C.L.; Jauert, P.A.; Kirkpatrick, D.T. Analysis of base excision and nucleotide excision repair in *Candida albicans*. *Microbiology* **2008**, *154*, 2446–2456. [CrossRef] [PubMed]

8. Wyatt, M.D.; Pittman, D.L. Methylating agents and DNA repair responses: Methylated bases and sources of strand breaks. *Chem. Res. Toxicol.* **2006**, *19*, 1580–1594. [CrossRef] [PubMed]

9. Valenti, A.; Napoli, A.; Ferrara, M.C.; Nadal, M.; Rossi, M.; Ciaramella, M. Selective degradation of reverse gyrase and DNA fragmentation induced by alkylating agent in the archaeon *Sulfolobus solfataricus*. *Nucleic Acids Res.* **2006**, *34*, 98–108. [CrossRef] [PubMed]

10. Lindahl, T.; Andersson, A. Rate of chain breakage at apurinic sites in double-stranded deoxyribonucleic acid. *Biochemistry* **1972**, *11*, 3618–3623. [CrossRef] [PubMed]

11. Lundin, E.; North, M.; Erixon, K.; Walters, K.; Jenssen, D.; Goldman, A.S.; Helleday, T. Methyl methane sulfonate (MMS) produces heat-labile DNA damage but no detectable in vivo DNA double-strand breaks. *Nucleic Acids Res.* **2005**, *33*, 3799–3811. [CrossRef] [PubMed]

12. Ma, W.; Panduri, V.; Sterling, J.F.; Van Houten, B.; Gordenin, D.A.; Resnick, M.A. The transition of closely opposed lesions to double-strand breaks during long-patch base excision repair is prevented by the coordinated action of DNA polymerase delta and Rad27/Fen1. *Mol. Cell. Biol.* **2009**, *29*, 1212–1221. [CrossRef] [PubMed]

13. Symington, L.S.; Rothstein, R.; Lisby, M. Mechanisms and regulation of mitotic recombination in *Saccharomyces cerevisiae*. *Genetics* **2014**, *198*, 795–835. [CrossRef] [PubMed]

14. Begley, T.J.; Rosenbach, A.S.; Ideker, T.; Samson, L.D. Damage recovery pathways in *Saccharomyces cerevisiae* revealed by genomic phenotyping and interactome mapping. *Mol. Cancer Res.* **2002**, *1*, 103–112. [PubMed]

15. Vázquez, M.V.; Rojas, V.; Tercero, J.A. sMultiple pathways cooperate to facilitate DNA replication fork progression through alkylated DNA. *DNA Repair* **2008**, *7*, 1693–1704. [CrossRef] [PubMed]

16. Tercero, J.A.; Diffley, J.F. Regulation of DNA replication fork progression through damaged DNA by the Mec1/Rad53 checkpoint. *Nature* **2001**, *412*, 553–557. [CrossRef] [PubMed]

17. González-Prieto, R.; Muñoz-Cabello, A.M.; Cabello-Lobato, M.J.; Prado, F. Rad51 replication fork recruitment is required for DNA damage tolerance. *EMBO J.* **2013**, *32*, 1307–1321. [CrossRef] [PubMed]

18. Legrand, M.; Chan, C.L.; Jauert, P.A.; Kirkpatrick, D.T. Role of DNA mismatch repair and double-strand break repair in genome stability and antifungal drug resistance in *Candida albicans*. *Eukaryot. Cell* **2007**, *6*, 2194–2205. [CrossRef] [PubMed]

19. Drablos, F.; Feyzi, E.; Aas, P.A.; Vaagbø, C.B.; Kavli, B.; Bratlie, M.S.; Peña-Diaz, J.; Otterlei, M.; Slupphaug, G.; Krokan, H.E. Alkylation damage in DNA and RNA-repair mechanisms and medical significance. *DNA Repair* **2004**, *3*, 1389–1407. [CrossRef] [PubMed]

20. Larson, K.; Sahm, J.; Shenkar, R.; Strauss, B. Methylation-induced blocks to in vitro DNA replication. *Mutat. Res.* **1985**, *150*, 77–84. [CrossRef]

21. García-Prieto, F.; Gómez-Raja, J.; Andaluz, E.; Calderone, R.; Larriba, G. Role of the homologous recombination genes *RAD51* and *RAD59* in the resistance of *Candida albicans* to UV light, radiomimetic and anti-tumor compounds and oxidizing agents. *Fungal Genet. Biol.* **2010**, *47*, 433–445. [CrossRef] [PubMed]

22. Bellido, A.; Andaluz, E.; Gómez-Raja, J.; Álvarez-Barrientos, A.; Larriba, G. Genetic interactions among homologous recombination mutants in *Candida albicans*. *Fungal Genet. Biol.* **2015**, *74*, 10–20. [CrossRef] [PubMed]

23. Andaluz, E.; Ciudad, T.; Gómez-Raja, J.; Calderone, R.; Larriba, G. Rad52 depletion in *Candida albicans* triggers both the DNA-damage checkpoint and filamentation accompanied by but independent of expression of hypha-specific genes. *Mol. Microbiol.* **2006**, *59*, 1452–1472. [CrossRef] [PubMed]

24. Bärtsch, S.; Kang, L.E.; Symington, L.S. *RAD51* is required for the repair of plasmid double-stranded gaps from either plasmid or chromosomal templates. *Mol. Cell. Biol.* **2000**, *20*, 1194–1205. [CrossRef] [PubMed]

25. Guthrie, C.; Fink, G.R. *Guide to Yeast Genetics and Molecular Biology*; Academic Press: San Diego, CA, USA, 1991.

26. Gillum, A.M.; Tsay, E.Y.; Kirsch, D.R. Isolation of the *Candida albicans* gene for orotidine-5′-phosphate decarboxylase by complementation of S. cerevisiae ura3 and E. coli pyrF mutations. *Mol. Gen. Genet.* **1984**, *198*, 179–182. [CrossRef] [PubMed]

27. Fonzi, W.A.; Irwin, M.Y. Isogenic strain construction and gene mapping in *Candida albicans*. *Genetics* **1993**, *134*, 717–728.

28. Ciudad, T.; Andaluz, E.; Steinberg-Neifach, O.; Lue, N.F.; Gow, N.A.; Calderone, R.A.; Larriba, G. Homologous recombination in *Candida albicans*: Role of CaRad52p in DNA repair, integration of linear DNA fragments and telomere length. *Mol. Microbiol.* **2004**, *53*, 1177–1194. [CrossRef] [PubMed]

29. Andaluz, E.; Gómez-Raja, J.; Hermosa, B.; Ciudad, T.; Rustchenko, E.; Calderone, R.; Larriba, G. Loss and fragmentation of chromosome 5 are major events linked to the adaptation of rad52-DeltaDelta strains of *Candida albicans* to sorbose. *Fungal Genet. Biol.* **2007**, *44*, 789–798. [CrossRef] [PubMed]

30. Andaluz, E.; Bellido, A.; Gómez-Raja, J.; Selmecki, A.; Bouchonville, K.; Calderone, R.; Berman, J.; Larriba, G. Rad52 function prevents chromosome loss and truncation in *Candida albicans*. *Mol. Microbiol.* **2011**, *79*, 1462–1482. [CrossRef] [PubMed]

31. Walther, A.; Wendland, J. An improved transformation protocol for the human fungal pathogen *Candida albicans*. *Curr. Genet.* **2003**, *42*, 339–343. [CrossRef] [PubMed]

32. Noble, S.M.; Johnson, A.D. Strains and strategies for large-scale gene deletion studies of the diploid human fungal pathogen *Candida albicans*. *Eukaryot. Cell* **2005**, *4*, 298–309. [CrossRef] [PubMed]

33. Shi, Q.M.; Wang, Y.M.; Zheng, X.D.; Lee, R.T.; Wang, Y. Critical role of DNA checkpoints in mediating genotoxic-stress-induced filamentous growth in *Candida albicans*. *Mol. Biol. Cell* **2007**, *18*, 815–826. [CrossRef] [PubMed]

34. Legrand, M.; Chan, C.L.; Jauert, P.A.; Kirkpatrick, D.T. The contribution of the S-phase checkpoint genes MEC1 and SGS1 to genome stability maintenance in *Candida albicans*. *Fungal Genet. Biol.* **2011**, *48*, 823–830. [CrossRef] [PubMed]

35. Ciudad, T.; Hickman, M.; Bellido, A.; Berman, J.; Larriba, G. Phenotypic consequences of a spontaneous loss of heterozygosity in a common laboratory strain of *Candida albicans*. *Genetics* **2016**, *203*, 1161–1176. [CrossRef] [PubMed]

36. Covo, S.; Westmoreland, J.W.; Gordenin, D.A.; Resnick, M.A. Cohesin is limiting for the suppression of DNA damage-induced recombination between homologous chromosomes. *PLoS Genet.* **2010**, *6*, e1001006. [CrossRef] [PubMed]

37. Redon, C.; Pilch, D.R.; Rogakou, E.P.; Orr, A.H.; Lowndes, N.F.; Bonner, W.M. Yeast histone 2A serine 129 is essential for the efficient repair of checkpoint-blind DNA damage. *EMBO Rep.* **2003**, *4*, 678–684. [CrossRef] [PubMed]

38. Auerbach, P.A.; Demple, B. Roles of Rev1, Pol zeta, Pol32 and Pol eta in the bypass of chromosomal abasic sites in *Saccharomyces cerevisiae*. *Mutagenesis* **2010**, *25*, 63–69. [CrossRef] [PubMed]

39. Fu, D.; Calvo, J.A.; Samson, L.D. Balancing repair and tolerance of DNA damage caused by alkylating agents. *Nat. Rev. Cancer* **2013**, *12*, 104–120. [CrossRef] [PubMed]

Review

Transcriptomic and Genomic Approaches for Unravelling *Candida albicans* Biofilm Formation and Drug Resistance—An Update

Pei Pei Chong [1,*], Voon Kin Chin [1], Won Fen Wong [2,*], Priya Madhavan [3], Voon Chen Yong [1] and Chung Yeng Looi [1]

[1] School of Biosciences, Faculty of Health and Medical Sciences, Taylor's University Malaysia, Subang Jaya, 47500 Selangor, Malaysia; VoonKin.Chin@taylors.edu.my (V.K.C.); PhelimVoonChen.Yong@taylors.edu.my (V.C.Y.); ChungYeng.Looi@taylors.edu.my (C.Y.L.)

[2] Department of Medical Microbiology, Faculty of Medicine, University Malaya, 50603 Kuala Lumpur, Malaysia

[3] School of Medicine, Faculty of Health and Medical Sciences, Taylor's University Malaysia, Subang Jaya, 47500 Selangor, Malaysia; Priya.Madhavan@taylors.edu.my

[*] Correspondence: PeiPei.Chong@taylors.edu.my (P.P.C.); wonfen@um.edu.my (W.F.W.)

Received: 21 September 2018; Accepted: 30 October 2018; Published: 7 November 2018

Abstract: *Candida albicans* is an opportunistic fungal pathogen, which causes a plethora of superficial, as well as invasive, infections in humans. The ability of this fungus in switching from commensalism to active infection is attributed to its many virulence traits. Biofilm formation is a key process, which allows the fungus to adhere to and proliferate on medically implanted devices as well as host tissue and cause serious life-threatening infections. Biofilms are complex communities of filamentous and yeast cells surrounded by an extracellular matrix that confers an enhanced degree of resistance to antifungal drugs. Moreover, the extensive plasticity of the *C. albicans* genome has given this versatile fungus the added advantage of microevolution and adaptation to thrive within the unique environmental niches within the host. To combat these challenges in dealing with *C. albicans* infections, it is imperative that we target specifically the molecular pathways involved in biofilm formation as well as drug resistance. With the advent of the -omics era and whole genome sequencing platforms, novel pathways and genes involved in the pathogenesis of the fungus have been unraveled. Researchers have used a myriad of strategies including transcriptome analysis for *C. albicans* cells grown in different environments, whole genome sequencing of different strains, functional genomics approaches to identify critical regulatory genes, as well as comparative genomics analysis between *C. albicans* and its closely related, much less virulent relative, *C. dubliniensis*, in the quest to increase our understanding of the mechanisms underlying the success of *C. albicans* as a major fungal pathogen. This review attempts to summarize the most recent advancements in the field of biofilm and antifungal resistance research and offers suggestions for future directions in therapeutics development.

Keywords: *Candida albicans*; biofilm; antifungal resistance; transcriptomics

1. Introduction

Candida albicans is the leading etiological agent for fungemia and disseminated candidiasis, which are associated with high mortality rates. According to statistics provided by the Centre for Disease Control, *C. albicans* is the third most commonly isolated microbe from bloodstream infections among hospitalized patients in the US [1]. The success of this eukaryotic microbe in causing a myriad range of human infections from superficial skin and nail infections, oral and vaginal candidiasis, to the more serious invasive candidemia and deep organ infections, is in part due to its arsenal of virulence factors and its morphology switching capability. Unlike most other fungi, *C. albicans* is able to exist in yeast, pseudohyphal as well as hyphal forms depending on the in vivo surrounding environment or in vitro culture conditions.

This versatile fungus is able to grow in biofilms on medical devices such as intravenous catheters, urinary catheters, heart pacers and other equipment that is in contact with biological fluids or organs. A huge problem encountered by clinicians treating invasive candidiasis is the enhanced antifungal drug resistance displayed by *Candida* sp. biofilms. Indeed, *C. albicans* biofilm cells have been reported in multiple studies to display up to 1000-fold greater drug resistance than planktonic, non-biofilm cells [2–4]. Globally, the impact of medical device-related candidiasis is undeniably serious considering the high mortality and morbidity rates ascribed to these infections that are often recalcitrant to routine antifungal therapies.

In this review, we summarize the switch from commensalism to colonization and active infection for *C. albicans* in host cells and discuss the various stages, biochemical processes and molecular changes that are essential for biofilm development and pathogenesis. The intricate transcription regulatory networks that play a critical part in biofilm formation are discussed. Next, drug resistance associated with biofilm growth of *C. albicans* will be dissected. A section will be dedicated to the chief genomic differences observed between *C. albicans* and its relatively less virulent close relative, *C. dubliniensis*. This is in line with our efforts to understand the inherent genetic factors that contribute to the success of *C. albicans* as a human pathogen. Recent studies, which report the transcriptomic analysis of genes and metagenomic profiling of antifungal drug resistance related to biofilms, are also highlighted. The final section of this review focuses on the strategies for future research on targeted therapeutics that could combat *C. albicans* biofilm formation.

2. Morphology Switching and Pathogenesis of Biofilm Formation

Owing to its dimorphic switching property, *C. albicans* is able to switch from a yeast to a hyphal form thereby exiting the "harmless" commensal stage to become a pathogen. In addition, the fungus possesses the trait of biofilm development; another major contributor to its pathogenesis. Normally, in healthy hosts, *C. albicans* is a commensal microbe that inhabits mucosal surfaces especially in the intestines and is almost ubiquitous in the human microbiome. Factors such as the normal microbial flora, innate immunity and also epithelial barriers prevent *C. albicans* from overgrowing or invading the deeper layers of skin or penetrating the intestinal barrier. Constant interaction between the fungus and the host immune system is believed to take place during this commensal stage [5].

During the transition from commensalism to pathogenesis, three distinct yet dynamic stages are seen, namely (i) adhesion, (ii) invasion, and (iii) damage [6]. Wächtler and colleagues were the first to show that the three stages are mediated by distinct factors. In the adhesion stage, factors that play a crucial role include the adhesins from the Als family and the cell wall components Hwp1 and Als3 [6]. The adhesion factor Eap1 was separately shown by another study to be involved in adhesion [7]. Several of these adhesins are linked to the cell membrane whereas others are linked to the cell wall via glycosylphosphatidylinositol moieties. In a study by Wächtler, a *C. albicans* microarray from Eurogentec (Belgium) was used for transcriptional profiling of genes during the adhesion, invasion and damage stages in interactions with oral epithelial cells. Many hyphal associated genes including *ALS3, HWP1, ECE1, SOD5, PHR1, PRA1,* and *RBT1* were found to be upregulated when in contact with the oral epithelial cells [6].

The Hwp1 and also the Als3 proteins are produced predominantly during hypha formation, to allow for the adhesion of the fungal cells either to the host cells or to a substrate surface [8]. In the invasion stage, a different set of genes are expressed; although the common involvement of Als3, a multifunctional adhesin and invasin protein, is also present. The proteases such as secreted aspartyl proteinases (Saps) and phospholipases have long been known to be crucial players in the hyphal invasion of host cells. On the other hand, the onset of host cell damage is a key feature of pathogenesis. Tissue damage occurs when *C. albicans* hyphae penetrate deep into or through the epithelial layer (interepithelial layer), a process facilitated by secreted hydrolases [9]. A previous study by Wächtler [6] suggested that Icl1, Sod5 and Yhb1 are involved in the active infection of epithelial cells in particular during epithelial damage. In a recent study, Allert and coworkers screened libraries of C. albicans deletion mutants in a quest to delineate genes involved in host epithelial damage and translocation through the intestine barrier. They found that candidalysin, a peptide toxin of C. albicans, is crucial for this process [10].

Biofilms are complex three-dimensional structures that are composed of a core microbial cell community (either a single species or a mixed species) attached to host tissue or abiotic surfaces surrounded by an extracellular matrix (ECM) of polysaccharides that provide a shield or scaffold for the microbes beneath [11,12]. Not surprisingly, a biofilm provides protection against antimicrobials for the microbes associated with it, and thus biofilm-associated infections are notoriously difficult to treat. Among the clinical isolates of Candida species and even across different species, there is great heterogeneity in terms of the biofilm forming capability. The ability of clinical Candida isolates to form a biofilm has been shown by many studies to correlate with a higher virulence and hence an increased mortality [13–15].

In C. albicans biofilm formation, several stages are distinguishable: (i) adhesion, (ii) initiation, (iii) maturation, (iv) dispersal; which typically progress in sequence over a period of 24–48 h [16,17]. In the adhesion step, single C. albicans yeast cells adhere to the substrate to form a basal layer of yeast cells. The cell proliferation phase ensues followed by filamentation whereby the yeast cells begin to elongate and develop into filamentous hyphae. This is the initiation step whereby the cells change their morphology and invade either the host mucosal site or plastic or other polymer surfaces in inert medical devices. An arsenal of hydrolytic enzymes such as proteinases, haemolysins, and phospholipase are secreted by C. albicans, which enables the fungus to invade host tissue or other solid substrates. The secreted aspartyl proteinases (Saps), comprising a family of ten genes (SAP1–10), are the most well studied among the many secreted enzymes [16,17].

In the maturation step, the production of hyphae is a key feature accompanied by the secretion of ECM of polysaccharides. The biofilm ECM of C. albicans is complex, where the major polysaccharides include α-mannan, β-1,6 glucan and β-1,3 glucan [18]. Among these, although β-1,3 glucan is a minor constituent, it is the chief matrix polysaccharide linked to biofilm resistance to antifungals as it could block drug diffusion [19]. The hyphal invasion into tissues is driven by physical hydrostatic forces (turgor), which drive the cytoplasmic forces. The cells could communicate with other cells through quorum sensing mechanisms and one of the most studied quorum sensing molecules that could regulate biofilm formation is farnesol [20].

Finally, in the dispersal stage, lateral yeast cells are released from the matured biofilm and are then able to disseminate to distant sites to initiate a new cycle of biofilm formation. The dispersal stage of biofilm is of immense clinical relevance, as the newly released cells from the mature biofilm located in either indwelling catheter or an infectious nidus are able to not only initiate new rounds of biofilm formation but also enter into the bloodstream to establish a distant focus of infection. This is the reason why biofilm formation is closely associated with candidemia and disseminated invasive candidiasis clinically [21]. Importantly, a previous study had shown that the dispersed cells are predominantly yeast cells, with associated enhanced adherence, filamentation capacity, biofilm formation, increased resistance to azole drugs, and are more pathogenic than their planktonic counterparts [22]. Interestingly, in the latest study by Uppuluri and colleagues, they demonstrated that a portion (~33%) of dispersed yeast cells express the hypha-specific hyphal wall protein HWP1 gene, whereas the yeast wall protein YWP1 gene was expressed in ~64% of the dispersal yeast cells [23]. It is intriguing that up to a third of the lateral yeast cells express HWP1 and although the underlying cause is still yet unknown, we postulate that this may be a clever strategy for the fungus so that these HWP1-expressing cells are "primed" and ready to colonize and invade the host cells as soon as they land on another site. The same authors also found that PES1, which is essential for yeast cell growth, had upregulated expression in dispersal yeast cells compared to biofilm, presumably as an inducer for generating more lateral yeast cells for dispersal [23].

In previous studies, differential expression analysis of the genes involved in various stages of biofilm production in *C. albicans* was mostly assessed via qualitative reverse transcription PCR (RT-PCR) or sometimes quantitative real-time RT-PCR systems. More recent transcriptomic and genomic approaches for studying regulatory networks involved in biofilm formation and the associated antifungal drug mechanisms are discussed in a later section in this review.

3. Transcription Regulatory Network of Biofilm Formation

As discussed in the above section, there are still many unanswered questions pertaining to biofilm formation by *C. albicans* in terms of the pathogenesis and the virulence mechanism. In this section, we will further dissect *C. albicans* biofilm formation in the context of the transcriptional regulatory network involved in this phenomenon.

Based on earlier literature, a master transcriptional regulatory network that consists of six major transcription regulators Efg1, Bcr1, Brg1, Ndt80, Tec1 and Rob1 are involved in controlling the normal process of biofilm formation. These regulators were discovered via screening a mutant library and in vivo studies in animal models [24,25]. Efg1 and Tec1 are involved in cell morphology regulation while Ndt80 is involved in biofilm formation. Meanwhile, Brg1 and Rob1 are present only in a species closely linked with *C. albicans* whereas, for Bcr1, its functions are yet to be fully characterized. Nobile et al deduced that transcription regulators form a complex and interconnected network with more than one thousand genes in controlling biofilm formation [25], where most of the target genes were bound by at least two or more transcriptional regulators. The same authors also surmised that the complexity in the biofilm network could be due to a number of factors including environmental influences and formation of cell memory to coordinate the cooperation between cells in order to maintain the dynamism and stability of the biofilm over generations. The authors also postulate that the complexity in the architecture of regulatory network helps to control the gene expression more precisely. On the other hand, evolutionary analysis suggests that transcriptional circuitry for the biofilm network in *C. albicans* has just evolved recently, where broad changes in the cis-regulatory sequence and regulators such as Brg1 and Rob1 are necessary for this modernized biofilm circuit [25].

A few years later, by screening the expanded library for biofilm formation at four different time points (immediately after adherence, at 8, 24 and 48 h), Fox and his coworkers [26] have identified three new transcriptional regulators involved in the biofilm circuit. These new regulators include Rfx2, Gal4 and Flo8, which are imperative and have specific roles in the development of the biofilm over time. The authors suggest that Flo8 is the most critical regulator identified in addition to the six master regulators identified earlier by Nobile [25], as double deletion mutants of Flo8 had severe disruption of biofilm formation at all time-points and the biofilm formed was similar to those formed by strains that resulted from deletions of any one of the previously discovered master regulators. Additionally, Flo8 is speculated to be a biofilm-specific regulator, as the upregulation of its expression was not affected by the form of reference cells (either yeast or filamentous form). Meanwhile, the authors reported that Rfx2 and Gal4 are involved in the biofilms formed at intermediate time points. Additionally, the authors hypothesized that Rfx2 and Gal4 are negative regulators for biofilms, where the enhancement of biofilm formation was observed in both rfx2Δ/Δ and gal4Δ/Δ mutant strains. Furthermore, through chromatin immunoprecipitation studies, the authors reported that *FLO8* is bound by Efg1, Brg1 and Ndt80 while *RFX2* and *GAL4* are bound by Ndt80. These findings further prove that Rfx2, Gal4 and Flo8 are well integrated into the existing *C. albicans* transcriptional biofilm regulatory network [26].

The pathogenesis of a biofilm often begins with adherence and colonization of *C. albicans* on the cell surface. Thus, analyzing the regulatory processes that occur during *C. albicans* adherence could provide a new paradigm to further understand the initial stage of biofilm formation. Finkel et al first reported that adherence of *C. albicans* on a silicone surface is under the control of 29 transcription factors [27]. Amongst these transcription factors, only mutants for Ace2p, Arg81p, Bcr1p, and Snf5p exhibit the anti-adherence properties in vitro. Meanwhile, the deletion of Zcf28p, Zfu2p and Crz2p is only able to disrupt biofilm formation on catheters in animal models [27]. Lee et al adopted a functional genomic analysis approach to screen for a library consisting of 1481 double barcoded doxycycline-repressible conditional gene expression strains which encompass approximately 25% of the *C. albicans* genome [28]. The authors have identified five important adherence regulators namely *ARC18*, *PMT1*, *MNN9*, *SPT7*, and *orf19.831* where transcriptional repression of these genes impaired the adherence of *C. albicans*. Of all, transcriptional repression of *ARC18* results in the strongest adherence defect and cell wall physiology changes in *C. albicans*. Arc18 is one of the putative members of the Arp2/3 complex. The authors demonstrated that perturbation of the Arp2/3 complex reduces biofilm formation, impairs the adherence process, and increases the cell surface hydrophobicity. Additionally, the disruption of the Arp2/3 complex also leads to hyperactivation of small G-protein Rho1-mediated cell wall related stress pathways, where extensive fungal cell wall remodeling is taking place. Taken together, in this study, the authors have identified a novel molecular mechanism between Arp2/3 complex and Rho1 in regulating *C. albicans* adhesion and biofilm formation [28].

Previous findings from Nobile [25], which used genome-wide approaches, have revealed that the transcription regulatory network of biofilm formation is highly integrated. Nevertheless, the functional consequences of this integration and the interactions between transcriptional regulators in this transcriptional circuitry remain obscure. Additionally, the study by Nobile and other studies on transcriptional circuitry for biofilm formation usually employed double homozygous deletion mutants of particular transcriptional regulator genes to study its effect on biofilm formation [24–26]. This genetic analysis approach will eventually limit the potential to discover novel genes that contribute towards alterations in biofilm formation, particularly for a single homozygous strain. To address this limitation, Glazier and his workers [29] have adopted a simple haploinsufficiency genetic analysis approach to analyze the transcription regulatory network of biofilm formation. The authors generated and analyzed the interaction between all possible double heterozygous mutants of the transcriptional regulators (Efg1, Bcr1, Brg1, Ndt80, Tec 1 and Rob1). From the study, the authors reported that the biofilm network is remarkably susceptible to genetic perturbation where all of the six transcriptional regulator mutants showed changes in biofilm formation. Additionally, the double heterozygous mutants showed a comparable or more severe disruption in biofilm development than the double homozygous mutants. In this study, the authors also shed light on the functions and involvement of individual transcriptional regulators. The authors revealed that *TEC1* expression is highly sensitive to small disruptions by other transcriptional regulators while *NDT80* expression is under the influence of *TEC1*. Meanwhile, *ROB1* expression was found to be dependent on the cooperative interaction among transcriptional regulators and auto-regulation mechanism [29]. Figure 1 summarizes and highlights the involvement of different transcription factors, master regulators and effectors in the complex biofilm regulatory circuitry according to the distinct stages of biofilm development in *C. albicans*.

4. Drug Resistance in *Candida albicans* Biofilm

The five major groups of antifungal agents that are used in the treatment of *C. albicans* infections are azoles, allylamines, echinocandins, polyenes and nucleoside analogues. The azoles and allylamines inhibit the ergosterol synthesis by blocking different enzymes respectively, whereas echinocandins disrupt the cell wall integrity by inhibiting the enzyme β-1,3-glucan synthase. Polyenes bind to sterols and cause intracellular leakage, whereas nucleoside analogues perturb DNA/RNA synthesis thus inhibiting cell growth [30].

Treatment of *Candida* biofilm was shown to be effective with amphotericin B and echinochandins in some previous studies [31–37]. However, antifungal drug resistance in biofilms of *C. albicans* has been reported in the past few years. Furthermore, the pyrimidine analogs, allylamines and classic formulations of polyenes are not effective against biofilms of *C. albicans* [3,38–40]. Factors contributing to antifungal resistance in *C. albicans* have been described. One of the more prominent factors is the ability of this species to effectively pump out drugs via its efflux pumps mechanism [41]. Besides the drug efflux pumps mechanism, the biofilm architecture, which is a thick layer of matrix and its constituents such as the polysaccharide β-1,3-glucan, has the ability to bind to the antifungal agents, preventing them from reaching their targets, which eventually increases the resistance of *C. albicans* towards these drugs [41,42].

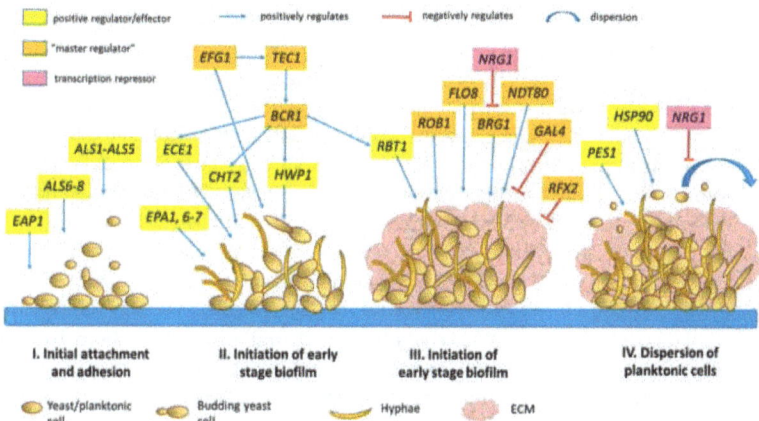

Figure 1. Schematic diagram depicting the stages of biofilm formation in *Candida albicans* and the transcription regulatory network involved in the process. The information on the "master regulators" originated from Nobile [25], Fox [26] and Glazier [29].

Persister cells that exist within the biofilms as metabolically inactive cells are also known to be resistant to many antifungal drugs such as amphotericin B [16,40]. Due to the dormancy of these persister cells, it is difficult for antifungal drugs to evoke an effect as these drugs normally target actively metabolizing cells. In this case, where persister cells exist within the biofilm matrix, much higher minimum inhibitory concentrations (MICs) of the drugs may be required to achieve the intended therapeutic goal. The acquisition of drug resistance in *C. albicans* has been studied using transcriptomes. The RNAseq analyses have been shown to be more accurate and sensitive when compared to microarray and, thus, provide a better platform to unravel the complexity of drug resistance genes in *C. albicans* [43,44]. Using RNA seq, more than 50 genes were found to be overexpressed in drug-resistant *C. albicans* [45]. The transcriptional factor encoded by the *CZF1* gene that is associated with hyphal transition and white/opaque switching was upregulated along with the *CDR1* and *CDR2* genes. This gene is also known to negatively control the expression of one of the three genes which encode β-1,3-glucan synthase, i.e., *GSL1*. Besides these genes, several other genes regulating adherence (*ALS1*), carbon metabolism (*CIT1*, *HGT10*, *GAL7*, *GUT1*, *FUM12* and *GDH3*), cell wall maintenance (*MNN4* and *CHR11*), drug transport (*YOR1*) and morphogenesis (*WHI1*, *ADAEC*, *SFL2*, *SCH9*, *CZF1*, *ECE1*, *DLD1*, *GPR1* and *SRR1*) of *C. albicans* were overexpressed. The study also revealed a few other genes that were repressed such as gene regulating adherence (*ALS2*), carbon metabolism (*HGT12*, *MAL2* and *MAE1*), copper and iron homeostasis (*HEM13*, *CRP1* and *SMF3*), drug transport (*CDR4* and *MFS*), extracellular proteins (*PLB1*, *ALS2* and *MAL2*), morphogenesis (*NAT4* and *PHHB*) and steroid binding (*EBP1* and *CBP1*). A new transcribed region was identified upstream of the *TAC1* gene, which encodes the major *CDR* transcriptional regulator, which is yet to be characterized [45].

5. Comparative Studies on *C. albicans* vs. "Avirulent" *C. dubliniensis*

C. albicans is the most commonly found opportunistic yeast that can exist as a normal microflora in healthy individuals or as an etiological agent in human candidiasis. *C. dubliniensis*, first identified in 1995, is a pathogenic species that is phylogenetically close to *C. albicans* [46,47]. Similar to *C. albicans*,

C. dubliniensis is able to form a chlamydospore and germ tube and to cause oral candidiasis, particularly in human immunodeficiency virus (HIV)-infected patients [48]. Despite the phylogenetic similarities between *C. dubliniensis* and *C. albicans*, the former exhibits a poorer virulence and lower prevalence rate due to the reduced capability to colonize the host and to form filaments [48–50]. Previous studies have focused on elucidating the phenotypical differences between these two closely related species to understand the fungal virulence mechanism. Some of the prominent findings from these studies indicate that *C. dubliniensis* exhibits impaired growth kinetics at 42 °C [51], has no β-glucosidase activity [52] and fails to produce true hyphae under *N*-acetylglucosamine stimulation or a nutrient-rich environment [46,53]. However, it shows a higher extracellular proteinase expression, adheres more strongly to buccal epithelial cells and is less susceptible to 5-flucytosine compared to *C. albicans* [54].

Using an in vivo animal infection model, Vilela demonstrated that mice intravenously infected with *C. dubliniensis* show higher survival rates than those infected with *C. albicans* and this is likely due to a more effective host inflammatory immune response to *C. dubliniensis* [48]. This is supported by in vitro culture experiment whereby *C. dubliniensis* had a lower survival rate in the presence of human polymorphonuclear leukocytes [55]. Furthermore, in an immunosuppressed mouse model, Stoke demonstrated that *C. dubliniensis* poses a weaker degree of expansion and dissemination to internal organs than *C. albicans* [50]. When co-cultured with oral reconstituted human epithelial cells, *C. dubliniensis* appears predominantly in yeast form that renders minimal effect to the cells; whereas *C. albicans* forms abundant hyphae that damage the epithelial cells in vitro [50].

Given that *C. albicans* and *C. dubliniensis* share a close phylogenetic relationship, the puzzle remains as to how these two species are varied in terms of virulence. Studies have used genomics and transcriptomics approaches to gain further insights into the different morphogenesis pathways in both species [53,55–57]. Moran has a utilized comparative genomic hybridization (CGH) method by co-hybridizing *C. albicans* microarrays with fluorescently labelled *C. albicans* and *C. dubliniensis* genomic DNA to assess the cross species genomic homology [55]. The outcome from this experiment shows a high genome similarity between the two species. Up to 95.6% of *C. dubliniensis* genomic DNA is homologous to, and hybridizes with, nucleotides from the *C. albicans* genome; while only a small proportion (4.4%, 247 genes) shows sequence divergence. The divergent genes include those encoding the hypha-specific human transglutaminase substrate *HWP1P*, which are important for hyphae formation. It is also noted that two of the *C. albicans* virulence factors implicated in invasion, the secreted aspartyl proteinase-encoding genes (*SAP5*, and either one of the *SAP4* or *SAP6*), are missing in the *C. dubliniensis* genome. Further, Jackson and coworkers sequenced the 14.6-megabase genome of *C. dubliniensis* and compared it to that of *C. albicans* using whole-genome shotgun sequencing at 11-fold average coverage [56]. Similar to results from Moran [55], that utilize the CGH method, a highly conserved sequence and synteny are shown throughout the genome of both species and only a total of 168 species-specific genes are identified. The absence of *SAP4* and *SAP5* genes in *C. dubliniensis* genome is reconfirmed in this study. Other genes reported in this study include the proposed invasin *ALS3* and a group of 115 pseudogenes that are orthologs of filamentous growth regulator (*FGR*) genes with predicted functions in fungal pathogenesis. A study by Butler compared eight *Candida* species (without *C. dubliniensis*) and shows some genes involved in mating and meiosis pathways are missing throughout evolution in certain species [58]. This suggests that distinct *Candida* species may have modified their genomes during evolutionary adaptation, which may contribute to the different virulence levels of each species. Figure 2 provides a summary of the differences between *C. albicans* and *C. dubliniensis* in different aspects, which contribute to the differences in virulence.

Compared to the genomic data, transcriptomic analysis provides more vital information of transactive genes under different stimuli or conditions. Transcriptomic analysis has been used to analyze the cross species gene expression between *Candida* versus other fungal species [57,59–61]. In 2013, Grumaz [57] used RNA-sequencing (RNA-Seq) on the Illumina (San Diego, CA, USA) next generation sequencing platform to compare the transcriptional landscapes between *C. dubliniensis* and *C. albicans* in both hyphal and yeast stages. A comparison of the differentially expressed orthologs was quantitatively determined. Recently, an RNA-Seq approach has also been used to compare the transcriptome of *C. albicans* versus *C. africana*, a biovariant with a low degree of virulence and inability to produce chlamydospores; which unveiled two novel transcriptionally active regions in both species [61]. A recent study by Caplice [53] uses a microarray meta-analysis to compare the transcriptional response of *C. dubliniensis* and *C. albicans* to different stimuli such as pH and temperature [53]. Interestingly, *C. dubliniensis* displays no or minimal expression of several Efg1-regulated, hypha-induced genes, such as the extent of cell elongation 1 (*ECE1*) and *HWP1* in response to 37 °C incubation. Other genes that are induced in *C. albicans* but not *C. dubliniensis* include those involved in the cell cycle, cytoskeleton organization, and the maintenance of hyphal growth and DNA replication. Hence, -omics approaches to compare between a less virulent and less versatile pathogen *C. dubliniensis* (or *C. africana*) to the highly virulent *C. albicans* provide important clues for the intricate gene regulatory network of the virulence process.

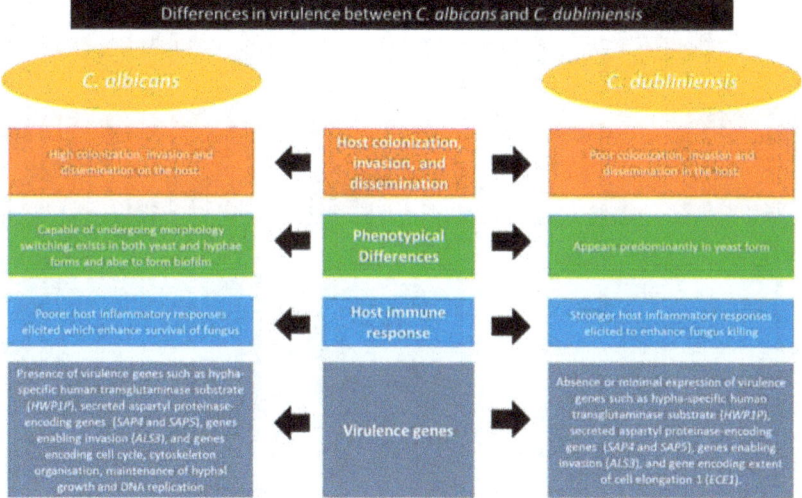

Figure 2. Comparative differences in the virulence determinants between *C. albicans* and *C. dubliniensis*. Differences in (i) host colonization, invasion, and dissemination; (ii) morphology; (iii) host immune response; and (iv) virulence genes are highlighted.

6. Exploitation of Transcriptomic and Genomic Technologies for Dissecting *C. albicans* Biofilm Formation and Drug Resistance

The *C. albicans* genome was sequenced and annotated via the efforts of the Stanford Genome Technology Centre as well as the European Galar Fungail network. The CandidaDB genome database for *C. albicans* pathogenomics was first launched in January 2002. Later, the release of CandidaDB launched in June 2004 represented an up-to-date annotation of Assembly 19 of the *C.albicans* genome sequence [62]. The genome database provided a strong impetus for and an indispensable base, which spurred numerous whole-genome related transcriptomics and proteomic studies. Table 1 below is a summary of the key publications within the last decade, which helped to advance our knowledge in *C. albicans* biofilm formation and drug resistance, particularly through those that describe the use of new -omics platforms. The list is in no way exhaustive but merely a snapshot of the studies that have contributed to this field.

Table 1. Gene expression studies on *C. albicans* biofilm and drug resistance.

Methodology/Platform	Biological Issue Studied	Major Findings	Reference
Microarrays with probes for *C. albicans* genes/Eurogentec (Seraing, Belgium)	Transcription profiles of biofilm cells vs. planktonic cells under different conditions of flow, oxygenation, and glucose concentration	Gcn4p, a regulator of amino acid metabolism, is required for biofilm growth	[63]
Microarrays/Eurogentec SA (Ivoz-Ramet, Belgium) in collaboration with European Galar Fungail Consortium (www.pasteur.fr)	Genome-wide expression profile of *C. albicans* to polyene, pyrimidine, azole, and echinocandin antifungal agents	Different expression profile signatures obtained in exposure to different classes of antifungals with various genes overexpressed	[64]
Oligonucleotide microarray (Agilent Technologies, Santa Clara, CA, USA) with *C. albicans* Assembly 21 genome (http://www.candidagenome.org/). Full Genome Chromatin Immunoprecipitation Tiling Microarray (ChIP-chip), RNA Sequencing (RNA-seq)	Comparative transcriptional analysis of *C. albicans* biofilm and planktonic cells, with *C. albicans* transcription regulator (TR) deletion mutants that are deficient in biofilm formation	Six master regulators Bcr1, Tec1, Efg1, Ndt80, Rob1, and Brg1 are essential for biofilm formation.	[25]
NanoString expression profiling and nCounter platform (NanoString Technologies, Inc., Seattle, US) with ~150 probes from cell wall-related genes, ~50 host-pathogen interaction genes, ~100 genes highly regulated during hypha development or biofilm formation, oxidative or osmotic stress	Expression profiling of genes involved in *C. albicans* adherence to substrate (silicone), an early step in medical device-related infections	Biofilm regulators Bcr1 and Ace2 have a role in adherence. A large regulatory network of 11 adherence regulators, the zinc-response regulator Zap1, and approximately 25% of the predicted cell surface protein genes known as Cell Surface Targets of Adherence Regulators (CSTARs) are involved in adherence.	[27]
RNA-sequencing (mRNA-Seq 8, Illumina) and Genome Analyzer (Illumina Inc.)	Transcriptome analysis of a *Candida* Drug Resistance (CDR) strain against its isogenic drug susceptible counterpart	Identified ~50 genes overexpressed in CDR strain: *CZF1* which is involved in transcription regulation of white/opaque switching and hypha formation is upregulated with *CDR1* and *CDR2, 5'UTR* region of *TAC1*	[45]
Gene expression microarrays (Agilent Technologies), Chromatin immunoprecipitation quantitative real-time PCR (ChIP-qPCR)	Genome-wide expression analysis of biofilm formation at different intervals, immediately after adherence, at 8, 24 and 48 h	Identified Flo8, Gal4, and Rfx2 to be involved in different time points of biofilm formation	[26]
High-throughput next-generation sequencing/Hi-Seq 2500 platform (Illumina)	Use pooled Gene Replacement and Conditional Expression (GRACE) library conditional expression strains to identify novel regulators of cell-to-surface adherence	Novel functional relationship between the Arp2/3 complex and Rho1 important for modulating actin cytoskeleton, endocytosis and cell wall remodeling.	[28]
RNA-seq with TruSeq RNA v2 kit/HiSeq2500 platform (Illumina)	Transcriptomic profiling of 124 mutant *C. albicans* strains in 10 in vitro conditions for filamentation ability	Genes encoding cell wall/membrane proteins, adhesins, alcohol dehydrogenases, and iron uptake and utilization genes were common genes upregulated across different conditions	[65]
Candida genome microarray (CapitalBio Corp., Beijing, China) with *C. albicans* genome database (http://www.candidagenome.org/)/CapitalBio BioMixer II hybridization station	Gene expression profiling of fluconazole-resistant *C. albicans* strain treated with osthole (a natural coumarin) in synergy with fluconazole	Genes in oxidation-reduction process (e.g., catalase encoded by *CAT1* and mitochondrial glycosylase encoded by *OGG1*); *CTN1* (carnitine acetyltransferase) and *ICL1* (isocitrate lyase) were upregulated	[66]
RNA-seq with BIOO Scientific NEXTflex Directional RNA-seq kit/HiSeq2000 platform (Illumina)	To decipher transcriptional gene expression patterns of dispersal cells versus core biofilm cells and planktonic cells	Transcription pattern of dispersal cells mostly similar to parent biofilm, *YWP1* expression ~2-fold higher in dispersal > biofilm > planktonic cells; ~33% of dispersal yeast cells express *HWP1*	[23]

7. Future Directions of Research on New Antifungal Drugs Targeted at *Candida* Biofilm

Undoubtedly, biofilm formation by *C. albicans* is a complicated process leading to life-threatening infections, which are difficult to eradicate. Additionally, current antifungal therapies have minimal effects on biofilm formation and there is no effective solution to solve this problem. Moreover, the development of resistance and toxicity further hindered the efficacy of antifungal drugs. Nonetheless, there are some promising strategies that could be carried out to tackle biofilm formation. These strategies include lock therapies for an infected catheter, catheter coatings, natural products and synthetic products screening, photodynamic inactivation, and targeting of the molecular pathways related to *C. albicans* biofilm formation.

The "lock therapy" is where high concentrations of antifungal drugs are directly administered into the catheter lumen for a period of time, such as several hours to days, prior to contact with patients. This approach allows the antifungal agent to eliminate biofilm formation in the catheter and also to avoid undesirable systemic toxicity build-up in patients as the high dosage of the antifungal agent only acts in the catheter [67]. This therapy has been tested on silicone catheters infected with different *C. glabrata* and *C. albicans* strains, using caspofungin, micafungin and posaconazole as antifungal agents. The outcomes were positive where all antifungal agents used in the study have successfully reduced biofilm formation, with micafungin showing the most promising result [68]. Meanwhile, another lock therapy using amphotericin B lipid formulation (L-AMB) was not effective to eradicate *Candida* biofilms [69]. There are several concerns when considering lock therapy in catheter-related candidemia including (1) the higher failure rate for biofilm infection on the outer surface or on the catheter tip; (2) the possibility of the development of antifungal drug resistance and (3) concomitant systemic antifungal treatment may need to be considered for disseminated infection. To counter these concerns, the choice of non-antifungal agents or antiseptics such as ethanol, EDTA or a high dose of minocycline are more preferable in the lock therapy against *C. albicans* and non-*albicans* biofilms [70].

Another alternative to counter *C. albicans* biofilm formation is through the modification of the catheter coating. A study has shown that the modification of the catheter coating with chlorhexidine, minocycline-rifampin or silver sulfadiazine could reduce the incidence of bloodstream infections caused by central venous catheters in the intensive care unit [71]. Meanwhile, a novel silane system coated on the implant surfaces of titanium and zirconia has been shown to inhibit *C. albicans* biofilm formation [72]. On the other hand, Karlsson fabricated and coated antifungal β-peptide-containing multilayered polymer films onto surfaces and demonstrated a profound inhibition of the antifungal-containing polymer films on the growth and proliferation of *C. albicans*. The authors suggest that this approach could be used to suppress the biofilm formation caused by *C. albicans* on film-coated surfaces, which could be further applied onto the surface of medical devices to inhibit *C. albicans* biofilm in clinical settings [73]. Hoque also demonstrated that by coating the surface of medical devices with water-insoluble and organo-soluble polymeric materials inhibition on the growth and proliferation for a number of bacteria and fungi including *C. albicans* was remarkable [74].

Exploiting the efficacy and synergistic effect of combinations of antifungal therapies with other drug classes could be another strategy for new antifungal development against *C. albicans* biofilm formation. For example, synergistic effects between cyclosporine A with fluconazole, caspofungin, voriconazole, nystatin and amphotericin B against *C. albicans* biofilm formation were observed [75]. Additionally, our previous work also demonstrated that allicin, a pure compound from garlic extract, when combined with fluconazole reduced the *C. albicans* biofilm formation and altered the expression of biofilm-related genes in vitro [76]. Besides that, the combination of Hsp90 inhibitors and the non-steroidal anti-inflammatory drugs (NSAIDs) with antifungal drugs also showed a promising synergistic activity against *C. albicans* biofilm formation [77,78]. The additive effects of different drugs such as that between fluconazole and doxycycline were also effective against *C. albicans* biofilms [79]. Nevertheless, the possibility of the development of cell toxicity has to be considered when adopting such an approach.

On the other hand, formulation of new drugs based on existing antifungal drugs could be another approach against *C. albicans* biofilm formation. Hitherto, several new formulated drugs, including an amphotericin B lipid complex and liposomal amphotericin B, showed efficacy against *C. albicans* biofilm in a bioprosthetic model [32]. Caspofungin, echinocandin, and micafungin also showed similar effects against *C. parapsilosis* and *C. albicans* biofilms in the same model [32]. Recent studies also documented the efficacy of amphotericin B lipid and echinocandin against *Candida* biofilm formation both in vitro (34) and in vivo (35,37). Meanwhile, aspirin was shown to inhibit *C. albicans* filamentation and biofilm formed by *C. albicans*, *C. parapsilosis* and *C. glabrata* [80]. Taken together, future work on the discovery of new antifungal drugs through in silico modeling or structural modification of existing antifungal drugs can help us in reducing the emergence of antifungal drug resistance.

Targeting the biofilm-related pathways in *C. albicans* could serve as a promising strategy in combating *C. albicans* biofilm formation. One instance is targeting the quorum-sensing pathway. Farnesol is a quorum-sensing molecule involved in facilitating the communication between *Candida* cells during cell proliferation. Previous studies have shown that farnesol is able to inhibit *C. albicans* biofilm formation and augment the efficacy of azoles [22,81–83]. Further evaluation of the efficacy of farnesol in vivo and the discovery of more quorum sensing molecules could shed light on the possibility of targeting the quorum-sensing pathway for *C. albicans* biofilm treatment. Meanwhile, targeting the biofilm ECM of *C. albicans*, such as β-1,3 glucan and extracellular DNA (eDNA) could be another potential approach for anti-biofilm therapy. As such, studies have shown fluconazole activity is enhanced upon digestion of β-1,3 glucan while amphotericin B activity is enhanced by the degradation of eDNA [11,42,84]. More studies identifying potential inhibitors for targeting the biofilm matrix and pathways could help us in designing better antifungal drugs for anti-biofilm therapy.

On-going and intense research has explored natural products or synthetic peptides against *C. albicans* biofilm formation. Compounds under investigation include terpenoids, polyphenols and phenylpropanoids from plant and tea extracts [85,86] and also phenazines produced by *P. aeruginosa* [87]. These compounds can suppress *C. albicans* biofilm formation and inhibit yeast-to-hyphal transition. Additionally, some synthetic peptides, such as KSL-W have significant effects on biofilm formation, growth and yeast-to-hypha transition of *C. albicans* [88]. On the other hand, several high-throughput screenings aimed at identifying small-molecule inhibitors against *C. albicans* filamentation and biofilm formation have been conducted. Siles screened for 1200 off-patent drugs approved by the Food and Drug Administration within the Prestwick Chemical Library and identified 38 bioactive compounds with ability in suppressing *C. albicans* biofilm [89]. Subsequently, Wong screened for 50,240 small molecules from a library and identified SM21 as a potent inhibitor for *C. albicans* yeast-to-hypha transition [90]. Pierce identified a series of diazaspiro-decane structural analogs which inhibit the filamentation and biofilm formation of *C. albicans* from a chemical library (NOVACore™) of 20,000 small molecules [91]. Meanwhile, Romo screened for 30,000 small-molecules within ChemBridge's DIVERSet chemical library and identified N-[3-(allyloxy)-phenyl]-4-methoxybenzamide as the leading compound for preventing *C. albicans* filamentation and biofilm formation [92].

Another potential approach to eliminate *C. albicans* biofilm is through photodynamic inactivation. This technique adopts the use of a nontoxic dye (photosensitizer) and visible light to produce reactive oxygen species, which are able to destroy the DNA, cell membrane or proteins of microbial cells and subsequently kill them. Several photosensitizers have been tested on *Candida* biofilm formation including methylene blue and toluidine blue [93,94]. A study has shown that toluidine blue (0.1 mg/mL) is able to reduce *C. albicans* biofilm at up to 60% [93]. In addition to that, when combined with chitosan, a greater reducing effect on *C albicans* biofilm formation was observed [95]. This non-toxic and yet cost-effective technique will definitely serve as a future expansion field in reducing biofilm formation of *Candida* spp.

In recent years, nanoparticles with antifungal properties have been described. There is rife interest on the anti-biofilm properties of silver (Ag) nanoparticles, which have been shown to damage the cell wall and membrane of *C. albicans* biofilm cells and inhibit filamentation [96]. A recent study also showed that Ag nanoparticles have altered multiple cellular targets including ergosterol content, fatty acid composition, cell membrane integrity and ultrastructure [97]. Another biopolymer with promising anti-biofilm property is chitosan nanoparticles, which were demonstrated by a recent study to be effective in *C. albicans* biofilm inhibition on a resin denture surface [98]. Moreover, the efficacy of other nanoparticles namely gold nanoparticles [99], silica nanoparticles [100] and selenium nanoparticles either alone or as a carrier for antifungal drugs has also been explored [101].

Vaccination is another potential strategy in preventing invasive fungal infections, particularly on high-risk groups with identifiable risk factors [102,103]. The key protection lies in the ability of vaccines to boost host immunity, including pro-inflammatory, cell-mediated, Th1 or Th17 responses to enhance phagocytic killing of the fungus [103]. Dedicated studies have been conducted by researchers in recent years to develop safe and effective fungal vaccines [104,105]. Though no specific vaccine has been developed to prevent *Candida* biofilms, however, two promising fungal vaccines against invasive candidiasis have been developed and they are currently under clinical trials. The first *Candida* vaccine containing the rAls3p-N antigen is presently under a phase IIa clinical trial whereby the results indicated that this vaccine hinders fungal adhesion and invasion in immunized subjects [105,106]. The second *Candida* vaccine is a virosome-based vaccine comprising of Sap2 antigen/truncated recombinant Sap2 antigen, which confers protection for both systemic and mucosal candidiasis [107]. Concerted efforts should be given to unravel new compounds and molecules that can be applied in the prophylaxis and treatment of *Candida* infections.

In conclusion, the majority of the data generated on *C. albicans* biofilms has mainly relied on in vitro models, which pose limitations in translating the findings from bench to bedside. Each in vitro biofilm model could be limited by the species/strains used, the specific environmental conditions and the choices of biotic interphase. It is important to take into consideration the clinical relevance of the adopted model as well. The lack of in vivo studies also warrants the development of reliable and novel in vitro biofilm models that resemble the conditions in vivo, which can be utilized for long-term anti-biofilm and antimicrobial activity prediction. On the other hand, from the accumulated data derived from transcriptome expression profiles of *C. albicans* biofilm, various key regulators of biofilm development have been identified. Dispersal cells, in particular, should be targeted as they are programmed to survive in nutrient-starved niches and to infect new sites in the host. Using the latest molecular docking and in silico modeling software to screen libraries of small molecules and peptides for candidates that could bind to and inactivate selected CSTARs targets and key regulators such as *PES1*, *YWP1*, *HWP1* and the Arp2/3 complex, we might be able to identify potential anti-*C. albicans* molecules that could become the next marketed antifungal drugs.

Funding: PPC's research was funded by the University Putra Malaysia Research University Grants Scheme (RUGS). WFW was funded by University of Malaya internal research grant.

Acknowledgments: PPC's research was funded by the University Putra Malaysia Research University Grants Scheme (RUGS). WFW was funded by Institut Mérieux Young Investigator Award Grant to University of Malaya (UM.0000107/HIF.IF; IF039-2017).

Conflicts of Interest: The authors declare no conflict of interest.

References

1. Wisplinghoff, H.; Bischoff, T.; Tallent, S.M.; Seifert, H.; Wenzel, R.P.; Edmond, M.B. Nosocomial bloodstream infections in U.S. hospitals: Analysis of 24,179 cases from a prospective nationwide surveillance study. *Clin. Infect. Dis.* **2004**, *39*, 309–317. [CrossRef] [PubMed]
2. Baillie, G.S.; Douglas, L.J. Role of dimorphism in the development of *Candida albicans* biofilms. *J. Med. Microbiol.* **1999**, *48*, 671–679. [CrossRef] [PubMed]
3. Mukherjee, P.K.; Chandra, J.; Kuhn, D.M.; Ghannoum, M.A. Mechanism of fluconazole resistance in *Candida albicans* biofilms: Phase-specific role of efflux pumps and membrane sterols. *Infect. Immun.* **2003**, *71*, 4333–4340. [CrossRef] [PubMed]

4. LaFleur, M.D.; Kumamoto, C.A.; Lewis, K. *Candida albicans* biofilms produce antifungal-tolerant persister cells. *Antimicrob. Agents Chemother.* **2006**, *50*, 3839–3846. [CrossRef] [PubMed]

5. Mochon, A.B.; Ye, J.; Kayala, M.A.; Wingard, J.R.; Clancy, C.J.; Nguyen, M.H.; Felgner, P.; Baldi, P.; Liu, H. Serological profiling of a *Candida albicans* protein microarray reveals permanent host-pathogen interplay and stage-specific responses during candidemia. *PLoS Pathog.* **2010**, *6*, e1000827. [CrossRef]

6. Wächtler, B.; Wilson, D.; Haedicke, K.; Dalle, F.; Hube, B. From attachment to damage: Defined genes of *Candida albicans* mediate adhesion, invasion and damage during interaction with oral epithelial cells. *PLoS ONE* **2011**, *6*, e17046. [CrossRef] [PubMed]

7. Zordan, R.; Cormack, B. Adhesins in Opportunistic Fungal Pathogens. In *Candida and Candidiasis*, 2nd ed.; Calderone, R.A., Clancy, C.J., Eds.; ASM Press: Washington, DC, USA, 2012; pp. 243–259.

8. Liu, Y.; Filler, S.G. *Candida albicans* Als3, a multifunctional adhesin and invasin. *Eukaryot. Cell* **2011**, *10*, 168–173. [CrossRef] [PubMed]

9. Dalle, F.; Wächtler, B.; L'Ollivier, C.; Holland, G.; Bannert, N.; Wilson, D.; Labruere, C.; Bonnin, A.; Hube, B. Cellular interactions of *Candida albicans* with human oral epithelial cells and enterocytes. *Cell. Microbiol.* **2010**, *12*, 248–271. [CrossRef] [PubMed]

10. Allert, S.; Förster, T.M.; Svensson, C.M.; Richardson, J.P.; Pawlik, T.; Hebecker, B.; Rudolphi, S.; Juraschitz, M.; Schaller, M.; Blagojevic, M.; et al. *Candida albicans*-induced epithelial damage mediates translocation through intestinal barriers. *MBio* **2018**, *9*, e00915-18. [CrossRef] [PubMed]

11. Al-Fattani, M.A.; Douglas, L.J. Biofilm matrix of *Candida albicans* and *Candida tropicalis*: Chemical composition and role in drug resistance. *J. Med. Microbiol.* **2006**, *55*, 999–1008. [CrossRef] [PubMed]

12. Ghannoum, M.A.; Roilides, E.; Katragkou, A.; Petraitis, V.; Walsh, T.J. The role of echinocandins in *Candida* biofilm-related vascular catheter infections: In vitro and in vivo model systems. *Clin. Infect. Dis.* **2015**, *61*, S618–S621. [CrossRef] [PubMed]

13. Tumbarello, M.; Fiori, B.; Trecarichi, E.M.; Posteraro, P.; Losito, A.R.; De Luca, A.; Sanguinetti, M.; Fadda, G.; Cauda, R.; Posteraro, B. Risk factors and outcomes of candidemia caused by biofilm-forming isolates in a tertiary care hospital. *PLoS ONE* **2012**, *7*, e33705. [CrossRef] [PubMed]

14. Sherry, L.; Nile, C.J.; Sherriff, A.; Johnson, E.M.; Hanson, M.F.; Williams, C.; Munro, C.A.; Jones, B.J.; Ramage, G. Biofilm formation is a risk factor for mortality in patients with *Candida albicans* bloodstream infection-Scotland, 2012–2013. *Clin. Microbiol. Infect.* **2016**, *22*, 87–93.

15. Soldini, S.; Posteraro, B.; Vella, A.; De Carolis, E.; Borghi, E.; Falleni, M.; Losito, A.R.; Maiuro, G.; Trecarichi, E.M.; Sanguinetti, M.; et al. Microbiological and clinical characteristics of biofilm-forming *Candida parapsilosis* isolates associated with fungaemia and their impact on mortality. *Clin. Microbiol. Infect.* **2018**, *24*, 771–777. [CrossRef] [PubMed]

16. Mathé, L.; Van Dijck, P. Recent insights into *Candida albicans* biofilm resistance mechanisms. *Curr. Genet.* **2013**, *59*, 251–264. [CrossRef] [PubMed]

17. Tsui, C.; Kong, E.F.; Jabra-Rizk, M.A. Pathogenesis of *Candida albicans* biofilm. *Pathog. Dis.* **2016**, *74*, ftw018. [CrossRef] [PubMed]

18. Mitchell, K.F.; Zarnowski, R.; Andes, D.R. Fungal super glue: The biofilm matrix and its composition, assembly, and functions. *PLoS Pathog.* **2016**, *12*, e1005828. [CrossRef] [PubMed]

19. Taff, H.T.; Mitchell, K.F.; Edward, J.A.; Andes, D.R. Mechanisms of *Candida* biofilm drug resistance. *Future Microbiol.* **2013**, *8*, 1325–1337. [CrossRef] [PubMed]

20. Polke, M.; Leonhardt, I.; Kurzai, O.; Jacobsen, I.D. Farnesol signalling in *Candida albicans*—More than just communication. *Crit. Rev. Microbiol.* **2018**, *44*, 230–243. [CrossRef] [PubMed]

21. Tournu, H.; Van Dijck, P. *Candida* biofilms and the host: Models and new concepts for eradication. *Int. J. Microbiol.* **2012**, *2012*, 845352. [CrossRef] [PubMed]

22. Uppuluri, P.; Chaturvedi, A.K.; Srinivasan, A.; Banerjee, M.; Ramasubramaniam, A.K.; Köhler, J.R.; Kadosh, D.; Lopez-Ribot, J.L. Dispersion as an important step in the *Candida albicans* biofilm developmental cycle. *PLoS Pathog.* **2010**, *6*, e1000828. [CrossRef] [PubMed]

23. Uppuluri, P.; Zaldívar, M.A.; Anderson, M.Z.; Dunn, M.J.; Berman, J.; Ribot, J.L.L.; Köhler, J.R. *Candida albicans* dispersed cells are developmentally distinct from biofilm and planktonic cells. *mBio* **2018**, *9*, e01338-18. [CrossRef] [PubMed]

24. Fox, E.P.; Nobile, C.J. A sticky situation: Untangling the transcriptional network controlling biofilm development in *Candida albicans*. *Transcription* **2012**, *3*, 315–322. [CrossRef] [PubMed]

25. Nobile, C.J.; Fox, E.P.; Nett, J.E.; Sorrells, T.R.; Mitrovich, Q.M.; Hernday, A.D.; Tuch, B.B.; Andes, D.R.; Johnson, A.D. A recently evolved transcriptional network controls biofilm development in *Candida albicans*. *Cell* **2012**, *148*, 126e38. [CrossRef] [PubMed]

26. Fox, E.P.; Bui, C.K.; Nett, J.E.; Hartooni, N.; Mui, M.C.; Andes, D.R.; Nobile, C.J.; Johnson, A.D. An expanded regulatory network temporally controls *Candida albicans* biofilm formation. *Mol. Microbiol.* **2015**, *96*, 1226–1239. [CrossRef] [PubMed]

27. Finkel, J.S.; Xu, W.; Huang, D.; Hill, E.M.; Desai, J.V.; Woolford, C.A.; Nett, J.E.; Taff, H.; Norice, C.T.; Andes, D.R.; et al. Portrait of *Candida albicans* adherence regulators. *PLoS Pathog.* **2012**, *8*, e1002525. [CrossRef] [PubMed]

28. Lee, J.A.; Robbins, N.; Xie, J.L.; Ketela, T.; Cowen, L.E. Functional genomic analysis of *Candida albicans* adherence reveals a key role for the arp2/3 complex in cell wall remodelling and biofilm formation. *PLoS Genet.* **2016**, *12*, e1006452. [CrossRef] [PubMed]

29. Glazier, V.E.; Murante, T.; Murante, D.; Koselny, K.; Liu, Y.; Kim, D.; Koo, H.; Krysan, D.J. Genetic analysis of the *Candida albicans* biofilm transcription factor network using simple and complex haploinsufficiency. *PLoS Genet.* **2017**, *8*, e1006948. [CrossRef] [PubMed]

30. Fox, E.P.; Singh-babak, S.D.; Hartooni, N.; Nobile, C.J. Biofilms and antifungal resistance. In *Antifungals from Genomics to Resistance and the Development of Novel Agents*; Coste, A.T., Vandeputte, P., Eds.; Caister Academic Press: Poole, UK, 2015; p. 71e90.

31. Bachmann, S.P.; VandeWalle, K.; Ramage, G.; Patterson, T.F.; Wickes, B.L.; Graybill, J.R.; López-Ribot, J.L. In vitro activity of echinocandins against *Candida albicans* biofilms. *Antimicrob. Agents Chemother.* **2002**, *46*, 3591–3596. [CrossRef] [PubMed]

32. Kuhn, D.M.; George, T.; Chandra, J.; Mukherjee, P.K.; Ghannoum, M.A. Antifungal susceptibility of *Candida* biofilms: Unique efficacy of amphotericin B lipid formulations and echinocandins. *Antimicrob. Agents Chemother.* **2002**, *46*, 1773–1780. [CrossRef] [PubMed]

33. Ramage, G.; VandeWalle, K.; Bachmann, S.P.; Wickes, B.L.; Lopez-Ribot, J.L. In vitro pharmacodynamic properties of three antifungal agents against preformed *Candida albicans* biofilms determined by time-kill studies. *Antimicrob. Agents Chemother.* **2002**, *46*, 3634–3636. [CrossRef] [PubMed]

34. Ramage, G.; Jose, A.; Sherry, L.; Lappin, D.F.; Jones, B.; Williams, C. Liposomal amphotericin B displays rapid dose-dependent activity against *Candida albicans* biofilms. *Antimicrob. Agents Chemother.* **2013**, *57*, 2369–2371. [CrossRef] [PubMed]

35. Mukherjee, P.K.; Long, L.A.; Kim, H.G.; Ghannoum, M.A. Amphotericin B lipid complex is efficacious in the treatment of *Candida albicans* biofilms using a model of catheter-associated *Candida* biofilms. *Int. J. Antimicrob. Agents* **2009**, *33*, 149–153. [CrossRef] [PubMed]

36. Kucharíková, S.; Tournu, H.; Holtappels, M.; Van Dijck, P.; Lagrou, K. In vivo efficacy of anidulafungin against *Candida albicans* mature biofilms in a novel rat model of catheter-associated candidiasis. *Antimicrob. Agents Chemother.* **2010**, *54*, 4474–4478. [CrossRef] [PubMed]

37. Kucharicová, S.; Sharma, N.; Spriet, I.; Maertens, J.; Van Dijck, P.; Lagrou, K. Activities of systematically administered echinocandins against in vivo mature *Candida albicans* biofilms developed in a rat subcutaneous model. *Antimicrob. Agents Chemother.* **2013**, *57*, 2365–2368. [CrossRef] [PubMed]

38. Ramage, G.; Bachmann, S.; Patterson, T.F.; Wickes, B.L.; Lopez-Ribot, J.L. Investigation of multidrug efflux pumps in relation to fluconazole resistance in *Candida albicans* biofilms. *J. Antimicrob. Chemother.* **2002**, *49*, 973e80. [CrossRef]

39. Anderson, J.B. Evolution of antifungal-drug resistance: Mechanisms and pathogen fitness. *Nat. Rev. Microbiol.* **2005**, *3*, 547e56. [CrossRef] [PubMed]

40. Nobile, C.J.; Johnson, A.D. *Candida albicans* biofilms and human disease. *Annu. Rev. Microbiol.* **2015**, *69*, 71e92. [CrossRef] [PubMed]

41. Nett, J.; Lincoln, L.; Marchillo, K.; Massey, R.; Holoyda, K.; Hoff, B.; VanHandel, M.; Andes, D. Putative role of β-1,3 glucans in *Candida albicans* biofilm resistance. *Antimicrob. Agents Chemother.* **2007**, *51*, 510e20. [CrossRef] [PubMed]

42. Vediyappan, G.; Rossignol, T.; d' Enfert, C. Interaction of *Candida albicans* biofilms with antifungals: Transcriptional response and binding of antifungals to β-glucans. *Antimicrob. Agents Chemother.* **2010**, *54*, 2096e111. [CrossRef] [PubMed]

43. Nagalakshmi, U.; Wang, Z.; Waern, K.; Shou, C.; Raha, D.; Gerstein, M.; Snyder, M. The transcriptional landscape of the yeast genome defined by RNA sequencing. *Science* **2008**, *320*, 1344–1349. [CrossRef] [PubMed]

44. Xiong, Y.; Chen, X.; Chen, Z.; Wang, X.; Shi, S.; Wang, X.; Zhang, J.; He, X. RNA sequencing shows no dosage compensation of the active X-chromosome. *Nat. Genet.* **2010**, *42*, 1043–1047. [CrossRef] [PubMed]

45. Dhamgaye, S.; Bernard, M.; Lelandais, G.; Sismeiro, O.; Lemoine, S.; Coppée, J.Y.; Le Crom, S.; Prasad, R.; Devaux, F. RNA sequencing revealed novel actors of the acquisition of drug resistance in *Candida albicans*. *BMC Genom.* **2012**, *13*, 396. [CrossRef] [PubMed]

46. Gilfillan, G.D.; Sullivan, D.J.; Haynes, K.; Parkinson, T.; Coleman, D.C.; Gow, N.A. *Candida dubliniensis*: Phylogeny and putative virulence factors. *Microbiology* **1998**, *144*, 829–838. [CrossRef] [PubMed]

47. Schorling, S.R.; Kortinga, H.C.; Froschb, M.; Muhlschlegel, F.A. The role of *Candida dubliniensis* in oral candidiasis in human immunodeficiency virus-infected individuals. *Crit. Rev. Microbiol.* **2000**, *26*, 59–68. [CrossRef] [PubMed]

48. Vilela, M.M.; Kamei, K.; Sano, A.; Tanaka, R.; Uno, J.; Takahashi, I.; Ito, J.; Yarita, K.; Miyaji, M. Pathogenicity and virulence of *Candida dubliniensis*: Comparison with *C. albicans*. *Med. Mycol.* **2002**, *40*, 249–257. [CrossRef] [PubMed]

49. Sullivan, D.J.; Moran, G.P.; Pinjon, E.; Al-Mosaid, A.; Stokes, C.; Vaughan, C.; Coleman, D.C. Comparison of the epidemiology, drug resistance mechanisms, and virulence of *Candida dubliniensis* and *Candida albicans*. *FEMS Yeast Res.* **2004**, *4*, 369–376. [CrossRef]

50. Stokes, C.; Moran, G.P.; Spiering, M.J.; Cole, G.T.; Coleman, D.C.; Sullivan, D.J. Lower filamentation rates of *Candida dubliniensis* contribute to its lower virulence in comparison with *Candida albicans*. *Fungal. Genet. Biol.* **2007**, *44*, 920–931. [CrossRef] [PubMed]

51. Sullivan, D.; Coleman, D. *Candida dubliniensis*: Characteristics and identification. *J. Clin. Microbiol.* **1998**, *36*, 329–334. [PubMed]

52. Schoofs, A.; Odds, F.C.; Colebunders, R.; Ieven, M.; Goossens, H. Use of specialised isolation media for recognition and identification of *Candida dubliniensis* isolates from HIV-infected patients. *Eur. J. Clin. Microbiol. Infect. Dis.* **1997**, *16*, 296–300. [CrossRef] [PubMed]

53. Caplice, N.; Moran, G.P. *Candida albicans* exhibits enhanced alkaline and temperature induction of Efg1-regulated transcripts relative to *Candida dubliniensis*. *Genom. Data* **2015**, *6*, 130–135. [CrossRef] [PubMed]

54. McCullough, M.; Ross, B.; Reade, P. Characterization of genetically distinct subgroup of *Candida albicans* strains isolated from oral cavities of patients infected with human immunodeficiency virus. *J. Clin. Microbiol.* **1995**, *33*, 696–700. [PubMed]

55. Moran, G.; Stokes, C.; Thewes, S.; Hube, B.; Coleman, D.C.; Sullivan, D. Comparative genomics using *Candida albicans* DNA microarrays reveals absence and divergence of virulence-associated genes in *Candida dubliniensis*. *Microbiology* **2004**, *150*, 3363–3382. [CrossRef] [PubMed]

56. Jackson, A.P.; Gamble, J.A.; Yeomans, T.; Moran, G.P.; Saunders, D.; Harris, D.; Aslett, M.; Barrell, J.F.; Butler, G.; Citiulo, F.; et al. Comparative genomics of the fungal pathogens *Candida dubliniensis* and *Candida albicans*. *Genom. Res.* **2009**, *19*, 2231–2244. [CrossRef] [PubMed]

57. Grumaz, C.; Lorenz, S.; Stevens, P.; Lindemann, E.; Schock, U.; Retey, J.; Rupp, S.; Sohn, K. Species and condition specific adaptation of the transcriptional landscapes in *Candida albicans* and *Candida dubliniensis*. *BMC Genom.* **2013**, *14*, 212. [CrossRef] [PubMed]

58. Butler, G.; Rasmussen, M.D.; Lin, M.F.; Santos, M.A.; Sakthikumar, S.; Munro, C.A.; Rheinbay, E.; Grabherr, M.; Forche, A.; Reedy, J.L.; et al. Evolution of pathogenicity and sexual reproduction in eight *Candida* genomes. *Nature* **2009**, *459*, 657–662. [CrossRef] [PubMed]

59. Bruno, V.M.; Wang, Z.; Marjani, S.L.; Euskirchen, G.M.; Martin, J.; Sherlock, G.; Snyder, M. Comprehensive annotation of the transcriptome of the human fungal pathogen *Candida albicans* using RNA-seq. *Genom. Res.* **2010**, *20*, 1451–1458. [CrossRef] [PubMed]

60. Movahed, E.; Munusamy, K.; Tan, G.M.; Looi, C.Y.; Tay, S.T.; Wong, W.F. Genome-wide transcription study of *Cryptococcus neoformans* H99 clinical strain versus environmental strains. *PLoS ONE* **2015**, *10*, e0137457. [CrossRef] [PubMed]

61. Giosa, D.; Felice, M.R.; Lawrence, T.J.; Gulati, M.; Scordino, F.; Giuffre, L.; Lo Passo, C.; D'Alessandro, E.; Criseo, G.; Ardell, D.H.; et al. Whole RNA-sequencing and transcriptome assembly of *Candida albicans* and *Candida africana* under chlamydospore-inducing conditions. *Genom. Biol. Evol.* **2017**, *9*, 1971–1977. [CrossRef] [PubMed]

62. d' Enfert, C.; Goyard, S.; Rodriguez-Arnaveilhe, S.; Frangeul, L.; Jones, L.; Tekaia, F.; Bader, O.; Albrecht, A.; Castillo, L.; Dominguez, A.; et al. CandidaDB: A genome database for *Candida albicans* pathogenomics. *Nucleic Acids Res.* **2005**, *33*, D353–D357. [CrossRef] [PubMed]

63. García-Sánchez, S.; Aubert, S.; Iraqui, I.; Janbon, G.; Ghigo, J.M.; d' Enfert, C. *Candida albicans* biofilms: A developmental state associated with specific and stable gene expression patterns. *Eukaryot. Cell* **2004**, *3*, 536–545. [CrossRef] [PubMed]

64. Liu, T.T.; Znaidi, S.; Barker, K.S.; Xu, L.; Homayouni, R.; Saidane, S.; Morschhäuser, J.; Nantel, A.; Raymond, M.; Rogers, P.D. Genome-wide expression and location analyses of the *Candida albicans* Tac1p regulon. *Eukaryot. Cell* **2007**, *6*, 2122–2238. [CrossRef] [PubMed]

65. Azadmanesh, J.; Gowen, A.M.; Creger, P.E.; Schafer, N.D.; Blankenship, J.R. Filamentation involves two overlapping, but distinct, programs of filamentation in the pathogenic fungus *Candida albicans*. *Genes Genom. Genet.* **2017**, *7*, 3797–3808. [CrossRef] [PubMed]

66. Li, D.D.; Chai, D.; Huang, X.W.; Guan, S.X.; Du, J.; Zhang, H.Y.; Sun, Y.; Jiang, Y.Y. Potent in vitro synergism of fluconazole and osthole against fluconazole-resistant *Candida albicans*. *Antimicrob. Agents Chemother.* **2017**, *61*, e00436-17. [CrossRef] [PubMed]

67. Carratalà, J. The antibiotic-lock technique for therapy of 'highly needed' infected catheters. *Clin. Microbiol. Infect.* **2002**, *8*, 282–289. [CrossRef] [PubMed]

68. Cateau, E.; Berjeaud, J.M.; Imbert, C. Possible role of azole and echinocandin lock solutions in the control of *Candida* biofilms associated with silicone. *Int. J. Antimicrob. Agents* **2011**, *37*, 380–384. [CrossRef] [PubMed]

69. Toulet, D.; Debarre, C.; Imbert, C. Could liposomal amphotericin B (L-AMB) lock solutions be useful to inhibit *Candida* spp. biofilms on silicone biomaterials? *J. Antimicrob. Chemother.* **2012**, *67*, 430–432. [CrossRef] [PubMed]

70. Raad, I.; Hanna, H.; Dvorak, T.; Chaiban, G.; Hachem, R. Optimal antimicrobial catheter lock solution, using different combinations of minocycline, EDTA, and 25-percent ethanol, rapidly eradicates organisms embedded in biofilm. *Antimicrob. Agents. Chemother.* **2007**, *51*, 78–83. [CrossRef] [PubMed]

71. Bonne, S.; Mazuski, J.E.; Sona, C.; Schallom, M.; Boyle, W.; Buchman, T.G.; Bochicchio, G.V.; Coopersmith, C.M.; Schuerer, D.J. Effectiveness of minocycline and rifampin vs chlorhexidine and silver sulfadiazine-impregnated central venous catheters in preventing central line-associated bloodstream infection in a high-volume academic intensive care unit: A before and after trial. *J. Am. Coll. Surg.* **2015**, *221*, 739–747. [CrossRef] [PubMed]

72. Villard, N.; Seneviratne, C.; Tsoi, J.K.H.; Heinonen, M.; Matinlinna, J. *Candida albicans* aspects of novel silane system-coated titanium and zirconia implant surfaces. *Clin. Oral Implants Res.* **2015**, *26*, 332–341. [CrossRef] [PubMed]

73. Karlsson, A.J.; Flessner, R.M.; Gellman, S.H.; Lynn, D.M.; Palecek, S.P. Polyelectrolyte multilayers fabricated from antifungal β-peptides: Design of surfaces that exhibit antifungal activity against *Candida albicans*. *Biomacromolecules* **2010**, *11*, 2321–2328. [CrossRef] [PubMed]

74. Hoque, J.; Akkapeddi, P.; Yadav, V.; Manjunath, G.B.; Uppu, D.S.; Konai, M.M.; Yarlagadda, V.; Sanyal, K.; Haldar, J. Broad spectrum antibacterial and antifungal polymeric paint materials: Synthesis, structure-activity relationship, and membrane-active mode of action. *ACS. Appl. Mater. Interfaces* **2015**, *7*, 1804–1815. [CrossRef] [PubMed]

75. Shinde, R.B.; Chauhan, N.M.; Raut, J.S.; Karuppayil, S.M. Sensitization of *Candida albicans* biofilms to various antifungal drugs by cyclosporine A. *Ann. Clin. Microbiol. Antimicrob.* **2012**, *11*, 27. [CrossRef] [PubMed]

76. Khodavandi, A.; Harmal, N.S.; Alizadeh, F.; Scully, O.J.; Sidik, S.M.; Othman, F.; Sekawi, Z.; Ng, K.P.; Chong, P.P. Comparison between allicin and fluconazole in *Candida albicans* biofilm inhibition and in suppression of *HWP1* gene expression. *Phytomedicine* **2011**, *19*, 56–63. [CrossRef] [PubMed]

77. Robbins, N.; Uppuluri, P.; Nett, J.; Rajendran, R.; Ramage, G.; Lopez-Ribot, J.L.; Andes, D.; Cowen, L.E. Hsp90 governs dispersion and drug resistance of fungal biofilms. *PLoS Pathog.* **2011**, *7*, e1002257. [CrossRef] [PubMed]

78. Bink, A.; Kucharíková, S.; Neirinck, B.; Vleugels, J.; Van Dijck, P.; Cammue, B.P.; Thevissen, K. The nonsteroidal antiinflammatory drug diclofenac potentiates the in vivo activity of caspofungin against *Candida albicans* biofilms. *J. Infect. Dis.* **2012**, *206*, 1790–1797. [CrossRef] [PubMed]

79. Gao, Y.; Zhang, C.; Lu, C.; Liu, P.; Li, Y.; Li, H.; Sun, S. Synergistic effect of doxycycline and fluconazole against *Candida albicans* biofilms and the impact of calcium channel blockers. *FEMS Yeast Res.* **2013**, *13*, 453–462. [CrossRef] [PubMed]

80. Stepanovic, S.; Vukovic, D.; Jesic, M.; Ranin, L. Influence of acetylsalicylic acid (aspirin) on biofilm production by *Candida* species. *J. Chemother.* **2004**, *16*, 134–138. [CrossRef] [PubMed]

81. Ramage, G.; Saville, S.P.; Wickes, B.L.; Lopez-Ribot, J.L. Inhibition of *Candida albicans* biofilm formation by farnesol, a quorum-sensing molecule. *Appl. Environ. Microbiol.* **2002**, *68*, 5459–5463. [CrossRef] [PubMed]

82. Sharma, M.; Prasad, R. The quorum-sensing molecule farnesol is a modulator of drug efflux mediated by ABC multidrug transporters and synergizes with drugs in *Candida albicans*. *Antimicrob. Agents Chemother.* **2011**, *55*, 4834–4843. [CrossRef] [PubMed]

83. Yu, L.H.; Wei, X.; Ma, M.; Chen, X.J.; Xu, S.B. Possible inhibitory molecular mechanism of farnesol on the development of fluconazole resistance in *Candida albicans* biofilm. *Antimicrob. Agents Chemother.* **2012**, *56*, 770–775. [CrossRef] [PubMed]

84. Martin, M.; Henriques, M.; Lopez-Ribot, J.L.; Oliveira, R. Addition of DNase improves the in vitro activity of antifungal drugs against *Candida albicans* biofilms. *Mycoses* **2012**, *55*, 80–85. [CrossRef] [PubMed]

85. Raut, J.S.; Shinde, R.B.; Chauhan, N.M.; Karuppayil, S.M. Phenylpropanoids of plant origin as inhibitors of biofilm formation by *Candida albicans*. *J. Microbiol. Biotechnol.* **2014**, *24*, 1216–1225. [CrossRef] [PubMed]

86. Raut, J.S.; Shinde, R.B.; Chauhan, N.M.; Mohan Karuppayil, S. Terpenoids of plant origin inhibit morphogenesis, adhesion, and biofilm formation by *Candida albicans*. *Biofouling* **2013**, *29*, 87–96. [CrossRef] [PubMed]

87. Morales, D.K.; Grahl, N.; Okegbe, C.; Dietrich, L.E.; Jacobs, N.J.; Hogan, D.A. Control of *Candida albicans* metabolism and biofilm formation by *Pseudomonas aeruginosa* phenazines. *MBio* **2013**, *4*, 1–9. [CrossRef] [PubMed]

88. Theberge, S.; Semlali, A.; Alamri, A.; Leung, K.P.; Rouabhia, M. *C. albicans* growth, transition, biofilm formation, and gene expression modulation by antimicrobial decapeptide KSL-W. *BMC Microbiol.* **2013**, *13*, 246. [CrossRef] [PubMed]

89. Siles, S.A.; Srinivasan, A.; Pierce, C.G.; Lopez-Ribot, J.L.; Ramasubramanian, A.K. High-throughput screening of a collection of known pharmacologically active small compounds for the identification of *Candida albicans* biofilm inhibitors. *Antimicrob. Agents Chemother.* **2013**, *57*, 3681–3687. [CrossRef] [PubMed]

90. Wong, S.S.; Kao, R.Y.; Yuen, K.Y.; Wang, Y.; Yang, D.; Samaranayake, L.P.; Seneviratne, C.J. In vitro and in vivo activity of a novel antifungal small molecule against *Candida* infections. *PLoS ONE* **2014**, *9*, e85836. [CrossRef] [PubMed]

91. Pierce, C.G.; Chaturvedi, A.K.; Lazzell, A.L.; Powell, A.T.; Saville, S.P.; McHardy, S.F.; Lopez-Ribot, J.L. A novel small molecule inhibitor of *Candida albicans* biofilm formation, filamentation and virulence with low potential for the development of resistance. *NPJ Biofilms Microb.* **2015**, *1*, 15012. [CrossRef] [PubMed]

92. Romo, J.A.; Pierce, C.G.; Chaturvedi, A.K.; Lazzell, A.L.; McHardy, S.F.; Saville, S.P.; Lopez-Ribot, J.L. Development of anti-virulence approaches for candidiasis via a novel series of small-molecule inhibitors of *Candida albicans* Filamentation. *mBio* **2017**, *8*, e01991-17. [CrossRef] [PubMed]

93. Rosseti, I.B.; Chagas, L.R.; Costa, M.S. Photodynamic antimicrobial chemotherapy (PACT) inhibits biofilm formation by *Candida albicans*, increasing both ROS production and membrane permeability. *Lasers Med. Sci.* **2014**, *29*, 1059–1064. [CrossRef] [PubMed]

94. Černáková, L.; Chupáčová, J.; Židlíková, K.; Bujdáková, H. Effectiveness of the photoactive dye methylene blue versus caspofungin on the *Candida parapsilosis* biofilm in vitro and ex vivo. *Photochem. Photobiol.* **2015**, *91*, 1181–1190. [CrossRef] [PubMed]

95. Chien, H.F.; Chen, C.P.; Chen, Y.C.; Chang, P.H.; Tsai, T.; Chen, C.T. The use of chitosan to enhance photodynamic inactivation against *Candida albicans* and its drug-resistant clinical isolates. *Int. J. Mol. Sci.* **2013**, *14*, 7445–7456. [CrossRef] [PubMed]

96. Lara, H.H.; Romero-Urbina, D.G.; Pierce, C.; Lopez-Ribot, J.L.; Arellano-Jiménez, M.J.; Jose-Yacaman, M. Effect of silver nanoparticles on *Candida albicans* biofilms: An ultrastructural study. *J. Nanobiotechnol.* **2015**, *13*, 91. [CrossRef] [PubMed]

97. Radhakrishnan, V.S.; Reddy Mudiam, M.K.; Kumar, M.; Dwivedi, S.P.; Singh, S.P.; Prasad, T. Silver nanoparticles induced alterations in multiple cellular targets, which are critical for drug susceptibilities and pathogenicity in fungal pathogen (*Candida albicans*). *Int. J. Nanomed.* **2018**, *13*, 2647–2663. [CrossRef] [PubMed]
98. Gondim, B.L.C.; Castellano, L.R.C.; de Castro, R.D.; Machado, G.; Carlo, H.L.; Valença, A.M.G.; de Carvalho, F.G. Effect of chitosan nanoparticles on the inhibition of *Candida* spp. biofilm on denture base surface. *Arch. Oral. Biol.* **2018**, *94*, 99–107. [CrossRef] [PubMed]
99. de Alteriis, E.; Maselli, V.; Falanga, A.; Galdiero, S.; Di Lella, F.M.; Gesuele, R.; Guida, M.; Galdiero, E. Efficiency of gold nanoparticles coated with the antimicrobial peptide indolicidin against biofilm formation and development of *Candida* spp. clinical isolates. *Infect. Drug Resist.* **2018**, *11*, 915–925. [CrossRef] [PubMed]
100. Baigorria, E.; Reynoso, E.; Alvarez, M.G.; Milanesio, M.E.; Durantini, E.N. Silica nanoparticles embedded with water insoluble phthalocyanines for the photoinactivation of microorganisms. *Photodiagn. Photodyn. Ther.* **2018**, *23*, 261–269. [CrossRef] [PubMed]
101. Mosallam, F.M.; El-Sayyad, G.S.; Fathy, R.M.; El-Batal, A.I. Biomolecules-mediated synthesis of selenium nanoparticles using *Aspergillus oryzae* fermented Lupin extract and gamma radiation for hindering the growth of some multidrug-resistant bacteria and pathogenic fungi. *Microb. Pathog.* **2018**, *122*, 108–116. [CrossRef] [PubMed]
102. Perlroth, J.; Choi, B.; Spellberg, B. Nosocomial fungal infections: Epidemiology, diagnosis, and treatment. *Med. Mycol.* **2007**, *45*, 321–346. [CrossRef] [PubMed]
103. Spellberg, B. Vaccines for invasive fungal infections. *F1000 Med. Rep.* **2011**, *3*, 13. [CrossRef] [PubMed]
104. Santos, E.; Levitz, S.M. Fungal vaccines and immunotherapeutics. *Cold Spring Harb. Perspect. Med.* **2014**, *4*, a019711. [CrossRef] [PubMed]
105. Shahid, S.K. Newer patents in antimycotic therapy. *Pharm. Pat Anal.* **2016**, *5*, 115–134. [CrossRef] [PubMed]
106. Edwards, J.E. Fungal cell wall vaccines: An update. *J. Med. Microbiol.* **2012**, *61*, 895–903. [CrossRef] [PubMed]
107. Vecchiarelli, A.; Pericolini, E.; Gabrielli, E.; Pietrella, D. New approaches in the development of a vaccine for mucosal candidiasis: Progress and challenges. *Front. Microbiol.* **2012**, *3*, 294. [CrossRef] [PubMed]

Article

Generation of A *Mucor circinelloides* Reporter Strain—A Promising New Tool to Study Antifungal Drug Efficacy and Mucormycosis

Ulrike Binder [1,*], Maria Isabel Navarro-Mendoza [2], Verena Naschberger [1], Ingo Bauer [3], Francisco E. Nicolas [2], Johannes D. Pallua [4], Cornelia Lass-Flörl [1] and Victoriano Garre [2,*]

1   Division of Hygiene and Medical Microbiology, Medical University Innsbruck, Schöpfstrasse 41,
    6020 Innsbruck, Austria; verena.naschberger@i-med.ac.at (V.N.); cornelia.lass-floerl@i-med.ac.at (C.L.-F.)
2   Departamento de Genética y Microbiología, Facultad de Biología, Universidad de Murcia,
    30100 Murcia, Spain; mariaisabel.navarro3@um.es (M.I.N.-M.); fnicolas@um.es (F.E.N.)
3   Division of Molecular Biology, Biocenter, Medical University of Innsbruck, Innrain 80-82,
    6020 Innsbruck, Austria; ingo.bauer@i-med.ac.at
4   Institute of Pathology, Neuropathology and Molecular Pathology, Medical University of Innsbruck,
    Müllerstraße 44, 6020 Innsbruck, Austria; johannes.pallua@i-med.ac.at
*   Correspondence: ulrike.binder@i-med.ac.at (U.B.); vgarre@um.es (V.G.)

Received: 5 November 2018; Accepted: 5 December 2018; Published: 7 December 2018

Abstract: Invasive fungal infections caused by Mucorales (mucormycosis) have increased worldwide. These life-threatening infections affect mainly, but not exclusively, immunocompromised patients, and are characterized by rapid progression, severe tissue damage and an unacceptably high rate of mortality. Still, little is known about this disease and its successful therapy. New tools to understand mucormycosis and a screening method for novel antimycotics are required. Bioluminescent imaging is a powerful tool for in vitro and in vivo approaches. Hence, the objective of this work was to generate and functionally analyze bioluminescent reporter strains of *Mucor circinelloides*, one mucormycosis-causing pathogen. Reporter strains were constructed by targeted integration of the firefly luciferase gene under control of the *M. circinelloides* promoter Pzrt1. The luciferase gene was sufficiently expressed, and light emission was detected under several conditions. Phenotypic characteristics, virulence potential and antifungal susceptibility were indifferent to the wild-type strains. Light intensity was dependent on growth conditions and biomass, being suitable to determine antifungal efficacy in vitro. This work describes for the first time the generation of reporter strains in a basal fungus that will allow real-time, non-invasive infection monitoring in insect and murine models, and the testing of antifungal efficacy by means other than survival.

Keywords: *Mucor circinelloides*; mucormycosis; firefly luciferase; reporter strain; bioluminescence

1. Introduction

Mucor circinelloides, a member of the Mucoromycota, is ubiquitously found in the environment. It is thermotolerant, able to grow on a wide range of organic substrates and sporulates fast and abundantly [1,2]. It can cause mucormycosis—a severe animal and human disease. In recent decades, the incidence of mucormycosis has increased all over the world, becoming the second most common fungal disease in patients with haematological malignancies and transplant recipients [3–5]. Infections with mucormycetes are highly aggressive and destructive, resulting in tissue necrosis, invasion of blood vessels and subsequent thrombosis. The rapid progression, linked with shortcomings in diagnosis and therapy, results in high mortality rates which are estimated to range between 40–>90%, depending on the site of infection, the condition of the host and the therapeutic interventions [3,5–7]. The different types of mucormycosis are classified according to the anatomic site of infection, such as rhino-orbital-cerebral, pulmonary, cutaneous, gastrointestinal and disseminated infections [8]. Antifungal therapy is complicated by the limited treatment options that comprise lipid amphotericin B (AMB) as first-line therapeutic and posaconazole (POS) or isavuconazole (ISA) as salvage treatment [9,10].

The most common genera associated with human disease next to *Mucor* are *Rhizopus*, and *Lichtheimia* (formerly *Absidia*) and infections are associated with severe graft-versus-host disease, treatment with steroids, neutropenia, iron overload, diabetes and malnutrition [10]. *M. circinelloides* isolates have been associated with outbreaks of mucormycosis in the US, the UK and Europe and it poses a threat to public health by contaminating food and producing 3-nitropropionic acid [11–15].

Despite the growing relevance of mucormycetes in public health, little is known about the physiology and virulence factors associated with this group of fungi. The heterogeneity of this group and the difficulties in genetic manipulation are reasons thereof. However, *M. circinelloides* stands out among the rest of basal fungi offering the opportunity to carry out genetic manipulation by the development of an increasing number of molecular tools [16,17]. The intrinsic resistance of mucormycetes to drugs used as resistant markers in other fungi, leaves only the use of auxotrophic markers [18,19].

Bioluminescence imaging is a very useful technique to track microorganisms in living animals and has provided novel insights into the onset and progression of disease. The great advantage is the real-time monitoring of infection in one individual organism over time. Different enzymes exist in living organisms, which use different substrates and different cofactors to emit light. The most prominently used are the firefly (*Photinus pyralis*) and the copepod (*Gaussia princeps*) luciferase [20]. Both have already been successfully transformed into opportunistic fungal pathogens e.g., *Candida albicans* [21,22], *Aspergillus fumigatus* [23–25] and *A. terreus* [26]. Bioluminescence imaging with these strains has significantly enhanced our understanding of fungal infection. It revealed unexpected host sites in disseminated candidiasis, showing persistence of *Candida* cells in the gallbladder, even after antifungal treatment. Studies comparing *A. fumigatus* and *A. terreus* revealed delayed onset of disease in *A. terreus* infected mice and survival in 50% of *A. terreus* infected mice, although progression of disease was similar to those that died.

In this study, we generated bioluminescent strains in the opportunistic human pathogen *M. circinelloides* based on the expression of firefly luciferase and controlled under a highly expressed *M. circinelloides* promoter, for the first time. Light emission was correlated to fungal growth and concentration of the substrate luciferin. Phenotypic analysis and virulence potential in the alternative host *Galleria mellonella* revealed no differences to parental strains. Antifungal efficacy was determined successfully by the use of the obtained reporter strains. The strains generated in this study will be a useful tool to test novel antifungal agents both in vitro and in murine and insect models, in addition to shed light on the onset and progression of mucormycosis in animal models.

2. Material and Methods

2.1. Fungal Strains, Plasmids, Media and Growth Conditions

The strains and plasmids used in this study are listed in Table 1. All fungal strains used were *M. circinelloides f. lusitanicus*, referred to in this work as *M. circinelloides* for simplicity. To obtain spores, strains were grown on YPG (yeast peptone glucose agar; 3 g/L yeast extract, 10 g/L peptone, 20 g/L glucose, pH 4.5) medium at 26 °C in the light for 4–5 days. Spores were collected by scraping the plates with sterile spore solution buffer (0.9% NaCl, 0.01% Tween 80) and spore concentration was determined by hemocytometer. Media and growth conditions for the individual assays are given below, for most assays YNB (yeast nitrogen base; 1.5 g/L ammonium sulfate, 1.5 g/L glutamic acid, 0.5 g/L yeast nitrogen base (*w/o* ammonium sulfate and amino acids, Sigma-Aldrich, Steinheim, Germany, cat. no. Y1251), 10 g/L glucose, thiamine 1 µg/mL and niacin 1 µg/mL) was used [16]. All chemicals used were purchased from Sigma-Aldrich, Germany, unless otherwise stated.

Table 1. List of strains and plasmids used in this study.

	Name	Genotypes/Characteristics	Reference
Mucor circinelloides	R7B	*leuA1* (mutant alelle of *leuA* gene)	Roncero et al., 1984 [27]
	R7B_luc	*carRP::leu*	Obtained in this study
	R7B_luc1	*carRP::leu*	Obtained in this study
	MU402	*leuA1, pyrG⁻*	Nicolas et al., 2007 [18]
	MU402_luc	*pyrG⁻, carRP::leuA*	Obtained in this study
	MU402_luc1	*pyrG⁻, carRP::leuA*	Obtained in this study
	DH5α	ampicillin resistance	Thermo Fisher (Germering, Germany)
Escherichia coli	pMAT1477	*Pzrt1, leuA*	Rodriguez-Frometa et al., 2013 [19]
	pGL3 basic vector	firefly luciferase reporter vector	Commercially available, Promega (Fitchburg, WI, USA)
Plasmids	pMAT1903	pMAT1477 + luciferase	Obtained in this study

Escherichia coli DH5α was used as a host for plasmid propagation and grown in lysogeny broth (LB) medium, supplemented with 0.1 mg/mL ampicillin if needed. Bacterial cultures were incubated at 37 °C overnight.

2.2. Cloning Procedures

2.2.1. Amplification of the Firefly Luciferase Gene

The plasmid pGL3 basic vector (Promega, Fitchburg, WI, USA) served as a template to amplify firefly luciferase by PCR using primers luc-FOWXhoI (AAACTCGAGATGGAAGACGCCAAAAAC ATAAAGAAAGG), luc-REVSacII (CGCCCCGCGGCTAGAATTACACGGCGATCTTTCC) and Herculase II fusion DNA polymerase (Agilent, Santa Clara, CA, UAS). This luciferase is an optimized version for the use in mammalian cells and does not contain the peroxisomal target sequence of the native firefly luciferase.

2.2.2. Plasmid Construction

The pMAT1477 plasmid carries the strong promoter of the *M. circinelloides zrt1* gene (Pzrt1) and a functional *leuA* gene as a selective marker [19]. Amplified luciferase gene was digested by XhoI and SacII and ligated into pMAT1477 to obtain pMAT1903 (Figure S1). By the targeted integration of the whole construct in the *M. circinelloides carRP* gene, which is involved in carotenoid biosynthesis, identification of clones with integrated gene was facilitated, as they formed albino colonies, while those without integration had a yellow phenotype [28].

2.3. Transformation of Mucor circinelloides *and Initial Screening*

Strain R7B (*leuA⁻*) and MU402 (*leuA⁻, pyrG⁻*) were chosen as recipient strains. MU402 is derived from R7B. Transformation of *M. circinelloides* was performed by electroporation of protoplasts as described previously [16]. In brief, freshly harvested spores were incubated in YPG media (pH 4.5) for 2–4 h until spores were germinated and then transferred to a fresh tube for digestion of the cell wall by lysing enzymes (L-1412, Sigma-Aldrich, St. Louis, MO, USA) and chitosanase (C-0794, Sigma-Aldrich). Linearized DNA (5 µg) was used in each transformation reaction. Protoplasts were incubated on YNBS agar (pH 3.2; containing 0.2 g/L uridine for MU402 strains) and checked daily for colonies [16]. Colonies formed by protoplasts with correct gene targeting appeared white, because of disruption of the *carRP* gene by targeted integration. Albino colonies were repeatedly transferred to fresh selective agar plates (3–4 cycles) to obtain homokaryons. Several clones in each background were then chosen and tested for light emission. Therefore, spores were incubated in YPG in 6-well plates (Nunc GmbH, Langenselbold, Germany) overnight, then D-luciferin (10 mM, Synchem, Felsberg, Germany) was added to the cultures and light emission detected by a monochrome scientific grade CCD camera (BIO-Vision 3000 imaging system, Golden, CA, USA (Figure S3)). Clones that showed highest light emission were chosen for further experiments.

2.4. Genomic DNA Extraction and Southern Analysis

For the preparation of genomic DNA, lyophilized mycelia were ground to powder using a tungsten carbide ball in a Retsch MixerMill 400 and resuspended in 1 mL of DNA isolation buffer (50 mM Tris-HCl, 250 mM NaCl, 100 mM EDTA (Ethylenediaminetetraacetic acid), 1% (w/v) SDS (Sodium dodecyl sulfate), pH 8.0 at 25 °C) and 300 µL of PCI (Phenyl-Chloroform-Isoamylalcohol; Carl Roth, Karlsruhe, Germany). After incubation at room temperature for 5 min, the mixture was centrifuged for 10 min at 20,000× g and 4 °C. RNase A (10 µL; 10 mg/mL) were added to the supernatant and incubated for 10 min at 65 °C and further 30 min at 37 °C. After RNase digestion, DNA was extracted by addition of 1/3 volume of PCI, centrifuged and the resulting supernatant was precipitated by addition of 1 volume of isopropanol. The DNA pellet was washed with 180 µL of 70% ethanol, briefly air-dried and solubilized in 50 µL of a.d. Concentration and quality were determined by agarose gel electrophoresis. 2 µg of genomic DNA were digested overnight with 10–20 units of either *Bgl*II or *Pst*I and separated on 0.8% agarose gels. DIG-labeled marker VII (Roche, Basel, Switzerland) served as a marker for fragment size estimation. Capillary transfer of DNA onto nylon membrane was performed overnight. Membranes were hybridized with a probe for the luciferase coding sequence. Probe labeling with DIG-dUTP was performed by PCR amplification using primers luc1f (5′-TCGCATGCCAGAGATCCT) and luc1r (5′-CGCCCGGTTTATCATCCC).

2.5. Luciferase Activity

Luciferase activity was tested in vitro by measuring light emission of bioluminescent *M. circinelloides* strains in dependency of inoculum density and growth conditions by luminometer Tecan infinite 200 PRO plate reader (Tecan Group AG, Männedorf, Switzerland). First, two transformants of each background were chosen, grown in YNB (2×10^5 spores/mL) in 24-well microtiter plates (Nunc) for 24 h, 100 µL of luciferin (Roche, Luciferase Reporter Gene Assay) were added to the cultures and light emission was detected immediately (2 min after addition of luciferase), or 30 min after addition of substrate to check for stability of light emission.

To determine correlation to inoculum size and subsequent fungal growth/biomass, different spore concentrations were incubated (2×10^2–2×10^6/mL) and light emission was detected as previously described.

A dilution series of substrate luciferin (Roche)—undiluted, 1:2, 1:5 and 1:10—was tested to check for the optimal concentration needed for further experiments.

To determine if light emission is dependent upon growth conditions, strains (2×10^5 spores/mL) were pre-grown in YNB to the same amount of biomass overnight, before the medium was replaced by fresh YNB, YPG or RPMI$_{1640}$ (Sigma-Aldrich, Spittal/Drau, Austria), respectively. Substrate addition and measurement of emitted light were carried out as described above.

2.6. Phenotypic Analysis in Different Growth Conditions

Growth on different media of the recipient strains and the resulting luciferase expressing clone was compared on different media (YNB, YPG, RPMI$_{1640}$ and supplemented minimal agar, SUP [29]; with supplements added when necessary). 10^4 spores of the individual strains were dotted onto the respective agar plates and incubated at 30 °C and 37 °C, respectively. After 24 h colony diameters were measured and growth documented visually. Experiments were carried out with 3 parallels and repeated twice. For growth assays in hypoxic conditions cultures were grown at 1% O_2 (Biospherix, C-Chamber & Pro-Ox controller, Parish, NY, USA).

2.7. Antifungal Susceptibility Testing

Minimum inhibitory concentration (MIC) of AMB, POS, ISA, and itraconazole (ITRA) were determined for all strains according to the European Committee on Antimicrobial Susceptibility Testing (EUCAST) guidelines 9.2 [30]. MIC was defined as the lowest concentration that completely inhibited growth. Additionally, MICs were also determined in YNB medium, as this was used for luciferin activity assays.

To determine antifungal drug efficacy by correlation of fungal growth with light emission, R7B_luc was grown in YNB overnight in the presence of AMB (0.25 μg/mL and 1 μg/mL) or POS (2, 16 and 32 μg/mL), respectively, in 96 well plates. To test the growth inhibiting activity of AMB and POS on *M. circinelloides* hyphae, cultures were pre-grown in YNB overnight before antifungal drugs were added. After 4 h of drug exposure, luciferin was added and light emission detected as described above. All experiments were carried out in parallels and repeated twice.

2.8. Virulence Assay in Galleria mellonella

Sixth instar larvae of *G. mellonella* (SAGIP, Bagnacavallo, Italy), weighing 0.3–0.4 g, were selected for experimental use. Larvae, in groups of twenty, were injected through the last pro-leg into the hemocoel with 1×10^6 spores in a volume of 20 μL as described previously and incubated at 30 °C in the dark [29,31,32]. Untouched larvae and larvae injected with sterile insect physiological saline (IPS) served as controls. Survival was determined every 24 h over a period of 144 h. Experiments were repeated three times and the average survival rate was calculated. Significance was determined with log-rank (Mantel-Cox) test, utilizing GraphPad Prism 7 software (GraphPad Software, San Diego, CA, USA). Differences were considered significant at p-values < 0.05.

2.9. Histology of Larvae

Specimen were fixed in formalin for at least 15 days before being embedded in paraffin. Longitudinal tissue sections were carried out with a microtome at 3.0 μm thickness and stained with Grocott for histological validation. Slides were digitalized using a Pannoramic SCAN digital slide scanner (3DHISTECH, Budapest, Hungary) with plan-apochromat objective (magnification: 20×, Numerical aperture: 0.8). The histological evaluation and the scoring of the fungal infection were done by using the Pannoramic Viewer software (3DHISTECH).

3. Results and Discussion

3.1. Generation of Firefly Luciferase-Producing Mucor circinelloides Strains Resulted in Detectable Light Emission

For the generation of luciferase-producing *M. circinelloides* strains, we cloned the firefly (*P. pyralis*) luciferase gene, optimized for use in mammalian cells, under the control of a strong *M. circinelloides* zrt1 promoter into plasmid pMAT1477 that contained the *leuA* gene, which was used as a selective marker in transformations. In the resulting plasmid, the luciferase gene and *leuA* are flanked by sequences of the *carRP* locus to favor targeted integration of the whole construct. The plasmid was linearized and then used to transform the leucine auxotroph strain R7B and the leucine and uridine auxotroph strain MU402. The double auxotroph was chosen to facilitate subsequent disruption or introduction of other genes in the bioluminescent strain. In both backgrounds, more than 50 colonies were obtained on selective transformation plates. Integration in the *carRP* locus renders albino colonies, hence fifteen independent transformants of each background with white appearance were selected. Due to the multinucleated nature of the protoplasts, they were repeatedly inoculated on selective agar until the transformants produced only white colonies, an indication that they were homokaryons. Five of these transformants per background were randomly selected and checked for luciferase production by observing light emission with the naked eye in the dark and visualization of light production (Figure S3). Two strains in each background, showing high light emission, were selected for Southern blot analysis using digoxigenin-labeled probes directed against the luciferase coding region (Figure S2). Restriction with *Bgl*II or *Pst*I confirmed correct insertion and single integration of the luciferase gene. The same four strains were chosen for further luminescence detection by microplate reader and CCD camera (Figure S3). As shown in Table 2, all transformants emitted more light than the parental strains, indicating that our approach of expressing firefly luciferase for the first time in *M. circinelloides* was successful. Measurements taken 30 min after substrate addition indicated that light emission was moderately stable and still significantly detectable after this time, which is essential for further use of the reporter strains. Furthermore, luciferase-harboring strains were stable over several generations because of the site- directed insertion and the selection for homokaryons. Light signals were lower for transformants in the MU402 background, which correlates to slower growth and less biomass of these strains compared to R7B. Therefore, mainly R7B_luc was used for further experiments. Regarding the difficulties with genetic engineering of basal filamentous fungi and in particular the opposition of mucormycetes to express foreign genes, this is an achievement that will be advantageous for further experimental work (e.g., in optimizing luciferase expression in *M. circinelloides* and other mucormycetes).

Table 2. Detection of luminescence signal by microplate reader. 10^5 spores/mL of the respective strains were grown in YNB medium (containing supplements where needed) for 24 h. Light emission was induced by the addition of D-luciferin (10 mM) and detected with a microplate reader (Tecan Group AG, Männedorf, Switzerland). Ten seconds were set as integration time. Measurements were carried out 2 min and 30 min after substrate addition. RLUs (relative light units) present the average of three experiments; SD represents standard deviation.

Strains	RLUs (2 min)	SD	RLUs (30 min)	SD
R7B	14	4	15	1
R7B_luc	4167	112	1444	129
R7B_luc1	3117	73	1194	95
MU402	10	1	7	3
MU402_luc	174	50	185	0
MU402_luc1	20	2	15	1

3.2. In Vitro Characterization of Bioluminescent Mucor circinelloides Reporter Strains

3.2.1. Radial Growth Is Not Altered by Insertion of the Luciferase Gene

To check growth ability of luciferase containing strains compared to their recipient strains, radial growth was determined on different media at 30 °C and 37 °C. Based on results shown in Table 2, R7B_luc and MU402_luc were chosen for the radial growth assays, because they showed higher light emission. R7B_luc is prototrophic, while MU402_luc is still auxotrophic for uridine. The aim to generate a reporter strain in the MU402 background was to have a tool in hand that can be used for further genetic manipulation, such as deletion of genes essential for virulence. Here, to rule out phenotypes resulting from luciferase integration, also the MU402_luc strains were used for growth characterization and light emission assays. However, these strains will not be used per se in future animal models. None of the strains displayed an obvious abnormal growth phenotype. At both temperatures, R7B_luc exhibits same growth and average colony diameter on each of the media tested compared to the parental strain R7B (Figure 1, Table S1). MU402_luc exhibited significantly smaller colony diameters on YNB (containing uridine) at both temperatures and on $RPMI_{1640}$ at 37 °C, but still, no differences were detected compared to the parental strain. As expected, colonies were smaller at 37 °C at this early time point, indicating difficulties of *M. circinelloides* with adaptation to high temperatures. Because oxygen levels are expected to be very low on site of infection in the human and animal body [33], and we aim to use the luciferase containing strains in animal models, growth was further evaluated in hypoxic conditions (1% oxygen). Even at this low oxygen concentration, all strains were able to grow and form hyphae, a pre-requisite of tissue invasion. Growth was reduced in all samples compared to normoxic conditions, especially on minimal media (YNB and RPMI). Surprisingly, MU402 and MU402_luc seemed to adapt better to the combination of low oxygen and elevated temperature than R7B_luc and its parental strain on SUP and YPG. At 48 h growth in hypoxia was restored at 30 °C and partially at 37 °C (Figure S4), indicating that *M. circinelloides* spores showed delayed germination in hypoxic conditions, but are able to adapt to low oxygen. This ensures that the luciferase strains will be suitable for use in infection models at a later time point.

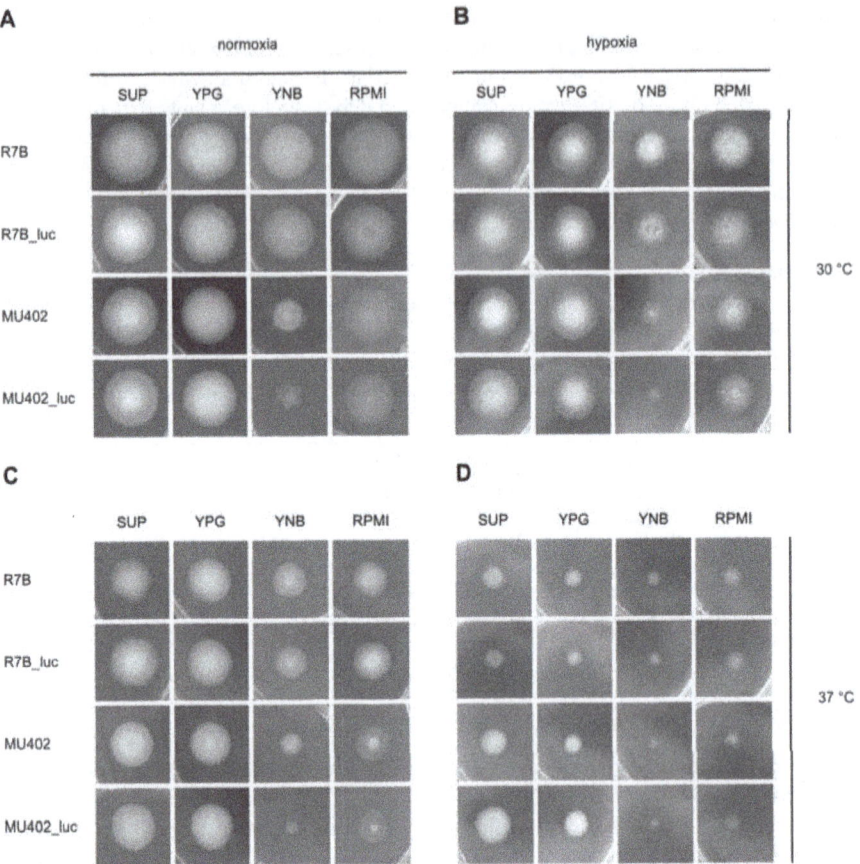

Figure 1. Growth phenotypes of recipient and luciferase-expressing strains grown for 24 h on different media at 30 °C (panels (**A,B**)), 37 °C (panels (**C,D**)) under normoxic (panels (**A,C**)) and hypoxic conditions (panels (**B,D**)). Hypoxia was induced by reducing the oxygen concentration in the incubator to 1%. SUP: supplemented minimal agar; YPG: yeast peptone glucose; YNB: yeast nitrogen base; RPMI: RPMI$_{1640}$.

3.2.2. Light Emission Correlates with Fungal Biomass and Amount of Available Luciferase Substrate

An important parameter for the use of bioluminescent reporter strains in animal infection is the detection limit of emitted light, which was determined by cultivation of R7B_luc spores at different inoculum densities and assessment of light emission after 24 h of growth. Light was detected in cultures inoculated with as low as 2×10^3 spores/mL compared to the controls without luciferin and increased with the number of spores used. Highest RLUs were observed at a spore concentration of 2×10^5/mL (Figure 2A, upper panel), the spore concentration that also led to the highest density of mycelia (Figure 2A, lower panel). All other spore concentrations led to significantly lower RLU measurement ($p < 0.05$). The highest inoculum concentration used (2×10^6/mL) did not result in highest light emission, which can be explained by lower growth rate and non-homogeneity of in the culture. Spores probably germinate but face a lack of nutrients at this high density. For all further in vitro experiments, 2×10^5 spores were used.

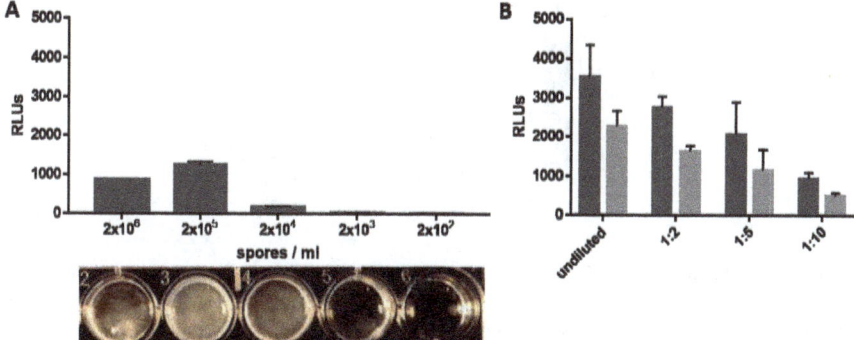

Figure 2. (**A**) Light emission in dependency of inoculum density. YNB medium was inoculated with different concentrations of R7B_luc spores, light emission was induced by addition of luciferin after 24 h and detected by plate reader (upper panel). Relative light units (RLUs) represent the average of three independent measurements. Error bars indicate standard deviation. The lower panel shows fungal growth at the various spore concentrations after 24 h of incubation. (**B**) Light emission in dependency of substrate concentration. 2×10^5 spores/mL were inoculated in YNB and light emission was induced by addition of different concentrations (undiluted, 1:2, 1:5; 1:10) of luciferin dissolved according to the manufacture's protocol (Roche, Basel, Switzerland). Light was detected immediately (dark grey bars) and 10 min after substrate addition (light grey bars). Error bars indicate standard deviation.

Different concentrations of the substrate luciferin were tested to evaluate the minimum amount necessary for R7B_luc to emit detectable light. As expected, light emission clearly correlated with amount of substrate added to the cultures (Figure 2B). As shown before, light emission decreased with time, with significantly reduced light emission at the later time point (*t*-test, $p < 0.05$), but was still detectable 10 min after substrate addition.

To evaluate the effect of growth media on luciferase expression, cultures were pre-grown in YNB overnight and the medium was replaced by fresh YNB, YPG or RPMI$_{1640}$ 3 h before addition of substrate and light detection. Measurement of RLUs revealed highest levels of emitted light in YNB medium and lowest in YPG (Figure 3). One possible explanation is the nature of the P*zrt1* promoter used for our construct. The gene *zrt1* codes for a zinc transporter whose expression is induced by reduced availability of zinc as is the case in minimal media such as YNB. In mammalian tissues such as human lung or blood, the concentration of zinc is very low; therefore, expression of luciferase driven by P*zrt1* should be high in vivo. Rich media, such as YPG contain higher concentrations of zinc, hypothetically resulting in downregulation of luciferase expression.

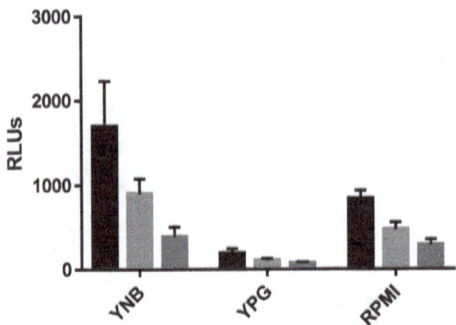

Figure 3. Light emission of R7B_luc in different growth media. YNB medium was inoculated with 2×10^5 spores/mL of R7B_luc, incubated for 16 h, and then replaced by fresh YNB, YPG, or RPMI, respectively. Light emission was induced by addition of substrate (luciferin 1:5) 3 h after medium exchange and detected by using a plate reader. RLUs were determined immediately after addition of substrate (black bars), 10 min (light grey bars) and 30 min (dark grey bars) after the addition of substrate. Error bars indicate standard deviation. RLUs emitted in each media were significantly different from the other media tested (Two-way analysis of variance (ANOVA), $p < 0.05$).

3.3. Antifungal Susceptibility Testing

3.3.1. Genetic Manipulation Does Not Influence Antifungal Susceptibility Patterns of *Mucor circinelloides* Strains

The expression of the luciferase gene does not affect the susceptibility patterns of *M. circinelloides* to commonly used antifungal agents, since strains expressing the luciferase gene showed the same susceptibility pattern as the recipient strains (Table 3). To better compare MIC results with results from light emission studies, MICs were also determined in YNB in addition to $RPMI_{1640}$ medium. All strains showed moderate susceptibility to AMB, resulting in complete growth inhibition at concentrations between 0.5 and 2 µg/mL in both media tested. This correlates well to other studies, and thus all 4 strains tested could be classified as susceptible to AMB according to the epidemiological cut-off value determined for *M. circinelloides* [34]. Despite of reducing growth, the azole concentrations applied were below the MIC in $RPMI_{1640}$, nevertheless, in YNB a posaconazole MIC could be determined for R7B and R7B_luc. Although POS is regarded as second-line treatment, it has been shown before that *M. circinelloides* isolates very often also exhibited resistance to this azole [35,36]. The fact that susceptibility patterns of luciferase-harboring and recipient strains were very similar, assures that the luciferase-harboring strains are suitable for the assessment of antifungal drug efficacy in vitro and in vivo.

Table 3. Minimal inhibitory concentrations (MICs; µg/mL) determined for amphotericin B (AMB) and azoles (posaconozale: POS; itraconazole: ITRA; isavuconazole: ISA) according to European Committee on Antimicrobial Susceptibility Testing (EUCAST) guidelines. MICs were determined after 24 h of incubation at 37 °C, except for MU402 and MU402_luc, where MICs were read after 48 h of growth in yeast nitrogen base (YNB).

	MIC (µg/mL)							
	$RPMI_{1640}$				YNB			
Strains	**AMB**	**POS**	**ITRA**	**ISA**	**AMB**	**POS**	**ITRA**	**ISA**
R7B	2	>16	>8	>4	1	2	>8	>4
R7B_luc	1	>16	>8	>4	0.5	4	>8	>4
MU402	2	>16	>8	>4	1	>16	>8	>4
MU402_luc	2	>16	>8	>4	1	>16	>8	>4

3.3.2. Bioluminescent Strains Can Be Used to Evaluate Efficacy of Antifungal Drugs

To test if the luciferase producing strain R7B_luc is suitable for monitoring the efficacy of antifungal substances, we used AMB and POS in two different experimental setups. First, the respective antifungal agent was added directly to spores of R7B_luc, mimicking the EUCAST protocol and light emission was determined after 24h of incubation. No light was detected in wells containing either AMB or POS; respectively (Figure 4A,B). Results obtained with 1 µg/mL AMB correlate well with the MIC determined (Table 3). At this concentration no growth was evident in the presence of AMB prior to germination and consequently no light is emitted. Although some spores could grow in the presence of 0.25 µg/mL AMB, also at this concentration biomass was too little to produce sufficient luciferase. Similarly, of all three POS concentrations tested, none resulted in the emission of detectable light units. Regarding a MIC of 4 mg/mL determined in YNB medium—the same medium that was used for the light emission assays—evidence of a correlation to standard MIC testing is observed. Even at a concentration of 2 µg/mL, growth (or at least fungal metabolism) was inhibited sufficiently to prevent production of luciferase and consequently, light.

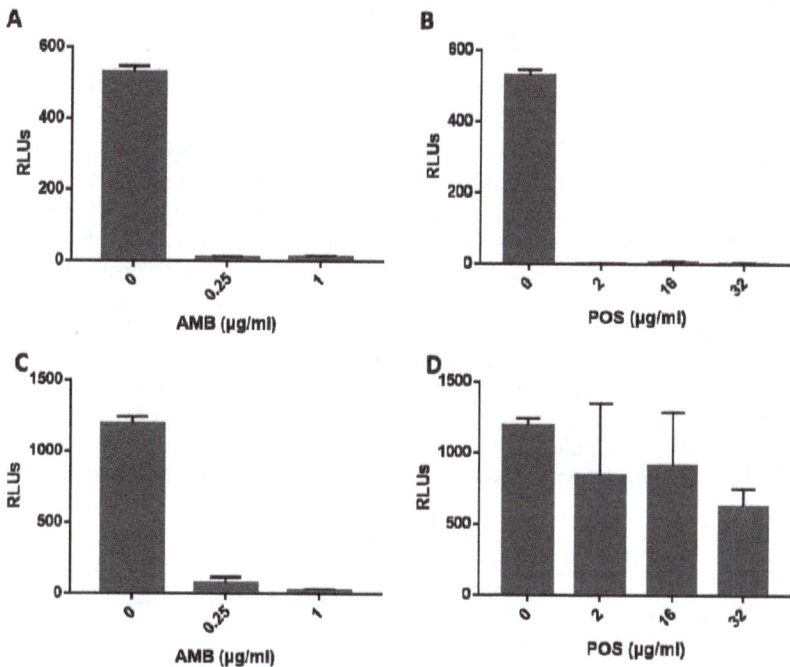

Figure 4. Graphical analysis of drug efficacy by detection of light emission. Luminescence was measured by a plate reader in wells containing 2×10^5 spores/mL of R7B_luc grown in YNB in the presence of amphotericin B (AMB) (**A**) or posaconazole (POS) (**B**), and with AMB (**C**) or POS (**D**) added to hyphae, respectively. Light emission was detected after 24 h of incubation (**A,B**) and subsequent luciferin (1:5, Roche) addition. Pre-grown cultures (16 h) were further incubated for 4 h once antifungals were added (**C,D**). Average values from three independent wells are given, error bars represent standard deviation.

The second approach was to test to what extent the production of luciferase is affected by the addition of antifungal agent, AMB and POS, to R7B_luc hyphae. AMB showed a tremendous effect on luciferase activity, resulting in no light emission when hyphae were incubated in 1 µg/mL AMB for 4 h, and only marginal light emission at 0.25 µg/mL (Figure 4C). This suggests a strong effect of AMB on hyphal metabolism, presumably resulting in decreased ATP levels within the hyphae, which consequently leads to reduced activity of the ATP-dependent luciferase. Another possible effect could be inhibition of substrate uptake due to AMB-induced metabolic changes or membrane dysfunction. Hyphae confronted with POS for 4 h exhibited reduced light emission, but with only a significant difference at 32 µg/mL compared to the untreated controls (Figure 4D). This correlated to previous data that showed reduced effect of POS on hyphae compared to AMB [37]. Furthermore, poor in vitro and in vivo POS efficacy against several *M. circinelloides* strains, due to a rather fungistatic than a fungicidal activity of POS, was shown by Salas et al. [38]. The results obtained with the luciferase-producing strains are in agreement with the MIC data shown before and studies undertaken by others [34,35]. Therefore, we can conclude that screening for antifungal drug efficacy is possible by using luciferase-expressing *M. circinelloides* strains and presents a valuable tool for testing novel antifungal drugs. For *M. circinelloides*, or Mucorales in general, this is of great importance, because many laboratories face difficulties applying and interpreting standard susceptibility test procedures, such as microbroth dilutions methods (EUCAST or clinical and laboratory standards institute (CLSI)) and especially Etest®, with this group of fungi. Often, results obtained with different methods do not correlate [39–41]; therefore, the use of bioluminescent strains will be an additional possibility to test the efficacy of (novel) antifungal drugs or combinations thereof.

3.4. Luciferase-Harboring Strains Exhibit Similar Virulence Potential as Recipient Strains in the Alternative Host Galleria mellonella

In order to test whether the integration of the luciferase gene influenced the virulence potential of the *M. circinelloides* strains, infection studies in the invertebrate host model *G. mellonella* were carried out. All strains were able to cause death to the larvae and no significant difference ($p > 0.05$) was detected between the luciferase-containing strains and the recipient strains (Figure 5). Lower mortality rates seen for MU402 and MU402_luc are most likely due to uridine auxotrophy, suggesting limited availability of uridine or uracil in *Galleria* hemolymph. This correlated to data obtained with *A. fumigatus*, that showed attenuated virulence potential of uridine or uracil auxotrophic strains in murine models [42]. Ability to cause disease in *Galleria* larvae and similarities in survival rates of luciferase-harboring and recipient strains confirms suitability of generated strains for future use in in vivo models. Fungal elements were found in tissue sections of larvae infected with R7B_luc (Figure S5), indicating larval killing by active fungal growth within the larval body. This is important for further studies, in which we aim to use this model system for in vivo bioluminescent imaging.

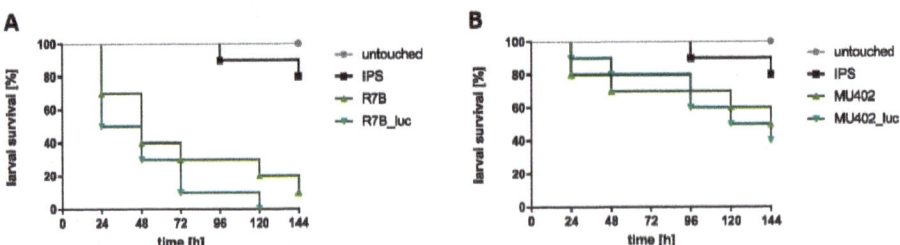

Figure 5. Survival of larvae infected with *M. circinelloides* strains. Larvae were infected with 10^6 spores of the respective strains and incubated at 30 °C. (**A**) represents Kaplan-Meier curves of larvae infected with R7B or R7B_luc strain. (**B**) represents Kaplan-Meier curves of larvae infected with MU402 or MU402_luc strain. Survival was monitored every 24 h up to 144 h. Untouched larvae and larvae injected with IPS buffer served as controls. Results are expressed as the mean of three independent experiments (60 larvae in total).

For other fungi, such as *A. fumigatus* and *C. albicans*, codon-optimized luciferase gene sequences were used and resulted in increased light emission, successfully detected in murine models [22,23,43].

Codon optimization of luciferase in *M. circinelloides* would probably increase luciferase expression and result in higher levels of detectable light, which would specifically be important in murine models, as light detection is quenched by tissue. Further, correlation between the copy number of luciferase and light emission was shown in *A. terreus* and *A. fumigatus* strains [23,26]. Although *M. circinelloides* is one of the few genetically tractable species among the basal fungi, it is not yet a robust genetic system, such as *A. nidulans* or *A. fumigatus*, and knowledge of other native, strong promoters besides *zrt1* is scarce. Nevertheless, using alternative promoters and/or integration of additional luciferase gene copies could further improve our model system.

4. Conclusions

The construction of bioluminescent reporter strains in the basal fungus *M. circinelloides* was successful and resulted in the first *M. circinelloides* strains expressing firefly luciferase, evident by detectable light emission. When comparing our newly generated reporter strains with their respective recipients, we obtained similar results regarding growth, antifungal susceptibility patterns, and virulence potential in the insect model *G. mellonella*. Further, luciferase-containing strains proved to be suitable for the evaluation of antifungal agents. The reporter strains obtained in this study represent a valuable tool for studies investigating the efficacy of novel antifungal agents and monitoring disease in a spatial and temporal manner in animal models in the future.

Supplementary Materials: The following are available online at http://www.mdpi.com/2073-4425/9/12/613/s1, Figure S1: Plasmid pMAT1903 containing luciferase gene without peroxisomal target sequence under the strong zrt1 promoter and functional *leuA*, Figure S2: Gene targeting strategy and confirmation of homologous recombination. (A) Schematic presentation of gene targeting and Southern analysis strategy. For generation of the targeting cassette, a leucine auxotrophic marker followed by the gene coding for firefly luciferase were cloned in between 1 kb each of the 5′ UTR and the 3′ UTR of carRP, which were used as flanks for homologous recombination after linearization of the plasmid by *Cfr*9I (*Xma*I) digestion. Restriction sites (*Bgl*II, *Pst*I) and the site for probe hybridization are shown. Expected fragment length to be identified upon proper integration at the carRP locus are 4.6 kb and 5.3 kb. (B) Southern analysis. Genomic DNA of individual transformants was digested with *Bgl*II and *Pst*I, respectively, separated on a 0.8% agarose gel (lower panels) and blotted onto nylon membranes. To determine fragment size, DIG-labelled marker VII (Roche; M) was used; the length of selected marker bands is indicated, Figure S3: Visualisation of bioluminescence from *M. circinelloides* cultures. The *M. circinelloides* transformants (2 per strain) and the respective parental strains are shown. 105 spores/ml were inoculated in YNB medium and grown for 24 h. Light emission was induced by the addition of D-luciferin (10 mM) to the medium, and bioluminescence images of the cultures were acquired by a monochrome scientific grade CCD camera (BIO-VISION 3000 imaging system, right panel), Figure S4: Growth phenotypes of recipient and luciferase expressing strains grown for 48 h on different media at 30 °C (panels A and B) and 37 °C (panels C and D) under normoxic (panels A and C) and hypoxic conditions (panels B and D) are shown. Hypoxia was induced by reducing the oxygen concentration in the incubator to 1 %, Figure S5: Histological examination of *Mucor circinelloides* infected *Galleria mellonella* larvae. Specimen were fixed in formalin 72 h after infection with 106 spores of R7B_luc and embedded in paraffin. Tissue sections were prepared at a thickness of 3.0 μm and stained with Grocott silver stain to optimize visualisation of fungal elements, Table S1: Colony diameter of *M. circinelloides* strains on various growth media. Colony diameter was determined in triplicates after 24 h of incubation at 30 °C and 37 °C. Numbers given represent the average of two independent experiments. Significance was determined by calculating standard deviation (SD).

Author Contributions: Conceptualization, U.B., V.G.; Methodology, U.B., M.I.N.-M., I.B., V.N. and F.E.N.; Formal Analysis, U.B., M.I.N.-M., I.B., V.N.; Investigation, U.B., M.I.N.-M., I.B., V.N., J.D.P.; Resources, U.B., V.G. and C.L.-F.; Data Curation, U.B., M.I.N.-M., I.B., V.N.; Writing—Original Draft Preparation, U.B., V.G., I.B.; Writing—Review & Editing, U.B., M.I.N.-M., I.B., V.G., F.E.N., J.D.P., C.L.-F.; Supervision, U.B., C.L.-F.; V.G. and F.E.N.; Project Administration, U.B., C.L.-F.; Funding Acquisition, U.B., C.L.-F., V.G.

Acknowledgments: This work was financially supported by the EMBO short-term fellowship 6856 to U.B., the "Christian Doppler Forschungsgesellschaft" (CD-Labor Invasive Pilzinfektionen) to C.L.-F., The "Ministerio de Educación, Cultura y Deporte, Spain" (FPU-14/01832) to M.I.N.-M., "Ministerio de Economía y Competitividad, Spain" (RYC-2014-15844) to F.E.N., and "Ministerio de Economía y Competitividad, Spain" (BFU2015-65501-P co-financed by FEDER) and "Fundación Séneca-Agencia de Ciencia y Tecnología de la Región de Murcia, Spain" (19339/PI/14) to M.I.N.-M., F.E.N. and V.G. The authors are grateful to Fabio Gsaller for useful discussion regarding light imaging and Carmen Kandelbauer, Ines Brosch, Inge Jehart and Sabine Jöbstl for technical assistance.

Conflicts of Interest: C.L.-F. has received grant support from the Austrian Science Fund (FWF), MFF Tirol, Astellas Pharma, Gilead Sciences, Pfizer, Schering Plough, and Merck Sharp & Dohme. She has been an advisor/consultant to Gilead Sciences, Merck Sharp & Dohme, Pfizer, and Schering Plough. She has received travel/accommodation expenses from Gilead Sciences, Merck Sharp & Dohme, Pfizer, Astellas, and Schering Plough and has been paid for talks on behalf of Gilead Sciences, Merck Sharp & Dohme, Pfizer, Astellas, and Schering Plough. All other authors declare no conflict of interest.

References

1. Richardson, M. The ecology of the Zygomycetes and its impact on environmental exposure. *Clin. Microbiol. Infect.* **2009**, *15*, 2–9. [CrossRef] [PubMed]
2. Ingold, C.T. *The Biology of Mucor and Its Allies*; Studies in biology no. 88; E. Arnold: London, UK, 1978.
3. Kontoyiannis, D.P.; Lionakis, M.S.; Lewis, R.E.; Chamilos, G.; Healy, M.; Perego, C.; Safdar, A.; Kantarjian, H.; Champlin, R.; Walsh, T.J.; et al. Zygomycosis in a tertiary-care cancer center in the era of *Aspergillus*-active antifungal therapy: A case-control observational study of 27 recent cases. *J. Infect. Dis.* **2005**, *191*, 1350–1360. [CrossRef] [PubMed]
4. Lanternier, F.; Sun, H.Y.; Ribaud, P.; Singh, N.; Kontoyiannis, D.P.; Lortholary, O. Mucormycosis in organ and stem cell transplant recipients. *Clin. Infect. Dis.* **2012**, *54*, 1629–1636. [CrossRef] [PubMed]
5. Lewis, R.E.; Kontoyiannis, D.P. Epidemiology and treatment of mucormycosis. *Future Microbiol.* **2013**, *8*, 1163–1175. [CrossRef] [PubMed]
6. Chamilos, G.; Marom, E.M.; Lewis, R.E.; Lionakis, M.S.; Kontoyiannis, D.P. Predictors of pulmonary zygomycosis versus invasive pulmonary aspergillosis in patients with cancer. *Clin. Infect. Dis.* **2005**, *41*, 60–66. [CrossRef]
7. Hammond, S.P.; Baden, L.R.; Marty, F.M. Mortality in hematologic malignancy and hematopoietic stem cell transplant patients with mucormycosis, 2001 to 2009. *Antimicrob. Agents Chemother.* **2011**, *55*, 5018–5021. [CrossRef]
8. Spellberg, B.; Edwards, J., Jr.; Ibrahim, A. Novel perspectives on mucormycosis: Pathophysiology, presentation, and management. *Clin. Microbiol. Rev.* **2005**, *18*, 556–569. [CrossRef]
9. Caramalho, R.; Tyndal, J.D.A.; Monk, B.C.; Larentis, T.; Lass-Flörl, C.; Lackner, M. Intrinsic short-tailed azole resistance in mucormycetes is due to an evolutionary conserved aminoacid substitution of the lanosterol 14α-demethylase. *Sci. Rep.* **2017**, *7*, 15898. [CrossRef]
10. Skiada, A.; Pagano, L.; Groll, A.; Zimmerli, S.; Dupont, B.; Lagrou, K.; Lass-Florl, C.; Bouza, E.; Klimko, N.; Gaustad, P.; et al. Zygomycosis in Europe: Analysis of 230 cases accrued by the registry of the European Confederation of Medical Mycology (ECMM) working group on Zygomycosis between 2005 and 2007. *Clin. Microbiol. Infect.* **2011**, *17*, 1859–1867. [CrossRef]
11. Antoniadou, A. Outbreaks of zygomycosis in hospitals. *Clin. Microbiol. Infect.* **2009**, *15*, 55–59. [CrossRef]
12. Duffy, J.; Harris, J.; Gade, L.; Sehulster, L.; Newhouse, E.; O'Connell, H.; Noble-Wang, J.; Rao, C.; Balajee, S.A.; Chiller, T. Mucormycosis outbreak associated with hospital linens. *Pediatr. Infect. Dis. J.* **2014**, *33*, 472–476. [CrossRef] [PubMed]
13. Garcia-Hermoso, D.; Criscuolo, A.; Lee, S.C.; Legrand, M.; Chaouat, M.; Denis, B.; Lafaurie, M.; Rouveau, M.; Soler, C.; Schaal, J.-V.; et al. Outbreak of invasive wound mucormycosis in a burn unit due to multiple strains of *Mucor circinelloides* f. *circinelloides* resolved by whole-genome sequencing. *MBio* **2018**, *9*, e00573-18. [CrossRef] [PubMed]
14. Lee, S.C.; Billmyre, R.B.; Li, A.; Carson, S.; Sykes, S.M.; Huh, E.Y.; Mieczkowski, P.; Ko, D.C.; Cuomo, C.A.; Heitman, J. Analysis of a food-borne fungal pathogen outbreak: Virulence and genome of a *Mucor circinelloides* isolate from yogurt. *MBio* **2014**, *5*, e01390-14. [CrossRef] [PubMed]
15. Magan, N.; Olsen, M. *Mycotoxins in Food: Detection and Control*; Woodhead Publishing Ltd.: Cambridge, UK, 2004.
16. Torres-Martinez, S.; Ruiz-Vázquez, R.M.; Garre, V.; López-García, S.; Navarro, E.; Vila, A. Molecular tools for carotenogenesis analysis in the zygomycete *Mucor circinelloides*. *Methods Mol. Biol.* **2012**, *898*, 85–107. [PubMed]
17. Vellanki, S.; Navarro-Mendoza, M.I.; Garcia, A.; Murcia, L.; Perez-Arques, C.; Garre, V.; Nicolas, F.E.; Lee, S.C. *Mucor circinelloides*: Growth, maintenance, and genetic manipulation. *Curr. Protoc. Microbiol.* **2018**, *49*, e53. [CrossRef] [PubMed]

18. Nicolas, F.E.; de Haro, J.P.; Torres-Martínez, S.; Ruiz-Vázquez, R.M. Mutants defective in a *Mucor circinelloides dicer*-like gene are not compromised in siRNA silencing but display developmental defects. *Fungal Genet. Biol.* **2007**, *44*, 504–516. [CrossRef] [PubMed]

19. Rodriguez-Frometa, R.A.; Gutiérrez, A.; Torres-Martínez, S.; Garre, V. Malic enzyme activity is not the only bottleneck for lipid accumulation in the oleaginous fungus *Mucor circinelloides*. *Appl. Microbiol. Biotechnol.* **2013**, *97*, 3063–3072. [CrossRef] [PubMed]

20. Brock, M. Application of bioluminescence imaging for in vivo monitoring of fungal infections. *Int. J. Microbiol.* **2012**, *2012*, 956794. [CrossRef] [PubMed]

21. Delarze, E.; Ischer, F.; Sanglard, D.; Coste, A.T. Adaptation of a *Gaussia princeps* Luciferase reporter system in *Candida albicans* for in vivo detection in the *Galleria mellonella* infection model. *Virulence* **2015**, *6*, 684–693. [CrossRef]

22. Jacobsen, I.D.; Lüttich, A.; Kurzai, O.; Hube, B.; Brock, M. In vivo imaging of disseminated murine *Candida albicans* infection reveals unexpected host sites of fungal persistence during antifungal therapy. *J. Antimicrob. Chemother.* **2014**, *69*, 2785–2796. [CrossRef]

23. Brock, M.; Jouvion, G.; Droin-Bergère, S.; Dussurget, O.; Nicola, M.-A.; Ibrahim-Granet, O. Bioluminescent *Aspergillus fumigatus*, a new tool for drug efficiency testing and in vivo monitoring of invasive aspergillosis. *Appl. Environ. Microbiol.* **2008**, *74*, 7023–7035. [CrossRef] [PubMed]

24. Donat, S.; Hasenberg, M.; Schäfer, T.; Ohlsen, K.; Gunzer, M.; Einsele, H.; Löffler, J.; Beilhack, A.; Krappmann, S. Surface display of *Gaussia princeps* luciferase allows sensitive fungal pathogen detection during cutaneous aspergillosis. *Virulence* **2012**, *3*, 51–61. [CrossRef] [PubMed]

25. Ibrahim-Granet, O.; Jouvion, G.; Hohl, T.M.; Droin-Bergère, S.; Philippart, F.; Kim, O.Y.; Adib-Conquy, M.; Schwendener, R.; Cavaillon, J.M.; Brock, M.; et al. In vivo bioluminescence imaging and histopathopathologic analysis reveal distinct roles for resident and recruited immune effector cells in defense against invasive aspergillosis. *BMC Microbiol.* **2010**, *10*, 105. [CrossRef]

26. Slesiona, S.; Ibrahim-Granet, O.; Olias, P.; Brock, M.; Jacobsen, I.D. Murine infection models for *Aspergillus terreus* pulmonary aspergillosis reveal long-term persistence of conidia and liver degeneration. *J. Infect. Dis.* **2012**, *205*, 1268–1277. [CrossRef] [PubMed]

27. Roncero, M.I.G. Enrichment Method for the isolation of auxotrophic mutants of mucor using the polyene antibiotic N-glycosyl-polifungin. *Carlsberg Res. Commun.* **1984**, *49*, 685–690. [CrossRef]

28. Nicolas, F.E.; Navarro-Mendoza, M.I.; Pérez-Arques, C.; López-García, S.; Navarro, E.; Torres-Martínez, S.; Garre, V. Molecular tools for carotenogenesis analysis in the mucoral *Mucor circinelloides*. *Methods Mol. Biol.* **2018**, *1852*, 221–237.

29. Maurer, E.; Hörtnagl, C.; Lackner, M.; Grässle, D.; Naschberger, V.; Moser, P.; Segal, E.; Semis, M.; Lass-Flörl, C.; Binder, U. *Galleria mellonella* as a model system to study virulence potential of mucormycetes and evaluation of antifungal treatment. *Med. Mycol.* **2018**. [CrossRef]

30. Arendrup, M.C.; Hope, W.W.; Lass-Flörl, C.; Cuenca-Estrella, M.; Arikan, S.; Barchiesi, F.; Bille, J.; Chryssanthou, E.; Groll, A.; et al. EUCAST technical note on the EUCAST definitive document EDef 7.2: Method for the determination of broth dilution minimum inhibitory concentrations of antifungal agents for yeasts EDef 7.2 (EUCAST-AFST). *Clin. Microbiol. Infect.* **2012**, *18*, E246–E247. [CrossRef]

31. Fallon, J.; Kelly, J.; Kavanagh, K. *Galleria mellonella* as a model for fungal pathogenicity testing. *Methods Mol. Biol.* **2012**, *845*, 469–485.

32. Maurer, E.; Browne, N.; Surlis, C.; Jukic, E.; Moser, P.; Kavanagh, K.; Lass-Flörl, C.; Binder, U. *Galleria mellonella* as a host model to study *Aspergillus terreus* virulence and amphotericin B resistance. *Virulence* **2015**, *6*, 591–598. [CrossRef]

33. Grahl, N.; Shepardson, K.M.; Chung, D.; Cramer, R.A. Hypoxia and fungal pathogenesis: To air or not to air? *Eukaryot. Cell* **2012**, *11*, 560–570. [CrossRef] [PubMed]

34. Dannaoui, E. Antifungal resistance in mucorales. *Int. J. Antimicrob. Agents* **2017**, *50*, 617–621. [CrossRef] [PubMed]

35. Espinel-Ingroff, A.; Chakrabarti, A.; Chowdhary, A.; Cordoba, S.; Dannaoui, E.; Dufresne, P.; Fothergill, A.; Ghannoum, M.; Gonzalez, G.M.; Guarro, J. Multicenter evaluation of MIC distributions for epidemiologic cutoff value definition to detect amphotericin B, posaconazole, and itraconazole resistance among the most clinically relevant species of Mucorales. *Antimicrob. Agents Chemother.* **2015**, *59*, 1745–1750. [CrossRef] [PubMed]

36. Khan, Z.U.; Ahmad, S.; Brazda, A.; Chandy, R. *Mucor circinelloides* as a cause of invasive maxillofacial zygomycosis: an emerging dimorphic pathogen with reduced susceptibility to posaconazole. *J. Clin. Microbiol.* **2009**, *47*, 1244–1248. [CrossRef] [PubMed]

37. Perkhofer, S.; Locher, M.; Cuenca-Estrella, M.; Rüchel, R.; Würzner, R.; Dierich, M.P.; Lass-Flörl, C. Posaconazole enhances the activity of amphotericin B against hyphae of zygomycetes in vitro. *Antimicrob. Agents Chemother.* **2008**, *52*, 2636–2638. [CrossRef]

38. Salas, V.; Pastor, F.J.; Calvo, E.; Alvarez, E.; Sutton, D.A.; Mayayo, E.; Fothergill, A.W.; Rinaldi, M.G.; Guarro, J. In vitro and in vivo activities of posaconazole and amphotericin B in a murine invasive infection by *Mucor circinelloides*: poor efficacy of posaconazole. *Antimicrob. Agents Chemother.* **2012**, *56*, 2246–2250. [CrossRef]

39. Caramalho, R.; Maurer, E.; Binder, U.; Araújo, R.; Dolatabadi, S.; Lass-Flörl, C.; Lackner, M. Etest cannot be recommended for in vitro susceptibility testing of mucorales. *Antimicrob. Agents Chemother.* **2015**, *59*, 3663–3665. [CrossRef] [PubMed]

40. Vitale, R.G.; de Hoog, G.S.; Schwarz, P.; Dannaoui, E.; Deng, S.; Machouart, M.; Voigt, K.; van de Sande, W.W.; Dolatabadi, S.; Meis, J.F.; et al. Antifungal susceptibility and phylogeny of opportunistic members of the order mucorales. *J. Clin. Microbiol.* **2012**, *50*, 66–75. [CrossRef] [PubMed]

41. Rodriguez, M.M.; Pastor, F.J.; Sutton, D.A.; Calvo, E.; Fothergill, A.W.; Salas, V.; Rinaldi, M.G.; Guarro, J. Correlation between in vitro activity of posaconazole and in vivo efficacy against *Rhizopus oryzae* infection in mice. *Antimicrob. Agents Chemother.* **2010**, *54*, 1665–1669. [CrossRef]

42. D'Enfert, C.; Diaquin, M.; Delit, A.; Wuscher, N.; Debeaupuis, J.P.; Huerre, M.; Latge, J.P. Attenuated virulence of uridine-uracil auxotrophs of *Aspergillus fumigatus*. *Infect. Immun.* **1996**, *64*, 4401–4405.

43. Galiger, C.; Brock, M.; Jouvion, G.; Savers, A.; Parlato, M.; Ibrahim-Granet, O. Assessment of efficacy of antifungals against *Aspergillus fumigatus*: Value of real-time bioluminescence imaging. *Antimicrob. Agents Chemother.* **2013**, *57*, 3046–3059. [CrossRef] [PubMed]

MDPI

St. Alban-Anlage 66

4052 Basel

Switzerland

Tel. +41 61 683 77 34

Fax +41 61 302 89 18

www.mdpi.com

Genes Editorial Office

E-mail: genes@mdpi.com

www.mdpi.com/journal/genes

Printed in June 2019
by Rotomail Italia S.p.A., Vignate (MI) - Italy